Handbook of Acoustics

Handbook of Acoustics

Editor: Benjamin Walker

STATES
ACADEMIC PRESS
www.statesacademicpress.com

Published by States Academic Press
109 South 5th Street,
Brooklyn, NY 11249, USA
www.statesacademicpress.com

Handbook of Acoustics
Edited by Benjamin Walker

International Standard Book Number: 978-1-63989-251-8 (Hardback)

Cataloging-in-Publication Data

Handbook of acoustics / edited by Benjamin Walker.
p. cm.
Includes bibliographical references and index.
ISBN 978-1-63989-251-8
1. Sound. 2. Physics. I. Walker, Benjamin.
QC225.15 .H36 2021
534--dc23

Contents

Preface

This book has been an outcome of determined endeavour from a group of educationists in the field. The primary objective was to involve a broad spectrum of professionals from diverse cultural background involved in the field for developing new researches. The book not only targets students but also scholars pursuing higher research for further enhancement of the theoretical and practical applications of the subject.

The branch of physics which deals with the study of mechanical waves in liquids, gases, and solids which includes topics such as vibration, sound, ultrasound, and infrasound is known as acoustics. It comprises the production, transmission, and effects of sound including the biological and psychological effects, and the qualities of a room that determine its character with respect to auditory effects. The application of acoustics is present in almost all aspects of modern society with the most common being the audio and noise control industries. It has many subdisciplines such as archaeoacoustics, aeroacoustics, acoustic signal processing, architectural acoustics, bioacoustics, electroacoustics, etc. This book outlines the processes and applications of acoustics in detail. It strives to provide a fair idea about this discipline and to help develop a better understanding of the latest advances within this field. This book will provide comprehensive knowledge to the readers.

It was an honour to edit such a profound book and also a challenging task to compile and examine all the relevant data for accuracy and originality. I wish to acknowledge the efforts of the contributors for submitting such brilliant and diverse chapters in the field and for endlessly working for the completion of the book. Last, but not the least; I thank my family for being a constant source of support in all my research endeavours.

Editor

1

Acoustic Simulation of Julius Caesar's Battlefield Speeches

Braxton Boren

Audio Technology Program, Department of Performing Arts, College of Arts & Sciences, American University, Washington, DC 20016, USA; boren@american.edu

Abstract: History contains many accounts of speeches given by civic and military leaders before large crowds prior to the invention of electronic amplification. Historians have debated the historical accuracy of these accounts, often making some reference to acoustics, either supporting or refuting the accounts, but without any numerical justification. The field of digital humanities, and more specifically archaeoacoustics, seeks to use computational techniques to provide empirical data to improve historical analysis. Julius Caesar recalled giving speeches to 14,000 men after the battle of Dyrrachium and another to 22,000 men before the battle of Pharsalus during the Roman Civil War. Caesar's background and education are discussed, including his training in rhetoric and oratory, which would have affected his articulation and effective sound pressure level while addressing his troops. Based on subjective reports about Caesar's oratorical abilities, his effective Sound Pressure Level (SPL) is assumed to be 80 dBA, about 6 dB above the average loud speaking voice but lower than that of the loudest trained actors and singers. Simulations show that for reasonable background noise conditions Caesar could have been heard intelligibly by 14,000 soldiers in a quiet, controlled environment as in the speech at Dyrrachium. In contrast, even granting generous acoustic and geometric conditions, Caesar could not have been heard by more than about 700 soldiers while his army was on the march before the battle of Pharsalus.

Keywords: acoustics; history; Julius Caesar; digital humanities; archaeoacoustics; acoustic simulation; historical speeches; general's harangue; military history

---

1. Introduction

Throughout most of human history, electronic amplification of the spoken word was unavailable; therefore, all human gatherings were effectively limited in size by the acoustic range over which the person speaking could be heard intelligibly. Many of the largest reported crowds in history are records of generals delivering a speech or "harangue" to an army [1], since the army's numbers were counted and thus represent a relatively rare instance of a counted crowd in ancient history (although there is certainly a large range of error in all—ancient and modern—crowd-counting methods [2,3]). These accounts were generally accepted by modern historians (e.g., [4,5]) until a landmark paper by Mogens Herman Hansen in 1993, which cast doubt upon the entire historicity of a general's speech to the army [6]. Hansen made many historical and textual arguments that need not be reproduced here in detail, but among these, he made acoustical claims, such as the area over which an ancient phalanx stretched, or noise from rattling hoplite armor, leading him to conclude that

> Under such circumstances it must have been impossible for a general, even if he had had the voice of a Stentor, to deliver a speech that could be heard by all the soldiers simultaneously [6].

Though propagation distance and background noise are certainly valid acoustical criteria to examine, the sentence above contains all the acoustical analysis the author thought necessary to include in his paradigm-shattering thesis. Of course, the problem with vague allusions to scientific truth to prove a point is that the other side can allude to science just as vaguely, which is indeed what happened in this case: a year later, the historian W. Kendrick Pritchett replied with a 100-page rejoinder defending the authenticity of the general's harangue, listing in detail different accounts of speeches to large crowds from ancient Greece and Rome, and also including more recent accounts, from Henry V at Agincourt to George Washington [7]. Pritchett mentioned these military examples as well as another from the preacher John Wesley, arguing that it is acoustically possible for a single speaker to reach such a large assembly.

This tendency to enlist science as an ally without careful numerical consideration often leads to an abuse or neglect of the specific issues in question. For example, Hansen's example of rattling hoplite armor, to which he returns several times in his essay, is based on a single reference to Alexander the Great in the midst of a great battle, rather than hoplites standing at attention *before* battle ([8], 4.13.37). In turn, Pritchett's reference to John Wesley uncritically accepted Wesley's estimate using an assumed crowd density of five people per square yard [9], while modern crowd estimation methods show that in large gatherings the highest densities achieved are less than half that value [10]. In such cases, quantitative references are somewhat superfluous; they are enlisted not to elucidate the past, but because the historians have already made up their minds and want to give their argument a veneer of scientific credibility (as some have asserted,"There are no statistics in ancient sources, just rhetorical flourishes made with numbers" [11]).

Humanities disciplines (including History, although it sometimes presents itself as Social Science instead) primarily *interpret* empirical facts which in themselves are not contested. The arguments employed in this process of interpretation are not (and in a sense cannot be) quantitative. Different historians come to different conclusions on the basis of the same collection of facts. The field which primarily seeks to collect empirical facts to be interpreted is not history but archaeology. In the past, archaeology focused on the excavation of physical artifacts, but recently has embraced computational methods to expand the range of acceptable empirical data [12]. This is a specific application of the movement generally known as the digital humanities, which seeks to use computational methods to shed light on uncontroversial but not immediately obvious facts, which may themselves be the basis for further humanistic interpretation and inquiry. The sub-field variously known as acoustic archaeology or archaeoacoustics uses computational acoustic simulation to uncover facts about the nature of sound in history, which is necessary to address the issue raised by Hansen and Pritchett's disagreement.

Both papers make reference to Julius Caesar's speeches to his army, both during the Gallic Wars and the Roman Civil War. The speeches during the Gallic Wars are smaller and less contested, but during the Civil War Caesar records giving two major speeches to large gatherings of soldiers, one following his army's defeat at the battle of Dyrrachium, and one immediately before the battle of Pharsalus, both in 48 BC [13]. This paper examines the historicity of these accounts through archaeoacoustic simulations of the speeches as Caesar describes them. Using this framework, it can be shown that, even without every historical detail preserved exactly, the plausibility or implausibility of certain historical speeches may be known with a high degree of certainty once we examine the acoustical evidence.

2. Background

After the controversy between Hansen and Pritchett, other historians writing on these speeches referenced the debate, but generally avoided making strong claims about whether these speeches actually occurred [14–16]. Again, this reticence may be a combination of not feeling comfortable with quantitative acoustics, as well as a general feeling that sound as a transient phenomenon is more or less lost to history once it is silenced. As it happens, there is a long tradition of working backwards through known physical laws to study sounds of the past, motivated by this same question: How many listeners can hear a single human voice?

2.1. Benjamin Franklin's Experiment

In 1739, the Methodist revivalist preacher (and friend of John Wesley) George Whitefield drew large outdoor crowds in London that were estimated as high as 80,000 people [17]. Across the Atlantic, Benjamin Franklin, who was at that time the publisher of The Pennsylvania Gazette in Philadelphia, had stopped printing the estimates of Whitefield's crowds because he thought they were exaggerated. Franklin described in his autobiography how he carried out an experiment to measure Whitefield's intelligible distance when the preacher came to Philadelphia. Using a semicircular approximation for the crowd shape, Franklin calculated that

> · · · [Whitefield] might well be heard by more than Thirty Thousand. This reconciled me to the newspaper accounts of his having preached to twenty five thousand people in the fields and to the ancient histories of generals haranguing whole armies of which I had sometimes doubted [18].

While Franklin used Whitefield's example as a conduit to explore the vocal ranges of ancient generals, the data from his experiment also provide information about Whitefield's vocal level, as the preacher was known for having one of the loudest voices of his generation [19]. Data from Franklin's experiment have been used to infer noise characteristics based on known noise sources and site geometry [20], study the maximum sound pressure [21] and directivity patterns for trained vocalists [22], and to simulate Whitefield's own maximum SPL and crowd size based on Franklin's data [23]. This work estimated that Whitefield's time-averaged on-axis L_{eq} could be as great as 90 dB_A, about 16 dB greater than the ANSI standard [24] for "loud speech" (about 74 dB when the ANSI spectrum is A-weighted [25]), and still significantly higher than that for "shouted speech" (about 82 dB_A). The corresponding ISO standard uses more discrete vocal levels, but similarly has its highest value of "Very Loud" speech as 78 dB after A-weighting is applied [25,26].

High vocal levels allow for animal communication at very long distances, producing an evolutionary sexual selection effect [27], which may vestigially influence human vocal capacity today. The computer simulations predicted that Whitefield could be heard intelligibly by a crowd of over 20,000 people without assuming overly optimistic acoustic conditions, although the reported crowd of 80,000 is acoustically implausible even under very favorable conditions [23].

This example shows that there exist regions of plausibility between being naively accepting or close-mindedly skeptical of all historical accounts of speeches to large crowds. Using what is already known about the human voice, sound propagation, and speech intelligibility, we can give a good approximation of how many people could hear a speaker on a specific occasion. While we cannot affix a precise crowd size to every historical account, we can shed a good deal of light on the historical account, which may inform the way we interpret the original text.

Even in situations where we do not have data as convenient as that which Franklin recorded, we can investigate a range of possible acoustic conditions (e.g., background noise). We can simulate the extreme "optimistic" and "pessimistic" ends of this range of conditions to better understand the historical situation. If, even under pessimistic conditions, the intelligibility (measured by the speech transmission index (STI)) is still acceptable throughout the crowd, we may consider the historical account plausible even if we cannot know the precise noise condition without the benefit of a time machine. Conversely, if under optimistic conditions the intelligibility is still too low throughout the simulated crowd, we may consider the account acoustically implausible. To apply this method to Caesar's speeches, we need to first consider his voice, the sites of his speeches, and the background noise present.

2.2. Caesar the Orator

Vocal training is an important factor to the maximum pressure achievable by a speaker [28], and Caesar received extensive training from childhood on. Though most famous for his achievements as a general and later as dictator, Julius Caesar was born in 100 BC into a high family in the Roman

Republic and was trained from childhood to perform the public ceremonies required of the family's inherited priesthood, and Caesar would later assume the title of *pontifex maximus*, the high priest of all Rome (the Latin title was later also given to the Bishop of Rome, i.e., the Pope) [29]. In 70 BC, Caesar traveled to the island of Rhodes to study oratory with the noted rhetoritician Apollonius Molon, who also trained Cicero in oratorical delivery [30]. In addition to his smaller speeches to his centurions or smaller military gatherings, Caesar also gave a noted speech in the Roman Senate advocating mercy for the Cataline conspirators in 63 BC [31].

In addition to the many examples of Caesar's experience with oratory, no less an orator than Cicero himself (who also wrote an entire text on the subject [32]) testified to the quality of Caesar's delivery. In a letter to Cornelius Nepos, Cicero wrote

> Do you know any man who can speak better than Caesar, even if he has concentrated on the art of oratory to the exclusion of all else? [33]

Despite his natural talent for oratory, Caesar chose to pursue the military instead of oratory as his chief vocation, and thus might not be expected to have perfected his delivery to the extent reported for speakers such as Demosthenes, who is reported to have undertaken specific exercises to perfect his elocution, lung control, and overall level [34]. In addition to vocal training, youth is also correlated with maximum vocal output, the level decreasing with increasing age [35]. In this regard, Caesar was no longer so young (52) at the time of the civil war in 48 BC, and thus simulations of his battle speeches cannot assume that he was near the highest vocal levels possible (90 dB_A), as in the case of Whitefield, who was only 24 at the time of Franklin's experiment. Because of this, a more moderate averaged SPL range of 74–80 dB_A is assumed for Caesar in the computer models.

3. Simulation Method and Results

3.1. Dyrrachium

3.1.1. Environmental Factors

In 49 BC, fearing prosecution if he lost his military position, Caesar "crossed the Rubicon" into Italy without resigning his military command, initiating the Civil War between his followers, known as the Populares, and the Optimates, who followed Pompey the Great. Caesar followed Pompey's army into Greece, where the two battles both occurred. Pompey's forces managed to break through Caesar's lines at the city of Dyrrachium, leading to a rare decisive defeat for Caesar's army. After the battle he recounted

> Caesar was now forced to abandon his previous plans and believed that he needed to change his entire strategy for this war. Accordingly, he simultaneously withdrew his troops from all their fortified positions, thus ending the siege. He gathered his entire army in one place and addressed the troops in assembly, urging them not to be discouraged by what had happened: they should neither be frightened by their present experience nor consider this single setback–which, at any rate, was relatively minor–as equivalent to their many successful battles [13].

After the speech, Caesar reports that his army successfully withdrew from Pompey's larger force, setting up the later conflict at Pharsalus. The exact location of the speech is not recorded by Caesar or later historians, so it is unknown if there were any distinguishing surfaces or geometric properties which may have aided or impaired speech intelligibility. However, he had retreated from his former fortifications, so it is reasonable to assume it took place on the plain to the east, along the line of Caesar's retreat, a relatively flat area. According to Caesar's own account, he had about 15,000 troops at Dyrrachium, and his losses from the battle were about 1000, suggesting that his entire army at the time of his speech consisted of about 14,000 [13].

Hansen did not address Caesar's speech at Dyrrachium, as assemblies of armies not in formation do not meet his criteria for a "battlefield speech". He even noted that, since on such an occasion the troops could be drawn up closer together, such assemblies are at least plausible [6]. However, it is certainly a historical account of a single speaker addressing a crowd of 14,000, which is a difficult task for most untrained vocalists and thus merits further investigation.

Despite Hansen's repeated concerns about rattling armor, as mentioned before, the only text he cites refers to noise during battle, not before or after. In fact, Caesar's legions at Dyrrachium consisted of battle-hardened soldiers who were intensely loyal to him and adept at moving or being still in a coordinated fashion (in fact, this second quality would more or less win the war for Caesar at Pharsalus). Caesar mentioned that "much self-reproach came over the entire army because of the disaster" which suggests a sombre tone over the whole assembly as their general spoke to them [13]. Assuming that the soldiers could be still and attentive during such a speech, a moderate background noise level of 45 dB_A seems reasonable based on the subjective descriptors of the occasion.

3.1.2. Simulation

The speech at Dyrrachium was modeled in CATT-Acoustic v9.1b [36,37]. Exact environmental data are not available, but the speech occurred in July and the weather was described as quite hot [13], so the model was assigned an air temperature of 27 °C, with 50% humidity, near the mid-summer averages for central Greece today, though to the author's knowledge no more exact temperature data are available for this period. Although geometric acoustic simulations (e.g., CATT) do not fully account for some wave effects [38,39], in this case, the model was needed only for propagation attenuation and STI calculation, so the geometrical model is sufficient for the level of historical precision that is known about the speech. CATT's diffraction prediction algorithm was enabled, allowing the simulation to account for the acoustic ground effect [40].

The model included a large map of a flat crowd, larger than that reported (200 m by 200 m), with assumed density of 2.7 people per m^2, corresponding to "strong" crowd conditions but not the extreme of "mosh-pit" conditions [10]. The modeled crowd's absorption coefficients are given in Table 1 using values measured in [41], and corresponding scattering coefficients were predicted based on an assumed average roughness depth of 0.1 m in CATT. Here and at Pharsalus, the measured Sabine absorption values were greater than 1 for the highest frequencies, but this is an artifact of the Sabine calculation. Since geometric acoustical simulations cannot accept values greater than 1, all such values were truncated to 0.99, as indicated in [37]. Above this, a source representing Julius Caesar was situated 1 m above the crowd, with listeners on all sides of him, assuming that his army would crowd around all sides to be able to hear him better.

Caesar's voice was simulated using a standard male vocal directivity pattern in CATT, since more specific information about Caesar's head shape or directivity pattern probably cannot be known. Caesar's voice was modeled with a steady-state L_{eq} ranging from 74 dB_A (standard "loud" speech) to 80 dB_A (closer to "shouted" speech) at 1 m and a standard male vocal spectrum (which CATT calculates directly from the ANSI-1997 standard spectrum based on simulated level [37]). Background noise was modeled as 45 dB_A.

Table 1. Absorption coefficients by octave band center frequency (Hz) for standing soldiers in Dyrrachium model (2.7 persons per m^2, [41]).

Surface	125	250	500	1000	2000	4000
Audience area (dense)	0.24	0.47	0.94	0.99	0.99	0.99

After the simulations, CATT exported grid maps of the crowd, representing the Speech Transmission Index (STI) in each 2 m × 2 m square. From these, the minimum intelligible area (MIA), defined as the area over which STI $\geq$ 0.3 was calculated for each grid with a custom Matlab script. 0.3

was first used as a threshold of intelligibility in [23] because the phoneme-group score for various word groups drops precipitously below this value [42], leading STI measures below this to be characterized as "bad" intelligibility. Based on the source vocal level, the MIA is predicted to range from 2104 to 8096 m^2, as shown in Figure 1 (Caesar's position is shown by a yellow dot). Final predicted intelligible crowd sizes are 2.7 multiplied by the MIA, assuming a density of 2.7 persons per m^2, as shown in Table 2.

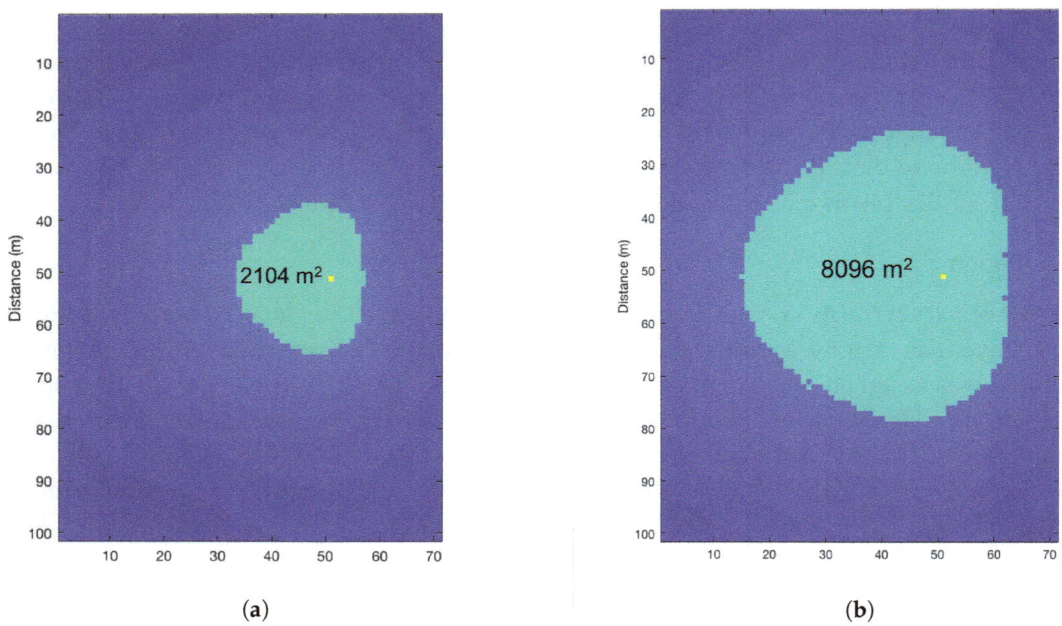

Figure 1. Minimally Intelligible Area (MIA) at Dyrrachium Based on Source Vocal Level: (**a**) 74 dB_A source; and (**b**) 80 dB_A source.

Table 2. Modeled crowd size, area, and density by vocal source level.

Vocal SPL (dB_A)	Background Noise Level (dB_A)	MIA (m^2)	Density (Persons/m^2)	Crowd Size
74	45	2104	2.7	5,681
80	45	8096	2.7	21,859

3.2. *Pharsalus*

3.2.1. Environmental Factors

After his speech at Dyrrachium, Caesar led his army across the Thessalian plain hoping to engage Pompey in the open field. Pompey, who had just won at Dyrrachium and had a much larger army (about 45,000 to Caesar's 22,000 at this point according to Caesar, who had received reinforcements from his commander Gnaeus Domitius Calvinus [13]), held higher ground and hoped for a defensive advantage. However, pressure from Pompey's associates to hurry up and finish the war led him to move out and engage Caesar's army. Caesar describes it thus:

> After··· the signal for departure had already been given and the tents struck, it was noticed that a little earlier Pompey's battle line, exceeding what it had been used to doing each day, had moved forward farther from their rampart, so that it appeared possible to fight on ground that would not be unfavorable. Then Caesar, who was with his men as they were about to march out of the gate in formation, said to them, "We need to put off our march for now and turn our thoughts to battle, which is what we have been demanding all this time.

> Let's take courage and be prepared to fight. It will not be easy to find another chance later." Then he immediately led out his troops, unencumbered and ready to fight [13].

Even Hansen the skeptic reckoned this speech may have actually happened, but argued that because it was so short it did not count as a true battlefield speech [6]. The exact site of the battle was unknown for nearly 2000 years, with many historians supposing that it happened south of the Epineus River, until F. L. Lucas suggested in 1921 that the battle be located on the north bank [43], which is now broadly accepted [44]. Because Pompey had come down out of the rougher hills to the west, it seems reasonable to model this as flat terrain as well.

However, unlike the controlled environment after Dyrrachium, the acoustic context of this speech is quite different. While in a gathering for the conveyance of information, we can assume Caesar's troops to be relatively quiet, in this instance, he specifically mentions that they were assembled and marching out the gate to challenge and exhaust Pompey. By Caesar's own account his troops were in battle formation and moving while he spoke to them, which would drastically increase the background noise. While we do not know the exact SPL this would produce, an optimistic figure is 55 dB_A, although the value could certainly be much greater.

3.2.2. Simulation

Caesar was simulated giving a speech to another 200 m × 200 m grid, with the same environmental factors as at Dyrrachium above. His vocal level was modeled as 80 dB (the quieter value of 74 dB being discarded for now to explore only the most optimistic scenario). The background noise was modeled as 55 dB_A to account for the louder noise of an army on the march. In addition, since Caesar's army was in formation, the greatest density that can be assumed for them is lower than at Dyrrachium, about 1 person per m^2 [45]. Thus, Caesar's army was modeled using measured absorption values for a standing crowd with density of 1.17 persons per m^2, as measured in [46] and shown in Table 3.

Table 3. Absorption coefficients by octave band center frequency (Hz) for soldiers in Pharsalus model (1.17 persons per m^2, [46]).

Surface	125	250	500	1000	2000	4000
Audience area (dense)	0.20	0.35	0.70	0.99	0.99	0.99

Again, the STI map from CATT was exported and all intelligible grid squares' areas were summed in Matlab, with the difference this time that no space was allowed behind Caesar since he was presumably in front of his army when he addressed them (in general, cutting off the space behind a speaker, given the shape of the human voice's directivity pattern, will reduce the MIA for that scenario by about 20%). The simulated MIA under these conditions is shown in Figure 2.

It can be seen that the predicted MIA drops drastically under these conditions, to 596 m^2, as shown in Table 4. Thus, the greatest number of men that could have heard Caesar intelligibly at once is closer to 500 (the size of a single cohort) rather than his entire army of 22,000.

Table 4. Modeled crowd size, area, and density by vocal source level.

Vocal SPL (dB_A)	Background Noise Level (dB_A)	MIA (m^2)	Density (Persons/m^2)	Crowd Size
80	55	596	1.17	697

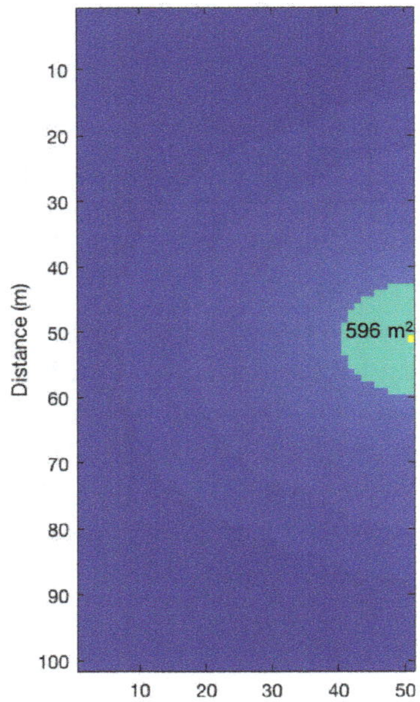

Figure 2. Minimally Intelligible Area (MIA) at Pharsalus.

4. Discussion

Based on these simulations, we may shed new light on the acoustical plausibility of these two speeches, which will affect our interpretation of Caesar's account. At Dyrrachium, we simulated rather pessimistic conditions—giving Caesar only a 6 dB boost above normal loud speech, while assuming background noise of 45 dB_A, over 10 dB greater than that measured at some ancient Greek amphitheaters [47]. It can be seen from the simulations that, even under these relatively pessimistic conditions, Caesar could have plausibly spoken to all his 14,000 soldiers at once (leaving room for more!), as long as he did not speak with a "normal" vocal level of 74 dB. Since the pessimistic case is plausible, we do not need to stretch the bounds of the simulation by probing how loud Caesar might have been, how quiet his men might have been, or assuming the existence of atypical environmental propagation, such as refraction patterns via a temperature inversion that would have carried his voice farther. The historical details must be sorted out by historians, but acoustically it must be said that this speech seems physically valid based on the descriptions we have of the event.

In contrast, before Pharsalus, even assuming the louder value for Caesar and a relatively quiet noise value for his army, his MIA value decreases by an order of magnitude. The maximum width of his intelligible area is less than 20 m, while the battlefront stretched up to 3 km once the armies were engaged [44]. Even assuming his army was somewhat more compact before he spoke to them, there does not appear to be any acoustically plausible scenario in which he could be heard intelligibly by his entire force if they were all in front of him, arranged less densely, and on the move. Since the optimistic case does not appear plausible, there is no need to simulate greater background noise values. In both cases, there is clearly some uncertainty about Caesar's exact STI value, vocal level, and background noise. However, by considering an implausible optimistic scenario or a plausible pessimistic scenario, it is possible to speak confidently about the general plausibility of an account even if we cannot know the precise intelligibility value at each point in the crowd.

Having made this acoustical point, however, it should not be inferred that Caesar's speech as written did not happen—it is quite possible that he did speak something similar to the soldiers and officers nearby when he realized Pompey's forces were extended far enough to engage them. The signal to engage could then have been conveyed by messengers, trumpet signal, or flag signals, all of which

Caesar used during the Civil War [13]. In fact, as Pompey's cavalry attempted to flank his right side, Caesar reacted as follows:

> However, once [Caesar] had a look at the enemy formation described above, he feared a flanking attack by the mass of enemy cavalry circling around his right wing; he therefore rapidly drew individual cohorts out of the third line of his formation, placed them as a fourth line to oppose Pompey's cavalry, and explained to them what he wanted them to do, making it clear that this day's victory would depend on the bravery of these cohorts. At the same time he commanded the third line not to move forward and engage with the enemy without explicit orders from himself: when he wanted this to happen, he would give the signal with a flag [13].

This action seems to be largely improvised, as Caesar recorded himself *explaining* and *commanding* his soldiers verbally of what they were about to do. Assuming that he could make himself heard by a single cohort of 500 men in formation, this seems reasonably in line with him riding behind his third line and speaking multiple times to multiple cohorts as he formed them into the fourth line which would counterattack Pompey's cavalry, resulting in Caesar's successful flank of Pompey, the route of Pompey's forces, and the end of the Civil War. The fact that Caesar ordered them all to coordinate this action by a flag signal again shows that he did not expect to be able to verbally communicate with his entire army, or even a portion of it, during the battle itself. However, his reported oratorical skill would have come in useful in this and many other noisy battles where Caesar's instincts and improvisational abilities led his forces to many such incredible victories and very few defeats.

Although this discussion has focused mainly on Caesar, clearly many such pitched battles between ancient armies occurred on fairly flat terrain without many significant reflecting surfaces, so these simulations may be seen as a proxy for any male speaker with the level and background noise conditions assumed here. This may be less useful for modern assemblies, which generally have electrical amplification, but it can be viewed as a lens through which to consider other ancient reported speeches before armies. However, for each situation, the orator's vocal ability, training, and age should be considered, along with the likely background noise of the crowd listening.

5. Conclusions

The simulations above show that Caesar's post-battle speech at Dyrrachium seems acoustically plausible, even without assuming optimistic conditions. In contrast, Caesar's speech before the battle at Pharsalus cannot plausibly have been heard intelligibly by most of his army in formation and on the march. This does not necessarily imply that we should adopt the extreme skepticism of Hansen, who doubted that commanders could even address 1000 soldiers at once in formation [6]. In other cases, when forces were drawn up to a static position before a battle, it may have been acoustically possible for their commanders to address all of them at once, depending on vocal level and background noise. Other cases such as accounts from Thucydides or Henry V's speech before Agincourt, seem to fall into this category, and require separate simulations to account for the particulars of each account. However, Caesar's speech at Pharsalus should be considered a local order to the troops immediately around him, rather than a grand address given to the entire army.

Funding: This research received no external funding.

Conflicts of Interest: The author declares no conflict of interest.

References

1. Thucydides. *History of the Peloponnesian War*; Bohn: London, UK, 1843.
2. Jacobs, H.A. To count a crowd. *Columbia Journal. Rev.* **1967**, *5*, 37–40.
3. Seidler, J.; Meyer, K.; Gillivray, L.M. Collecting Data on Crowds and Rallies: A New Method of Stationary Sampling. *Soc. Forces* **1976**, *55*, 507–519. [CrossRef]

4. Murphy, C.T. The Use of Speeches in Caesar's Gallic War. *Class. J.* **1949**, *45*, 120–127.
5. Miller, N.P. Dramatic Speech in the Roman Historians. *Greece Rome* **1975**, *22*, 45–57. [CrossRef]
6. Hansen, M.H. The battle exhortation in ancient historiography: Fact or fiction? *Historia* **1993**, *42*, 161–180.
7. Pritchett, W.K. The General's Exhorations in Greek Warfare. In *Essays in Greek History*; J. C. Gieben: Amsterdam, The Netherlands, 1994; pp. 27–109.
8. Rufus, Q.C. *History of Alexander*; Harvard University Press: Cambridge, MA, USA, 1946.
9. Wesley, J. *The Journal of John Wesley*; Moody Press: Chicago, IL, USA, 1951.
10. Watson, R.; Yip, P. How many were there when it mattered? *Significance* **2011**, *8*, 104–107. [CrossRef]
11. O'Neill, T. The Great Myths 5: The Destruction of The Great Library of Alexandria. 2017. Available online: https://historyforatheists.com/2017/07/the-destruction-of-the-great-library-of-alexandria/ (accessed on 8 February 2018).
12. Bordes, N.; Pailthorpe, B.; Hall, J.; Loy, T.; Williams, M.; Ulm, S.; Zhou, X.; Fletcher, R. Computational Archaeology. In Proceedings of the WACE 2004: Workshop on Advanced Collaborative Environments, Nice, France, 23 September 2004.
13. Raaflaub, K.A. (Ed.) *The Landmark Julius Caesar*; Pantheon Books: New York, NY, USA, 2017.
14. Ehrhardt, C.T.H.R. Speeches before Battle? *Hist. Z. Alte Gesch.* **1995**, *44*, 120–121.
15. Nordling, J.G. Caesar's Pre-Battle Speech at Pharsalus (B.C. 3.85.4): Ridiculum Acri Fortius ··· Secat Res. *Class. J.* **2005**, *101*, 183–189.
16. Zoido, J.C.I. The Battle Exhortation in Ancient Rhetoric. *Rhetorica* **2007**, *25*, 141–158. [CrossRef]
17. Dallimore, A. *George Whitefield: The Life and Times of the Great Evangelist of the Eighteenth–Century Revival*; Banner of Truth Trust: London, UK, 1970.
18. Franklin, B. *The Autobiography of Benjamin Franklin*, 2nd ed.; Yale University Press: New Haven, CT, USA; London, UK, 1964.
19. Stout, H. *The Divine Dramatist: George Whitefield and the Rise of Modern Evangelicalism*; William B. Eerdmans Publishing Company: Grand Rapids, MI, USA, 1991.
20. Boren, B.B.; Roginska, A. Analysis of noise sources in colonial Philadelphia. *Internoise* **2012**, *9*, 7543–7553.
21. Boren, B.B.; Roginska, A. Maximum Averaged and Peak Levels of Vocal Sound Pressure. In Proceedings of the 135th Audio Engineering Society Convention, New York, NY, USA, 17–20 October 2013.
22. Boren, B.B.; Roginska, A. Sound Radiation of Trained Vocalizers. In Proceedings of the Meetings on Acoustics: 21st International Congress on Acoustics, New York, NY, USA, 4 May 2013.
23. Boren, B.B. George Whitefield's Voice. In *George Whitefield: Life, Context and Legacy*; Jones, D.C., Hammond, G., Eds.; Oxford University Press: Oxford, UK, 2015.
24. American National Standards Institute. *American National Standard: Methods for Calculation of the Speech Intelligibility Index*; Acoustical Society of America: New York, NY, USA, 1997.
25. Rindel, J.H. *Odeon Application Note–Calculation of Speech Transmission Index in Rooms*; Odeon A/S: Lyngby, Denmark, 2014.
26. International Organization for Standardization. *ISO 9921: 2003 Ergonomics–Assesment of Speech Communication*; International Organization for Standardization: Geneva, Switzerland, 2003.
27. Forrest, T.G.; Raspet, R. Models of female choice in acoustic communication. *Behav. Ecol.* **1994**, *5*, 293–303. [CrossRef]
28. Mendes, A.P.; Rothman, H.B.; Sapienza, C.; Brown, W. Effects of vocal training on the acoustic parameters of the singing voice. *J. Voice* **2003**, *17*, 529–543. [CrossRef]
29. Kahn, A. *The Education of Julius Caesar*; Schocken Books: New York, NY, USA, 1986.
30. Plutarch. *The Parallel Lives*; Loeb Classical Library: Cambridge, MA, USA; London, UK, 1919.
31. Sallust. *Conspiracy of Catiline*; Harper and Brothers: New York, NY, USA; London, UK, 1899.
32. Cicero, M.T. *De Oratore*; Harvard University Press: Cambridge, MA, USA, 1967.
33. Suetonius, G. *The Twelve Caesars*; Penguin Books: London, UK, 1957.
34. Dobson, J.F. *The Greek Orators*; Methuen and Co.: London, UK, 1919.
35. Kent, R.; Kent, J.; Rosenbek, J. Maximum Performance Tests of Speech Production. *J. Speech Hear. Disord.* **1987**, *52*, 367–387. [CrossRef] [PubMed]
36. Dalenbäck, B.I.L. Room Acoustic Prediction Based on a Unified Treatment of Diffuse and Specular Reflection. *J. Acoust. Soc. Am.* **1996**, *100*, 899–909. [CrossRef]
37. Dalenback, B. *CATT-Acoustic v9*; CATT: Gothenburg, Sweden, 2011.

38. Rindel, J. The Use of Computer Modeling in Room Acoustics. *J. Vibroeng.* **2000**, *3*, 219–224.
39. Siltanen, S.; Lokki, T.; Savioja, L. Rays or Waves? Understanding the Strengths and Weaknesses of Computational Room Acoustics Modeling Techniques. In Proceedings of the International Symposium on Room Acoustics, Melbourne, Australia, 29–31 August 2010.
40. Forrest, T.G. From Sender to Receiver: Propagation and Environmental Effects on Acoustic Signals. *Am. Zool.* **1994**, *34*, 644–654. [CrossRef]
41. Adelman-Larsen, N.W.; Thompson, E.R.; Gade, A.C. Suitable reverberation times for halls for rock and pop music. *J. Acoust. Soc. Am.* **2010**, *127*, 247–255. [CrossRef] [PubMed]
42. Steeneken, H.J.M.; Houtgast, T. Basics of the STI measuring method. In *Past, Present and Future of the Speech Transmission Index*; van Wijngaarden, S.J., Ed.; TNO Human Factors: Soesterberg, The Netherlands, 2002; pp. 13–43.
43. Lucas, F.L. The Battlefield of Pharsalos. *Annu. Br. Sch. Athens Suppl.* **1921**, *24*, 34–53. [CrossRef]
44. Morgan, J.D. Palaepharsalus–The Battle and the Town. *Am. J. Archaeol.* **1983**, *87*, 23–54. [CrossRef]
45. James, S. 48 BC: The Battle of Pharsalus. In *Rome's Legions: Decoding Their Pythagorean Organization 753 BC to 410 AD*; Academia: Prague, Czechoslovakia, 2010.
46. Martellotta, F.; D'Alba, M.; Crociata, S.D. Laboratory measurement of sound absorption of occupied pews and standing audiences. *Appl. Acoust.* **2011**, *72*, 341–349. [CrossRef]
47. Bo, E.; Kostara-Konstantinou, E.; Lepore, F.; Shtrepi, L.; Puglisi, G.; Astolfi, A.; Barkas, N.; Mangano, B.; Mangano, F. Acoustic characterization of the ancient theatre of Tyndaris: Evaluation and proposals for its reuse. In Proceedings of the ICSV 2016–23rd International Congress on Sound and Vibration: From Ancient to Modern Acoustics, Atene, Greece, 10–14 July 2016.

2

Acoustical Impact of Architectonics and Material Features in the Lifespan of Two Monumental Sacred Structures

Zühre Sü Gül

Department of Architecture, Bilkent University, 06800 Ankara, Turkey; zuhre@bilkent.edu.tr

Abstract: Hagia Sophia and Süleymaniye Mosque, built in the 6th and 16th centuries, respectively, are the two major monuments of the İstanbul World Heritage Site. Within the context of this study, sound fields of these two sacred multi-domed monumental structures are analyzed with a focus on their architectonic and material attributes and applied alterations in basic restoration works. A comprehensive study is undertaken by a comparative analysis over acoustical field tests held in different years and over an extensive literature review on their material and architectural characteristics. Initially, the major features of Hagia Sophia and Süleymaniye Mosque are presented, and later, basic alterations in regard to function and materials are provided. The methodology includes the field tests carried both within the scope of this research as well as the published test results by other researchers. Acoustical simulations are utilized for comparison of unoccupied versus occupied conditions and also for discussion on original materials. The impact of historical plasters on the acoustics of domed spaces is highlighted. Common room acoustics parameters as of reverberation time and clarity are utilized in comparisons. The formation of multi-slope sound energy decay is discussed in light of different spiritual and acoustical needs expected from such monumental sacred spaces.

Keywords: Hagia Sophia; Süleymaniye Mosque; room acoustics; historical structures; restoration

---

1. Introduction

This research aims to investigate the effects of the main architectural features of monumental sacred structures on their interior sound fields. With their large volumes and multi-domed upper shell typology in the form of main dome spaces coupled to sub spaces sheltered either by domes or vaults, these historical structures have an outstanding and particular acoustical environment or so called "interior soundscape". In architectural literature, domed structures are mostly observed in religious buildings. Sacred spaces have a specific role of emphasizing spirituality. Acoustics is one means of recognizing the indoor environment and a critical element to augment the divine character of sacred mega-structures. The basic architectural features of such monuments i.e., the central dome and huge volumes, result in the long and shallow reverberant decay, contributing to the key role of holy spaces.

1.1. Acoustics of Sacred Spaces

The literature on the acoustics of sacred spaces is extensive. Within this framework the acoustics of mosques [1–6], churches [7,8], basilicas [9] and cathedrals [10–13] have always been an attractive research topic for acousticians. In relation to this research there are mainly two groups of studies. The first group includes virtual re-construction of historical structures [2,3,11], renovation of pre-designed and built structures [1] for enhancing their indoor acoustical qualities, and acoustical design of new generation mosques [6]. The prominent methods in sound field analysis are field tests, scale modelling, and acoustical simulations [14]. The second group of studies focuses on non-exponential decay formation in multi-volume mega-structures [7,9,10,15]. The acoustical studies on sacred spaces search for the desired interior sound field in relation to function, which differs slightly in different religions. Worshippers sometimes need solitude, while at other times they want to feel in absolute unity with the others in attendance.

There are three basic acoustical requirements for mosques in relation to clarity as well as the spiritual effects of sound; audibility of the namaz—prayer orders of the Imam (prayer leader)—, recognizable sermon of the preacher, and listening to or joining in the recital of the musical versions of the Holy Quran. In mosques, in order to overcome the large sound attenuation of his voice over the rows of worshippers on the floor, the imam delivers his sermon at minbar and müezzin delivers the commands of namaz from müezzin's mahfili. This scheme was specifically developed in the time of historical mosques when there was no electronic sound reinforcement. It was a positive effect that the sound source is above the receiver locations for providing more sound strength to the audience in mosques' original schemes. In the acoustical design of mosques the male voice must be augmented by the architectural features of the space for the envelopment and spaciousness. Thus, the architecture must be incorporated within the expected formal language, which satisfies the spiritual aspects of a worship space. The common use of carpet as a floor finish in mosques instead of the stone floors prevalent in Christian spaces results in faster sound decay rates, thus shorter reverberation times. In Christian sacred spaces, intelligibility of sermons by a priest is one acoustical necessity [14]. On the other hand, recitals of liturgical music are much more frequent in churches and cathedrals when music as an activity is compared to those in mosques. For that reason, it is much more appropriate that the reverberation times are longer in churches or cathedrals when compared to those in mosques.

1.2. Room Acoustics Coupling in Multi-Domed Superstructures

Most od the time, the historical sacred spaces with multi-domed upper-structures have even longer reverberation times than classical venues for music. At this point, the basic challenge in the design of sacred spaces is optimizing reverberance. It is well-known that very high reverberation negatively affects speech intelligibility, while very short reverberation causes a dry acoustical environment that would reduce the envelopment and spaciousness in a religious space. These two contradictory requirements, clarity and reverberance, are hard to reconcile with classical decay properties of a single space room. Some previous studies highlight that multi-domed or vaulted monumental spaces are composed of multiple sub-spaces that are connected to each other with apertures in the form of arches. These arches behave as coupling apertures and depending upon the size of the apertures and as well as the volume of individual sub-spaces, a convex decay curve is observed in large cathedrals, basilicas [7,10,11], and in mosques [15]. Non-exponential or multiple sound energy decay with regard to the inherent properties of early and late decays can maintain the sense of intimacy and clarity that is mostly observed in coupled spaces. This outcome was further discussed in a previous study within a broader context of this research [16] and acoustical indicators in multi-domed superstructures were re-evaluated [17].

In order to contribute to the previous literature on acoustics of sacred structures, this study investigates two historically significant monuments of the İstanbul World Heritage Site. The first monument is Hagia Sophia, which was originally built in the 6th century and functioned as a church and later converted to a mosque, while currently serving as a museum. The second monument Süleymaniye Mosque is a 16th century structure. Their historical significance and religious uses in relation to acoustics have motivated many researchers to test and discuss characteristics of their interior sound fields. A comprehensive study is necessitated to assess the acoustical conditions of both structures in relation to their architectural features and interior finish materials and to examine the acoustical changes that have occurred due to the repairs undergone within their lifetime. Initially, the major architectural features of Hagia Sophia and Süleymaniye Mosque are presented, and later, basic alterations with regards to function and materials are provided. The methodology includes the field tests carried both within the scope of this research as well as the published field test results by other researchers. Acoustical simulations are applied for comparison of unoccupied versus occupied conditions, reflecting major activity patterns. The simulations also provide basis for a discussion on the acoustical outcomes of original or historical plasters versus the current condition after restorations. Finally, the formation of multi-slope sound energy decay in case structures is discussed briefly in light of different spiritual and acoustical needs expected from such monumental sacred spaces.

2. Hagia Sophia

Hagia Sophia, the Holy Wisdom of Christ, was constructed as a church between 532 and 537 in Constantinople (Istanbul) during the reign of the Byzantine Emperor Justinian. Hagia Sophia was both the center of religious life and had been a legend for the new wave of church buildings in the West in the 12th and 13th centuries. After the Ottoman conquest of 1453, during the rule of Mehmet II, it was converted from a church to a mosque. In 1932 upon order from Atatürk, Hagia Sophia (Figure 1) started to function as a museum. Hagia Sophia is a masterpiece of not only Byzantine art but also of the world's historical heritage. Its architectural success and carried messages affected both the Ottoman Empire and architects such as Sinan the Architect [18–21].

Figure 1. Hagia Sophia exterior view (www.ayasofyamuzesi.gov.tr).

The influences of Hagia Sophia's form have revitalized many other domed spaces both in churches and mosques. On the other hand, the building physics aspects of the structure and materials has been a subject for many other researches [22]. The interior sound field of such an immense monumental building has also inspired acousticians. In some acoustical studies on Hagia Sophia, the room acoustical parameters were analyzed for its current condition, as further detailed in Section 5.1. By some other researchers, the acoustical history of the structure was virtually generated for different time periods and activity patterns [23]. One study focused on recreating Hagia Sophia's acoustics within other venues by the use of electronic architecture [24].

Hagia Sophia is a large space with many coupled sub-spaces. High reverberance within such a large interior, lends unique properties to the music and demands suitable repertoire and singing ability in terms of liturgical music. The sound field of such a volume with dominating geometric and material attributes also inspired this study in terms of acoustical coupling investigations. In a broader framework, the non-exponential energy decay formation in Hagia Sophia for its current state was previously discussed [17].

2.1. Major Architectural Features

The major figure of Hagia Sophia is an expanded dome basilica: a rectangular building covered by a central dome between two half domes and integrating longitudinal and centralized planning (Figure 2). This structure has an interior length of 73.5 m and a width of 69.5 m, excluding the narthex and the apse. The length of the entire interior from the exonarthex to the edge of the apse is 92.3 m. The central nave is built on an east-west axis, and a large dome is constructed right in the center to cover up the space. The central dome that rises 55 m above the pavement of the nave is not exactly round but slightly elliptical today, with a diameter of 31.2 m on one axis and 32.8 m on the other [18]. The domed central space is skirted by two large hemicycles covered by half-domes to the east and west. The diameter of these half-domes roughly equals to that of the central dome. From the crown of each of the main piers and between the four great arches, the pendentives fan out on the interior and rise to a roughly circular projecting cornice [25,26]. The side aisles are separated from central nave by columns and arches sheltered with vaults. There are galleries over the side aisles and inner narthex, creating a U-shape that reinforces the centralizing tendency of this expanded dome basilica. The central oval vessel of enclosed space is further expanded by barrel-vaulted spaces that terminate along the building's longitudinal axis at the east and the west ends of its nave [18]. Overall, Hagia Sophia has an approximate interior acoustical volume of 150,000 m^3, which creates an outstanding visual and acoustical environment.

Figure 2. On the left: section view of Hagia Sophia from the central axis by Salzenberg [25] (p.280), on the right: interior view—an old painting [25] (p.284).

Stone, brick and mortar are the main materials that compose the above-ground structure including the piers, columns, arches, vaults, and domes. The stone is either limestone or green stone—local granite. Above the springing level of the gallery vaults, most of the piers are built of brick. The surfaces of all the walls as well as the large supportive piers, bordered by sculptured moldings, are covered with polished slabs of veined marble and other colored stone. Internally most of the original marble revetment of the walls remains in place and most of the original non-figural mosaic decoration of the vaults has remained undamaged at ground level [18,20]. Some of the non-figural decorations in the galleries and above are distinguished partly by a much greater use of tesserae cut from natural stones and terracotta. All surviving figural mosaics are of a later date. The floor is paved today with large rectangular marble slabs, not the original one which was crushed in 1346 [20,26].

2.2. Basic Repairs and Alterations

Over the 1400 years of its existence, Hagia Sophia has suffered much damage essentially due to the major earthquakes. Three main phases of structural repair and strengthening are recorded. The first repairing phase was in 1317, in the reign of Andronicus Palaeologus; the second was in 1573 under the architect Sinan; and the third repair was in 1847 under the Swiss architect Gaspare Fossati, assisted by his brother Giuseppe. The principal work undertaken in 1317 was the construction of new buttresses. In 1573, Sinan built a new minaret in place of one that was to be demolished. In 1847 major works were the rectification of a number of columns in the gallery exedra, the installation of new ties for critical locations, and other repairment at the level of the dome base [20,21].

Hagia Sophia has also undergone many alterations due to the changes in its activity patterns. In conversion from a church to a mosque, some Christian elements were removed, and Islamic additions were introduced. Since pictorial representations are traditionally not permitted in Islam, after 1453 the mosaics were gradually covered up, whitewashed, or plastered over and hence preserved. After the Ottoman conquest, all the Christian furnishings were removed [20]. In 1847 the Sultan commissioned a pair of Swiss architects, Gaspare and Giuseppe Fossati, to restore both the fabric and the decoration of the building. During these works all the surviving mosaics were uncovered and copied in order to provide visual record. Unfortunately, many of the mosaics recorded by Fossati's had disappeared by this date, most probably lost in the great earthquake of 1894. The original altar, screen, and ambo were badly damaged in the first collapse, and rebuilt in the subsequent restoration [18].

Among the Islamic additions, apart from four minarets on the exterior, a mihrab was added on the kiblah axis (the church is oriented to the east) The minbar was constructed in the same direction. Prayer carpets and banners of victory were hung on the walls during the mosque period. The Müezzin's mahfili in the center of the structure and four more galleries in the narthex were added. On the left side of the central nave there is the preaching pulpit, that still exists. Imperial Pavilion and Imperial Loge (Figure 3) are some other additions which did not exist during the Christian period. Another significant addition of the Islamic period are the huge disks of inscriptions/calligraphies over wooden roundels with diameter of 7.6 m, which are acknowledged to be largest inscriptions in the entire Islamic world [26].

Figure 3. On the left: imperial pavilion and loge, on the right: preaching pulpit [25] (p.176,179).

In 1932 upon order from Atatürk, Prof. Thomas Whittemore—founder of the Byzantine Institute of America—had been given the permission to uncover and clean the mosaics [19]. In 1992, a major restoration and consolidation of the mosaics in the dome was started by the Central Laboratory for Restoration and Conservation of Istanbul in collaboration with an international team of experts funded by UNESCO [18]. Not all listed here but many other Hagia Sophia restorations were held in its lifespan. Even today, the scaffoldings, which are used for the repairment of current mosaic damages at the central dome due to the water leakage and humidity, occupy a huge place within the nave of Hagia Sophia.

3. Süleymaniye Mosque

Being the largest of the Ottoman building enterprises of the time, Süleymaniye Mosque and Complex (Figure 4) was sponsored by Kanuni Sultan Süleyman in his ruling and designed by Sinan-the architect laureate of the Ottoman Empire-. The main construction was held between 1550 and 1557, while the whole structure with the fine works of tombs was completed in 1568 [27]. The complex is currently located in the historical island of Istanbul [28]. Süleymaniye Complex has a plan of three adjacent squares: the mosque itself, the courtyard with the last prayers section, and backyard, where the tombs of Sultan and his wife Hürrem are located. The site plan of the complex takes its references from the topography. The complex is composed of 22 different structures with different functions surrounding an interior courtyard [29–31]. Süleymaniye Mosque, which is the central figure of the whole complex, has always been an inspiration source for many fields including architectural aesthetics, structure and construction, acoustics, and material science. The acoustics of the mosque in this respect is an important subject of research.

Figure 4. Süleymaniye complex exterior view [28].

3.1. Major Architectural Features

Süleymaniye Mosque's main structural elements are domes, arches, and flying buttresses. The mosque is covered centrally by a single dome which is supported on two sides by semi domes. The two semi domes align with the direction of the mihrab. Side aisles are sheltered by five smaller domes which complete the upper structure. Pendentives are utilized to smooth the central dome, secondary half dome, and arch connections. The inner plan of the mosque measures approximately 63 by 69 m. The main dome rests on four elephant feet and 32 footings on a circular wheel with a diameter of 26.2 m. The height from the foundation to the impost is 33.7 m. The inner rise of the dome is 14.0 m, and thus the height of the dome from the ground to the keystone is 47.8 m (Figure 5). The middle and corner smaller domes on aisles have diameters of 9.90 and 7.20 m [32,33]. Süleymaniye Mosque's interior acoustical volume is approximately 75,000 m^3.

Figure 5. Interior views of domes and pendentives in Süleymaniye Mosque (photos by the author).

The uniqueness of the building comes from abundant sources of stone supplies delivered from various ruins of ancient cities all over the world [34]. For its original state, other basic interior materials are plaster and paint on brick, tile, ceramic pots, glass, and wood. The interior walls are faced with stone revetments. The ceilings of the pulpits and the royal box, the domical superstructure, and the pendentives are painted (Figure 5). In contrast to lavishly painted domes and pendentives in lower zones the stone revetments left relatively bare. The prominent architectural features of the interior are historical columns, marble panels, porphyry discs, great arches, mihrab, minbar (pulpit) and royal box, stained glass windows, and inscriptions.

Limestone is the main structural stone as well as the facing stone for interior walls and wall footings. Piers carrying the main dome and suspension arches are of cut limestone while the inner faces are painted as of Hereke conglomerate and Proconnesian marble. Columns are of Egyptian porphyry (red sparrow eye). Different bricks were utilized as the core material of domes due to its lightness. Painted brick domes are then decorated with gold foiled pen paintings. The mihrab and the minbar are of carved white marble and have stained glass windows on the sides [35–37]. Lime, fine sand, gypsum, linen, and straw are the basic ingredients of plaster layers and seams. Linen is applied in dome plasters of Süleymaniye Mosque in its original state. Wood in the interior is mostly used for flat ceilings, doors, window frames and furniture. Only the two windows on each side of mihrab have kündekari wooden work shutters. The floor finish of the mosque is carpet with straw backing—which were collected from the finest straws grown in Nile delta—as stated in original documents. Carpets had originally been woven in Egypt and Aydın-Tire [35,36]. Another feature of Süleymaniye Mosque of its original state is the use of Sebu's (clay pots), that are essentially applied to lessen the weight of the dome, but also believed to function as cavity resonators for the control of excessive low frequency sound (Figure 6).

Figure 6. Muqarnas details (photos by the author).

In accounting registers of Süleymaniye Mosque construction (account book number 108), it is declared that for plaster finish of dome (Beray-i sıva-i kubbeha-i cami'-i şerif) 134 kantar of linen was purchased [35]. This is significant information in terms of acoustical assessment of historical plasters that is detailed in following sections. As a final acoustical point, within the mosque muqarnas/stalactites are used as a transition element in column heads, in the skirting of half domes. These elements do not have any structural function and are out of gypsum. Muqarnas helps to enhance the sound diffusion in mostly curvilinear and concave transition planes by fragmenting these surfaces (Figure 6).

3.2. Basic Repairs and Restorations

Since the 16th century, Süleymaniye Mosque has not experienced any major structural or volumetric/spatial changes. Aside from some small extensions, the structure as a whole has kept its integrity and original form. However, there have been a couple of major interior material modifications. The initial restorations were held in 19th century successively in 1840, 1844, 1845, 1847, 1870, and 1873.

In these interventions basically pen-carved paintings and plasters were modified, which ended up being substantially different—in terms of chemical and physical ingredients—interior finish characteristics. In the 1840s and 1880s restorations, held by Italian experts, it is recorded that the clay pots were covered and closed, and the original dome plasters were modified with gypsum plaster [36–38].

Since 1958, the restorations have been held within the control of T.R. Grand Commission of Memorial'. In 1959–1969 restorations basic interventions were the renewal of oil paints on elephant feet, removal of wooden cabinets at some locations, painting of door and window jambs, and renewal of dome and arch paintings. In these restorations, some of the 19th century paintings were removed in order to uncover the original paintings [38,39]. Prior to 2007–2011 restorations, on stone and wooden surfaces different types of material deteriorations were detected. In particular, the pen-paintings on dome facings and on ceilings of exterior side galleries were corroded due to moisture. Damage reported prior to 2007–2011 restorations also included structural damage, the damage due to cement-based plastering or seam fills, and damage due to some other inappropriate use of material [36,38]. The cement-based plasters and application of pen-wall paintings on these plasters are significant in assessing the changes in acoustical field of the Mosque. During 2011 restorations, the samples of original Horasan plasters in the dome were collected. Tests and analyses were held for obtaining cement free plasters that are compatible with original ones, then applied on renewed plastering and pen-wall paintings (nakış). The paintings within the Mosque took their final shape after 1957–1959 and 2007–2011 restorations [38]. Besides, it was also declared that the mouths of in total 256 clay pot (Sebu) voids were opened and cavities were repaired [40].

4. Methodology

4.1. Room Acoustics Measurements

In a broader scope of this study [16], acoustical field tests were held and results specifically in relation to multiple-decay formation were discussed previously for Hagia Sophia [17] and Süleymaniye Mosque [15,17]. This particular paper aims to discuss the findings in relation to different restorations together with acoustical measurements taken in different years by various research teams. As the impulse responses collected by other researchers are not available to this research for post-processing, results of T30 and C80, which are the common parameters available in all field tests, are presented in comparative analysis.

Field tests, within this study's scope, were held in Süleymaniye Mosque on 23 February 2013, in between the hours of 19:30 p.m. and 3:00 a.m. at the main prayer hall, and in Hagia Sophia on 25 August 2014, in between the hours of 09.00 a.m. and 12.00 a.m. at the ground floor, for both in unoccupied conditions [17]. In assessing room acoustic parameters, field measurements were held in accordance with ISO 3382-1,2009 [41]. B&K (Type 4292-L) standard dodecahedron omni-power sound source was used in acoustical signal generation with B&K (Type 2734-A,) power amplifier. The impulse responses at various measurement points were captured by B&K (Type 4190ZC-0032) microphone, incorporated into the hand held analyzer (B&K-Type 2250-A). Sampling frequency of the recorded multi-spectrum impulse was 48 kHz, covering the interval of interest between 100 and 8000 Hz. DIRAC Room Acoustics Software Type 7841 v.4.1 was used for both generating different noise signals and for post-processing of the measured impulse response data for each receiver position. For reliable decay parameter estimations, it was aimed to obtain signal that was at least 45–50 dB higher than the noise in all octaves. Tested signals were E-sweep, MLS, MLS-pink, balloon pop, and wood clap. Up to five pre-averages were applied over multiple measurements, with impulse response length of 21.8 s. E-sweep provided the highest PNR values, thus those samples were mostly utilized in post-processing. As an example, comparison of some sample source signals is provided in Figure 7.

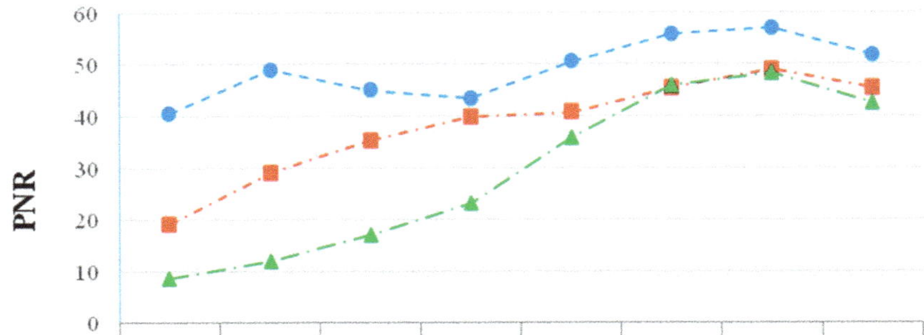

	63Hz	125Hz	250Hz	500Hz	1000Hz	2000Hz	4000Hz	8000Hz
e-sweep	41	49	45	43	50	56	57	52
baloon pop	19	29	35	40	41	45	49	45
wood clap	9	12	17	23	36	46	48	42

Figure 7. Hagia Sophia PNR measurements in 1/1 octave bands in Hz; comparison of e-sweep, balloon pop, and wood clap source signals.

The measurement positions of field tests are represented in Figure 8, in which the sizes of each structure in relation to one another can also be compared on the plan scheme. Major source locations in Süleymaniye Mosque are one in front of mihrab (S1) and the other at müezzin mahfili (S3). The positioning of sound source is important to assess the acoustics of Süleymaniye Mosque for its traditional use. For acoustical coupling or multi-decay rate investigation, additional source locations including one underneath the main dome (S4) and one underneath side corner dome (S2) were tested [42]. Eight receiver locations (R1–R8) were coupled with four source locations (S1–S4) providing measured source–receiver configurations. In order to observe multi-slope decay formation for different locations and to discuss the effects of spatial variations—such as main central space versus space underneath side galleries—in Hagia Sophia, three source (S1–S3) and six receiver (R1–R6) positions were tested in numerous configurations (Figure 8).

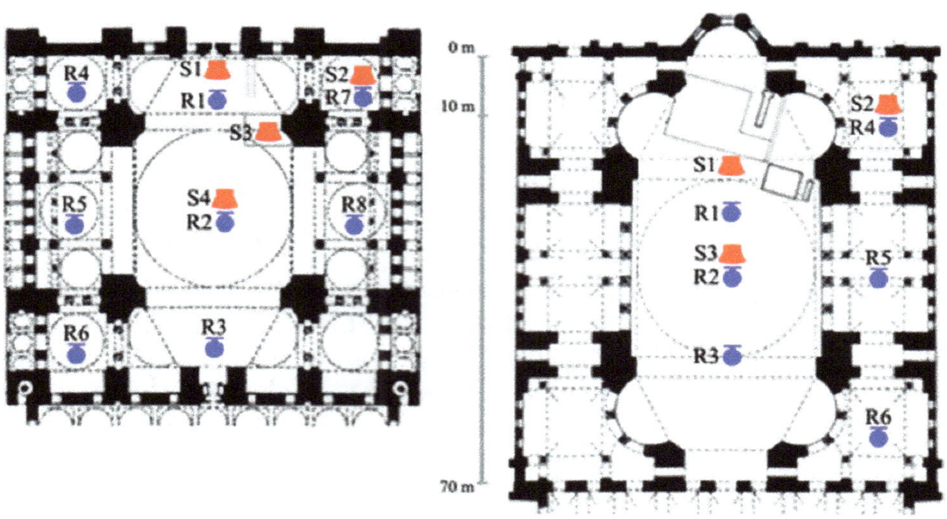

Figure 8. Sound source (**in red**) and receiver (**in blue**) locations used in field tests of Süleymaniye Mosque—METU-2013 tests (**on the left**) and Hagia Sophia—METU-2014 tests (**on the right**).

Other measurements following different restoration periods are also summarized in this section. The first set of measurement in Süleymaniye Mosque was held after 1959–1969 restorations by Gazi University [43] in 1988 (GU-1988); the measurements were taken in 1/3 octaves for the frequency spectrum in between 100 and 8 kHz. The following field tests were taken by Middle East Technical University (METU) in 1996 (METU-1996); broadband noise signal was emitted at two source locations including one in front of mihrab and one over müezzin's mahfili. Impulse responses were collected for six receiver locations at the prayer's area [44]. The final group of measurements after 1959–1969 restorations, as can be found in the academic literature, were held in 2000 within the scope of Conservation of the Acoustical Heritage by the Revival and Identification of the Sinan's Mosques CAHRISMA project [3]. Ferrara University (UNIFE-2000) and Denmark Technical University (DTU-2000) held two different measurements within that scope. Sweep signal was used as a source signal, and impulse responses were collected for different configurations of three source and six receiver locations. UNIFE used test signal possessing of the 80–18,000 Hz spectrum range and DTU used test signal possessing of the 35–11,500 Hz spectrum range. DTU recorded impulses for 10 s of capture length [3]. The first measurements after 1959–1969 restorations were held within the context of this thesis research [16] and abbreviated as METU-2013 in following discussion for the ease of comparisons.

Following the major alterations due to the functional changes, the basic restoration and consolidation of the mosaics in the dome of Hagia Sophia started in 1992, as previously summarized. There is no available measured acoustic data in the literature before that period. All of the available field data of Hagia Sophia were collected after 2000. Previous to Hagia Sophia measurements held in the scope of this research [16] in 2014 (METU-2014), two other set of data were recorded. Ferrara University (UNIFE-2000) and Denmark Technical University (DTU-2000) held two different measurements in the context of the CAHRISMA [3] project in 2000, similar to those explained for the Süleymaniye Mosque.

4.2. Acoustical Simulations

The acoustical simulations of historical structures are carried out in a broader frame of this research by two main methods that are ray-tracing simulations and diffusion equation model (DEM) analysis in a finite element scheme. For applying different estimation methods, the acoustical models of both structures were generated, utilizing the latest building survey registers (röleve) for Hagia Sophia provided by the Turkish Ministry of Culture and Tourism—General Directorate of Turkish Cultural Heritage—and for Süleymaniye Mosque provided by the General Directorate of Pious Foundations. For highlighting the different volumetric relations and major form of the encapsulated interior sound fields, in Figure 9 the ray-tracing models of both structures are presented.

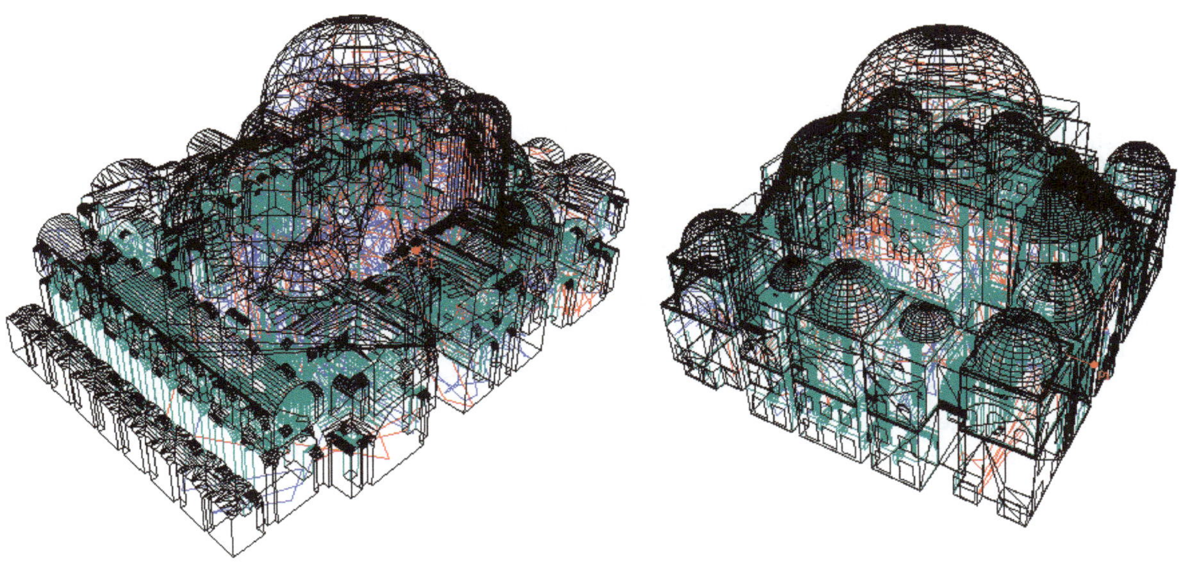

Figure 9. Ray tracing models of Hagia Sophia (on the left) and Süleymaniye Mosque (on the right).

Ray tracing simulations were carried out by ODEON Room Acoustics Software v.14.00 (Odeon A/S, Lyngby, Denmark) [45]. For field tests there were limited time within permitted dates, due to the historical and touristic significance of both structures. For that reason, before field tests, ray-tracing simulations were held for obtaining a general overview regarding to multi-slope decay formation for most probable positions. After field tests, the models were tuned and further utilized for experimentation of different materials of the original state and fully-occupied conditions. The DEM method was utilized for in depth-analysis and understanding of the probable reasons of multi-slope decay formation in such multi-domed monumental structures by energy flow vector, energy flow decay, and spatial sound energy distribution analysis and results are presented in previous papers [15,46,47].

In this study, main objective was to discuss the acoustics timeline of the case structures. For briefness, only the Süleymaniye Mosque's ray-tracing simulation results are presented for different materials and occupancy states to represent its original condition. The source and receiver positions applied in simulations are identical to those used in field tests. Yet again, similar to the field tests, omni-power sound sources are utilized in simulations. Materials are defined with associated sound absorption and scattering coefficients of the interior surfaces by tuning the model with the present field test results (Table 1). Later, the full occupancy of the Mosque during a Friday's sermon is estimated and compared to its unoccupied condition.

Table 1. Sound absorption coefficient and scattering data of applied materials in acoustical simulations.

Materials/Locations	Sound Absorption Coefficients over 1/1 Octave Bands						Scattering Coefficient
	125 Hz	250 Hz	500 Hz	1 kHz	2 kHz	4 kHz	
Ornamented stone piers, arches columns, and column heads	0.05	0.05	0.06	0.09	0.11	0.11	0.20
Current carpet [3]	0.02	0.12	0.25	0.38	0.55	0.58	0.10
Carpet with prayers [48]	0.18	0.22	0.41	0.58	0.69	0.72	0.70
Historical plasters tested in 30% humidity [49]	0.10	0.17	0.23	0.29	0.32	0.32	0.10
Current plasters on brick	0.13	0.09	0.07	0.05	0.03	0.04	0.10
Large pane of glass	0.18	0.06	0.04	0.03	0.02	0.02	0.05
Marble slabs	0.01	0.01	0.01	0.01	0.02	0.02	0.05
Wooden doors and furniture	0.10	0.07	0.05	0.04	0.04	0.04	0.10
Gypsum moldings for muqarnas and similar decorations	0.29	0.10	0.05	0.04	0.07	0.09	0.50
Copper decorative elements	0.12	0.08	0.02	0.01	0.01	0.01	0.50

The sound absorption coefficient data for standard materials as of single layer gypsum, marble on concrete, stone cladding, large pane or multi-layer glass surfaces are well defined in various sources that are more or less identical and mostly reflective as well. The critical issue in this case is assignment of comparatively absorptive materials to those listed in the former. The absorptive materials can significantly change the sound energy decay rates within such immense volumes depending upon their applied surface areas. For the optimization of sound absorption coefficients of current carpet two different sources were visited. Among those, one sample was tested under the scope of the CAHRISMA project [3] and the other is a previously tested carpet sample [50], which is widely used in Turkish Mosques. The sound absorption coefficient data of people on the prayer area applied in simulations—reflecting occupied state of the mosque—are taken from previous laboratory tests of mosque congregations [48]. The attained scattering coefficients on different surfaces are based on the amount of surface irregularities, in small or large scale. In the calibration process of acoustical models, field test data were utilized and T30 values were compared to simulations for different receiver locations. Just noticeable difference (JND) for T30 was kept under 5%, when simulated results were tuned to measured ones. Basically, the unknown sound absorption performances of plasters applied today -and modified in previous restorations-, has given direction to the tuning.

The original acoustical features of the Mosque were examined based on an assumption that the interiors, before restorations, were lime-based plasters with pozzolanic additives which are compatible with the historical ones belonging to the same era. Specifically, the sound absorption coefficients of interior plasters in current state were estimated to be in the range of 0.05 to 0.07 for the mid frequency range, while the sound absorption coefficients of the historical lime-based plasters were previously documented to be in between 0.20 to 0.30 in the mid frequency range [49]. Such a simulation analysis is expected to offer an insight into the probable acoustical conditions in respect to historical lime-based plasters versus today's repair plasters, as discussed in Section 5.3.

5. Results and Discussion

5.1. Acoustics of Hagia Sophia after the 1990s

The data presented for METU-2014 measurements were initially collected for different source signals including wood claps and popping of a balloon. Hagia Sophia was too large to be excited with either a balloon pop or wood clap, and measurements did not provide the sufficient signal (or peak) to noise ratio in the overall frequency spectrum, while neither of those source signals were perfectly omni-directional (Figure 7). Among MLS and e-sweep excitations, e-sweep provided the highest peak-to-noise (PNR) values ranging in between 45 to 60 in the overall spectrum, which were found more reliable to be used especially in multi-slope investigations. For that reason, in Figures 10 and 11 only the e-sweep generated impulse response results are presented for METU-2014 tests.

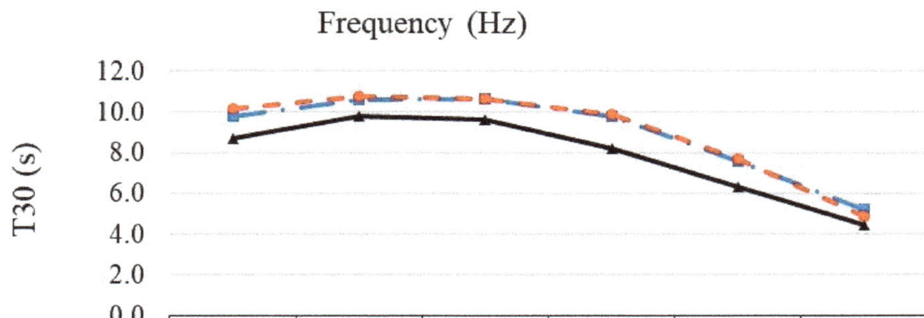

	125	250	500	1000	2000	4000
Hagia Sophia DTU-2000	9.8	10.6	10.7	9.8	7.6	5.2
Hagia Sophia UNIFE-2000	10.2	10.8	10.7	9.9	7.7	4.9
Hagia Sophia METU-2014	8.7	9.8	9.6	8.2	6.3	4.5

Figure 10. Comparison of T30 results over collected impulse responses in Hagia Sophia by Denmark Technical University (DTU)-2000, Ferrara University (UNIFE)-2000, and METU-2014 in field tests, in 1/1 octave bands from 125 to 4000 Hz.

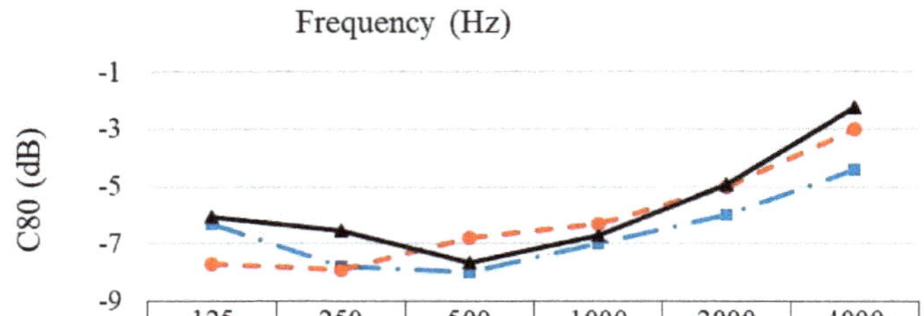

	125	250	500	1000	2000	4000
Hagia Sophia DTU-2000	-6.3	-7.8	-8.0	-7.0	-6.0	-4.4
Hagia Sophia UNIFE-2000	-7.7	-7.9	-6.8	-6.3	-5.0	-3.0
Hagia Sophia METU-2014	-6.1	-6.6	-7.7	-6.7	-4.9	-2.2

Figure 11. Comparison of C80 results over collected impulse responses in Hagia Sophia by DTU-2000, UNIFE-2000, and METU-2014 in field tests, in 1/1 octave bands from 125 to 4000 Hz.

In Figure 10, DTU and UNIFE indicate measurements taken in 2000 under the supervision of the CAHRISMA project [3]. Hagia Sophia field tests held in 2000, highlight that T30 values in overall are higher than field test results of METU-2014 measurements. The drop of T30 values in METU-2014 measurements below 250 Hz in Hagia Sophia could be the result of the barrier wall separating one side aisle from main volume over the whole longitudinal length, which was present within the space during field tests due to ongoing restorations. Out of single layer gypsum board connected to back studs/profiles in certain intervals, it is probable that the system behaved as an effective low frequency membrane absorber. As another outcome, the mid to high frequency drop in T30 values of METU-2014 measurements in comparison to pre-2000 restorations are one probable result of present scaffolding in on-going restorations that were re-built after 2012. Thus, the presence of additional architectural elements/constructions within the space during METU-2014 measurements resulted in a drop of 1 to 2 s, in the overall frequency spectrum, of reverberation times. On the other hand, the trend of the sound decay over the frequency spectrum is similar for all Hagia Sophia field tests.

C80 is a parameter that is very much dependent on source-receiver positions. Therefore, the deviations in average values of different field test data is expected. It is well known that, subjectively, clarity is the hearing or sensation of individual pieces within music or speech. In objective terms, clarity is the ratio of early to late energy and it is a competing or reverse parameter to the reverberation. Thus, the trend of C80 values for overall field tests is in the opposite manner of T30 results.

It is known that there were no major architectural or form modifications between 2000 and 2014 in Hagia Sophia restorations. Field recorded data is not available for before 1992 restorations, when the consolidation of the mosaics in the dome started. Thus, it is not possible to compare the basic alterations of Hagia Sophia, especially in regard to changes in its function, and to discuss on the success of cleaning of the mosaics, and their acoustical results through field tests.

Overall, in the current state of Hagia Sophia, the T30 results obtained in recent years by different research groups indicate very high reverberation times within the mega-structure. Mid and low frequency T30 averages are around 9 s and high frequency averages are around 5 s. This superb and extraordinary aural environment is unique and has the potential to provide acoustical field conditions in relation to coupled spaces for specific locations. Non-exponential sound energy decay is one outcome of Hagia Sophia's architectonic language, a brief discussion of which is presented in Section 5.4.

5.2. Acoustics of Süleymaniye Mosque for before and after 2007–2011 Restorations

The architectural and material modifications on Süleymaniye Mosque in different time periods are briefed in Section 3.2. With an aim of assessing the acoustical effects of previous restoration works, the field measurements taken in 2013 (METU-2013) are compared with previous field test data. The common acoustical parameter measured and assessed in all field tests for Süleymaniye

Mosque is the reverberation time (T30). Thus, measured T30 values over octave bands are the basis of comparisons (Figure 12). Some previous literature recommends reverberation times of 4-4,6 s for mid frequencies as an optimum range for mosques with similar volume [3]. According to Figure 12, all of the field tests indicate very long reverberation times, higher than recommended ranges for the mosques for unoccupied condition. Especially, 125 Hz is very problematic considering the intelligibility of speech; which will be even worse when the electro-acoustic system is on, as in today's applications. The T30 values over 2000 Hz are comparatively more suitable for the Mosque's function in comparison to mid to low frequencies. The main reason of this drop in higher bands is the absorption of air of such an immense volume.

In METU-2013 measurements below 100 Hz, reverberation times reaching 20 s are observed. It is also noted that there are some measurement spots reaching 20 s at 125 Hz in GU-1988 measurements that are excluded from the average as considered to be the highest deviation [43]. The rise in 250 Hz of METU-2013 measurements are due to the very high values of T30 at 200 Hz in the 1/3 octave band. One reason for the deviations among different measurements might also be due to the averaging of 1/3 octaves or only plotting 1/1 octaves. In 1996 and 2000 field tests, T30 values at 125 Hz were measured in between 9 and 11 s, which are much lower than 1988 and 2013 measurements. This result is most probably caused by the insufficient capture/impulse response lengths of the other field tests. For instance, the DTU-2000 measurement set-up uses 10 s of impulse response length, in the real case, which could easily have missed the chance of estimating a longer reverberation time. On the contrary, both of the first measurements taken by GU-1988 with analogue equipment and the final measurements taken by METU-2013 with digital equipment and with an impulse length of 22 s, indicate 15–17 s of T30 values at 125 Hz. Thus, it could be stated that 2007–2011 renovations have not significantly affected the T30 values at 125 Hz, which are still very high and above the acceptable limits for mosque function.

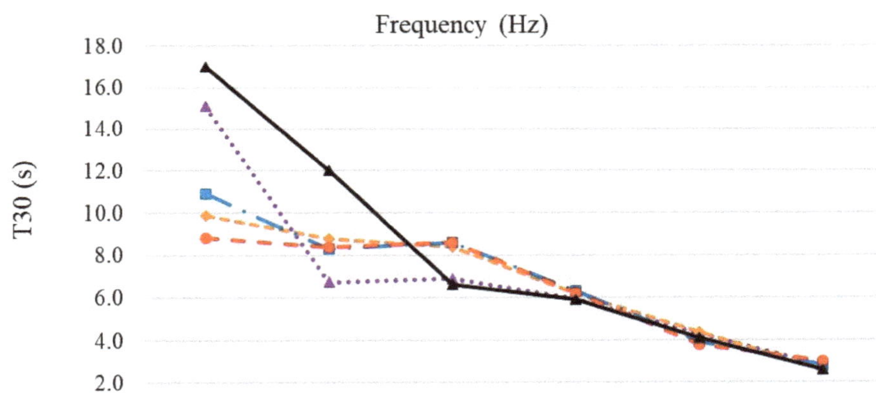

	125	250	500	1000	2000	4000
Süleymaniye GU-1988	15.1	6.7	6.9	5.9	4.3	2.8
Süleymaniye METU-1996	9.9	8.8	8.4	6.2	4.4	2.6
Süleymaniye DTU-2000	10.9	8.3	8.6	6.3	3.9	2.8
Süleymaniye UNIFE-2000	8.8	8.4	8.6	6.2	3.8	3.0
Süleymaniye METU-2013	17.0	12.0	6.6	5.9	4.1	2.6

Figure 12. Comparison of T30 results over collected impulse responses in Süleymaniye Mosque by Gazi University (GU)-1988, METU-1996, DTU-2000, UNIFE-2000, METU-2013, and AU-2013 in field tests, in 1/1 octave bands from 125 to 4000 Hz.

According to METU-2013 measurements, reverberation times for octave bands in between 500 and 4000 Hz have been lowered by different ratios compared to previous years' test results. The major change is the 2 s of drop at 500 Hz. When T30 values at 250 and 500 Hz are compared for 1988 measurements and 1996–2000 measurements, an increase of 1.5 to 2 s is observed in the latter. In order to be able to explain this increase, the probable undocumented minor-restorations or material changes in between the years of 1988 and 1996 should be known. Another point is that the measurement positions vary in different field tests, so only average values could be compared.

The comparative analysis of field test results indicate that the 2007–2011 restorations resulted in a slightly positive decrease in reverberation time at 500 Hz, whereas overall, the values are still higher than the recommended ranges. The attempts for removing cement-based plasters, and application of plasters that are compatible with historical lime-based plasters, as so declared, are constructive but not yet efficient acoustical interventions; especially in control of low frequency sound content. The effects of so stated intrusion on Sebu (clay pot) voids—opening the mouths and repair of inside cracks—are still vague in acoustical terms. In order to be able to scientifically comment on acoustical effects of Sebu repairs, the present geometrical and dimensional properties of these elements should be identified, which can only be possible by their systematic inspection on site.

Reverberation time (RT, T30, T20) is one of the key parameters affecting intelligibility of speech. Another important acoustical parameter that affects intelligibility is the level of background noise. High background noise can mask the speech subjectively and can drastically lower the speech transmission index (STI) values. Background noise is a frequency dependent parameter, so its spectrum differs for different noise sources. METU-2013 measurements were taken at night, when the Mosque is closed for visit or prayer and intriguing environmental/traffic noise was at minimum. The measured background noise level was LAeq 39.3 dBA, which is higher than the recommended—NC5 (35 dBA)—upper limit of background noise levels for religious spaces [51]. The high background noise levels detected are caused by the cooling fan units of electrical panels located within the Mosque at the first level right across the mihrab wall. It is evident that one of the basic reasons behind the claims for acoustical or intelligibility problems within Süleymaniye Mosque after the final restorations is the presence of fan noise, which can easily distract speech related activities such as hutbe's by imam.

5.3. Interpretation of Süleymaniye Mosque's Acoustics for Its Original State

Süleymaniye Mosque has gone through various restorations in years, and the acoustical conditions within the Mosque after specific restorations are compared and assessed in previous sections by the data out of field tests. However, none of the field test results reflect the acoustical conditions of the Mosque in its original state. On the other hand, the acoustical expectations from the Mosque in the years of its construction and expectations from a mosque in these contemporary years are quite different. All of the historical mosques today incorporate the use of sound-reinforcement systems. For historical mosques, which were designed for natural sound in their time, it is evident that problems would occur in application of electro-acoustic systems unless necessary precautions are taken.

One of the basic architectural parameters in room acoustics affecting natural speech/sound are the volume of the main space and its geometry. It is known that the dimensions and basic geometrical features of Süleymaniye Mosque have not been altered to date. The dominating form of the Mosque is the central dome which is supported by two semi-side domes. Acoustical focusing effects of the dome can be prevented to some extent, in the case that the lower end of the diameter/circumference of the dome section is located much higher than the receiver/prayer ear height [6]. In Süleymaniye Mosque, even the focusing zone of the biggest central dome is located 20 m above the prayer plane/floor. By the help of this, the first order reflections might have been minimized. However, the dome form causes sound foci at its focal points and negatively affects the sound scattering.

Sinan the Architect in his mosques applied the Sebu (clay pot) technique, which may have enabled acoustical asymmetry within the dome by scattering the sound and enlarging the dome reflection zone. By that, much even distribution of sound within the prayer zone could be provided. Sebu forms are similar to that of amphoras, which have a short neck and a backing volume. The Sebu technique in the structure enables to lessen the weight of the dome, while in acoustics they can easily function as Helmholtz resonators. These elements can scatter sound. Especially at low frequencies (63–250 Hz) they can behave as narrow band volume absorbers. In the account books (D.88. Yp. 19/a) of Süleymaniye Mosque's construction work, it is recorded that each for 2 akçe's (the chief monetary unit of the Ottoman Empire, a silver coin) total of 255 "Sebu" were purchased [35]. Several investigations state that 64 of these pots are located on a circular disk at the central dome [43]. On the other hand, after 2007–2011 restorations it was declared that 256 pots open towards the interior space were detected [40]. According to these declarations, the numerous applications of these clay pots (Sebu's) with various sizes could widen the frequency bandwidth that they are effective in. Thus, it could be predicted that such an application in the original state of the Mosque, to some extent, had healed the excessive low frequency sound content. Together with Sebu voids, the fragmentation of parallel surfaces in both sections and plan schemes of the Mosque by architectural elements such as mahfil's, niche's, and surface treatments such as muqarnas, kündekari, and glazed ceramics have provided sound scattering in a wide frequency spectrum, so that an even distribution of sound throughout the prayer zone is obtained.

In spaces with excessive volume as in Süleymaniye Mosque, the expected long reverberation times have to be controlled by increasing the sound absorptive surface area. For an even distribution of reverberation over frequency, the sound absorption of materials in octave bands should be well balanced. Carpet is an absorptive material for mid and high frequency range only, unless it has at least 5–10 cm height platform, or air space, underneath. In the records of construction documents of Süleymaniye Mosque such a platform is not mentioned [35]. This means that the original carpet's sound absorption performance would be similar to that of today's carpet for low frequencies, and not much of a difference is expected from high frequencies in reference to some other research on carpet effects' in mosques [50]. One other significant piece of information is that in its original state, straw is laid underneath carpet of Süleymaniye Mosque [35], which would have provided a slight improvement in absorption of mid frequency sound content. Carpet is still significant in terms of providing positive absorption area for intelligibility (of consonants). Apart from those, in a prayer scenario the presence of people and their compactness improve sound absorption as well as scattering within the Mosque, as observed in full-occupancy simulation results (Figure 13).

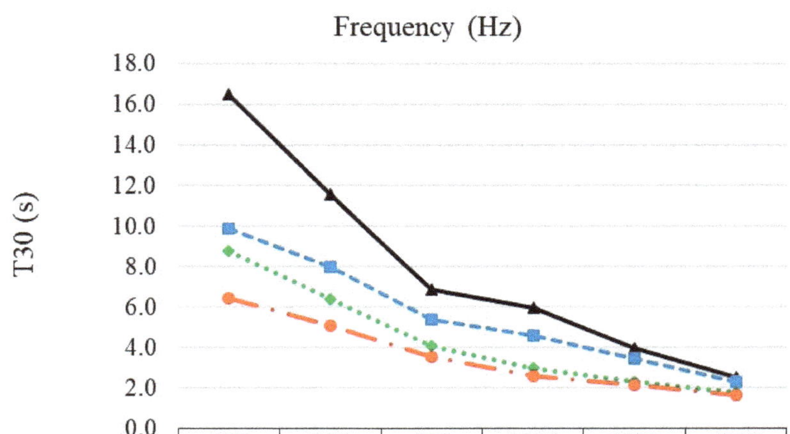

	125	250	500	1000	2000	4000
Current plasters (U)	16.5	11.6	6.9	6.0	4.0	2.5
Historical platers (U)	8.8	6.4	4.1	3.0	2.3	1.8
Current plasters (O)	9.9	8.0	5.4	4.6	3.5	2.3
Historical platers (O)	6.5	5.1	3.6	2.6	2.2	1.7

Figure 13. Simulated T30 results of Süleymaniye Mosque Hz in 1/1 octave bands from 125 to 4000 Hz for its current state and trial with historical plasters for unoccupied (**U**) and unoccupied (**O**) conditions.

The absorptive carpet floor surface is still not sufficient to tolerate the long reverberations of this immense volume. On the other side, in Süleymaniye Mosque shiny and tight stone walls, column, and elephant feet surfaces compose the reflective areas. The sound absorption coefficients of applied stones within Süleymaniye Mosque can be predicted for the current and original states in the range of 0.05–0.10. The unknown is the sound absorption coefficient data of original/historical plasters with or without paintings that are renewed in several restorations. These plasters are applied at dome, arch, and mostly upper wall surfaces, and compose a very large surface area of approximately 19,000 m^2. The influence of these plasters on reverberation times would be noteworthy [52].

In the account books (D.108) of Süleymaniye Mosque's construction work [35], it is recorded that 134 scale linen were purchased to be used in plastering of dome. In some other account books, it is also recorded that 524 kantars (Ottoman weight unit corresponding to 56.5 g) of linen were purchased for plastering of the structure in general. Use of linen within the composition of interior surface plasters in Süleymaniye Mosque is an important acoustical element. Linen is commonly used to improve the tensile strength of mortar and/or plaster materials; on the other hand, in acoustical terms it can also increase the sound absorption performance of the plaster at low frequency range. Linen can also augment the mechanical strength of plasters [53,54]. It can be predicted that by the use of linen Sinan the Architect increased sound absorption area serviced by plastered surfaces, which then provided a more controlled reverberation within the mosque in both low and mid frequencies.

Another significant point is that the previous literature highlights the acoustical significance of porous and soft horasan mortar with linen and hemp fiber ingredients that are applied on dome and wall surfaces in Sinan's mosques [43]. The replacement of historical mortar and plaster, which are good absorbers at low to mid frequency range, with tight and stiff cement-based plasters in specific restorations are accused of causing excessive reverberation times observed especially at low to mid frequencies at most of the historical mosques. The use of natural fibers as an ingredient in historical plasters, in fact, is considered to be significant information in terms of acoustical performance of historical plasters, and needs to be investigated with further studies.

In another study, the historical Turkish Baths of the same era were investigated in terms of the acoustical performances of plasters applied in their interior wall surfaces. The sound absorption coefficients of historical multi-layered pozzolanic lime plasters were tested by the impedance tube method [49]. According to that, historical Turkish Bath plasters have sound absorption coefficients of (α) 0.25–0.30 in the mid frequency range and (α) of 0.10–0.17 in the low frequency range, which are higher than the current plasters estimated to have (α) of 0.03–0.09 in the mid frequency range. In Süleymaniye Mosque the current plastered surfaces cover a significant area that should have a drastic impact on effective sound absorption within the Mosque.

For testing the above argument, the acoustical model of Süleymaniye Mosque (Figure 9) was utilized in a simulation study that compares today's plasters, after multiple restorations, with historical multi-layered pozzolanic lime-based plasters. The procedure of model tuning for current state with field tests are presented in Section 4.2. The results are initially attained for unoccupied condition as in field tests, later the results are obtained for occupied condition. The simulations are done for both current state of the Mosque and for modified acoustical model with historical plaster data [49]. Occupied condition represents the full-occupancy as in Friday's Sermon when clarity of speech is more important.

The plot of the simulations tuned by the acoustical field tests in Süleymaniye Mosque's current condition is quoted as "Current plasters (U)" in Figure 13. The measurement results used in this tuning are given in Figure 12 indicated by "Süleymaniye METU-2013." According to the simulation results for unoccupied condition, the replacement of cement-based repair plasters with historical ones ended up with a drop of 2.8 s in T30 value at 500 Hz and almost a drop of 8 s at 125 Hz (Figure 13). T30 values for the fully-occupied mosque in its current state is 6.6 s lower than that of its unoccupied condition for 125 Hz, and 3.6 s lower than unoccupied condition for 250 Hz. However, the low frequency reverberation times are very high and distracting for intelligibility in its current state.

This situation causes even worse acoustical defects, in fact, when the electro-acoustic system is on. In the fully-occupied state of the Mosque with the application of representative historical plasters, the reverberation times are lowered by 2 to 3 s in mid and low frequencies in comparison to the same occupancy state with current plasters. The reverberation time values in between 5 and 6.5 s in the low octaves and 2.6–3.6 s in the mid frequency range indicate that with this trial of historical plasters, Süleymaniye Mosque's interior sound field is much suitable and closer to the recommend limits for mosques with similar volume [3,55]. Although there were some recent attempts in removing cement-based plasters (2007–2011 restorations), the 2013 field measurements show that the problems are not thoroughly solved. Results indicate that if the historical plasters could have survived till now or if the Mosque underwent repairs with the plasters totally compatible with the historical ones, the acoustical conditions would be much more suitable for the function of a mosque today, as well.

As a final note, being located in a courtyard of a big complex (külliye) surrounded by walls together with a thick masonry wall construction—widths ranging in between 65 and 200 cm—and strong exterior shell of domes [29], Süleymaniye Mosque was very well isolated from any environmental noise that might be present in its time. Vehicle traffic, industrial noise, or mechanical equipment noise, as in today's, did not exist in the past. Thus, the intelligibility of speech within the Mosque, had been at least masked, or distorted, in its original state.

5.4. Interior Sound Fields of Hagia Sophia and Süleymaniye Mosque in Relation to Major Architectural Parameters

The aim of this section is initially to assess the dissimilar sound field of Hagia Sophia in comparison to Süleymaniye Mosque, due to volumetric, geometric, and material input. The joint interpretation of results over data obtained at field measurements in 2013 (or METU-2013) for Süleymaniye Mosque and in 2014 (or METU-2014) for Hagia Sophia are presented. The results are provided for T30 in Figure 14 and for C80 in Figure 15, over octave bands from 125 to 4000 Hz and with average, maximum, and minimum values for overall source-receiver positions (Figure 8). As can be observed from Figure 14, the deviation in between measured receiver positions are mostly less than 1 s, and reach up to 2 s in a few octave bands. It is evident that when different receiver positions are compared over minimum and maximum values, the difference is noticeable due to the immense volume. However, the discussion on subjective response evaluation should not rely on single decay metrics but rather be considered under the view of multi-slope sound energy decay formation.

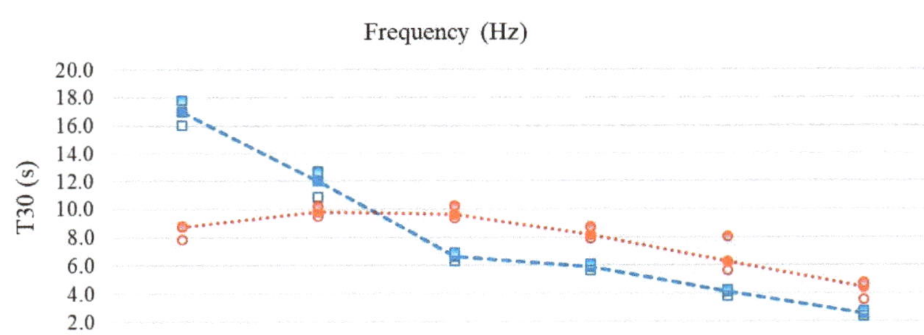

	125	250	500	1000	2000	4000
SM avr	17.0	12.0	6.6	5.9	4.1	2.6
SM max	17.8	12.7	6.9	6.1	4.3	2.8
SM min	16.1	10.9	6.4	5.6	3.8	2.4
HS avr	8.7	9.8	9.6	8.2	6.3	4.5
HS max	8.8	10.3	10.2	8.7	8.0	4.7
HS min	7.9	9.5	9.4	7.9	5.7	3.6

Figure 14. Comparison of T30 results over collected impulse responses in field tests of Süleymaniye Mosque by METU-2013 and in Hagia Sophia by METU-2014, in 1/1 octave bands from 125 to 4000 Hz.

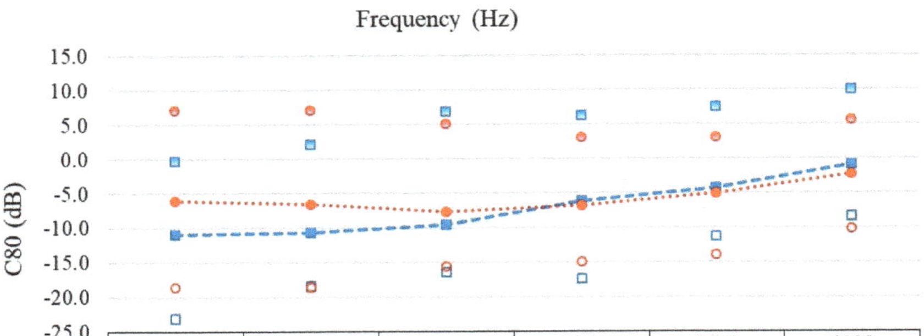

	125	250	500	1000	2000	4000
SM avr	-10.9	-10.6	-9.5	-6.1	-4.2	-0.9
SM max	-0.3	2.2	6.9	6.3	7.5	10.1
SM min	-23.1	-18.4	-16.4	-17.5	-11.2	-8.3
HS avr	-6.1	-6.6	-7.7	-6.7	-4.9	-2.2
HS max	7.2	7.2	5.1	3.1	3.2	5.7
HS min	-18.5	-18.5	-15.5	-14.9	-13.9	-10.1

Figure 15. Comparison of C80 results over collected impulse responses in field tests of Süleymaniye Mosque by METU-2013 and in Hagia Sophia by METU-2014, in 1/1 octave bands from 125 to 4000 Hz.

Due to volumetric differences, dimensional proportions and variable sound absorption characteristics of interior finish materials Süleymaniye Mosque have expectedly lower reverberation times in the frequency range over 250 Hz in comparison to Hagia Sophia. This result was expected considering the doubling of acoustical volumes, which is roughly around 75,000 m^3 for Süleymaniye Mosque, whereas roughly around 150,000 m^3 for Hagia Sophia. The plan schemes of two structures are compared in Figure 8, indicating the considerable differences in size, even in plan layout. Another significant aspect in relation to reverberation, aside from the interior volumes, is the interior finish materials. Marble floor surface in Hagia Sophia in comparison to the carpet floor finish in Süleymaniye Mosque is also very much effective in the increase of reverberation times, specifically over mid to high frequencies. Whereas, the carpet is not that effective under 250 Hz, and this time the reverberation is higher in Süleymaniye Mosque.

In Süleymaniye Mosque, T30 values at 125 Hz are observed to rise to very high levels, which are even much longer than T30 values obtained at 125 Hz in Hagia Sophia. That indicates a probable occurrence of resonances within the main space of Süleymaniye Mosque within that particular frequency band interval. Overall, the distribution of reverberation times over source-receiver positions are smooth and there are no big differences between the average and minimum-maximum values. This indicates a reasonably even distribution within both structures with such great volumes. The diffuseness is due to the surface fragmentations provided by efficient scattering of sound by means of arches, muqarnases or stalactites and mahfili's. It should also be noted that high reverberance inside Hagia Sophia provides suitable repertoire and singing ability in terms of liturgical music, while in Süleymaniye Mosque music is not practiced that much, but only ritual sermons. Thus, the interior sound field of Süleymaniye as a mosque and Hagia Sophia as a church are practically much more suitable for their authentic function of use.

As shown in Figure 15, the average trend of C80 values over frequency spectrum obtained for Hagia Sophia and Süleymaniye Mosque are similar, while values for both structures are much lower than those typically obtained in other venues such as concert halls. It is not surprising that such high reverberation times in overall frequency spectrum would cause low values of clarity. Another reason for very low clarity is the major geometric attributes of both mega-structures. The great sizes and heights of central domes and the lack of close-by surfaces to the receiver zone—such as overhead reflectors in concert halls—cause late reflections to dominate early reflections. The C80 parameter is very much dependent on receiver position, so the deviations in the data that can be observed from average, minimum, and maximum values obtained in both structures are as expected. The C80 information is provided not essentially for comparison but mostly for archiving the measured data obtained within the case historical structures over years.

Another significant finding is multi-slope decay formation as an outcome of the measurements in case structures through a detailed analysis by Bayesian decay parameter estimations. The number and level of decay rates/slopes and their locations were discussed in a previous study for both structures [17]. This discussion is not included in detail within the content of this paper, as the main objective is to discuss the material and geometric changes over years and analyze their possible acoustical effects from their past to present.

The available and comparable data in previous field tests are mostly for T30, and in a few for C80. The multi-slope decay analysis, on the other hand, do not solely rely on a 20 or 30 dB drop of energy levels, but rather investigates the sound energy decay curve for all portions within Bayesian probabilistic inference framework [15]. In regard to the non-exponential energy decay formation, it can be stated that the overall number of decay slopes is greater in Süleymaniye Mosque than in Hagia Sophia, especially in the main prayer zone underneath central domes. The absorption introduced by carpet floor finish in Süleymaniye Mosque versus a reflective upper structure provides un-diffused sound field. Consequently, the sound energy density fragmentation in Süleymaniye Mosque is much more obvious in comparison to Hagia Sophia, due to the relatively diffused sound field of Hagia Sophia.

On the other hand, the span of arches which separate the main prayer hall in Süleymaniye Mosque and the arches connecting the side galleries to the central nave in Hagia Sophia differ in dimension (Figure 16). Due to the larger proportions of arches in Süleymaniye Mosque, the side aisles behind those major arches supporting the central dome are not restricted as much as arcades separating side aisles of Hagia Sophia. While, the smaller arches of Hagia Sophia create much more defined coupling apertures and occasionally cause triple slope decays underneath side aisles. To summarize, interior finish material distribution and/or geometric attributes of the two mega-structures provide sound energy flows inside, among sub-volumes. All these geometric and material attributes cause sound energy concentration and fragmentation within the spaces that augment the non-exponential or multiple sound energy decay formation [17].

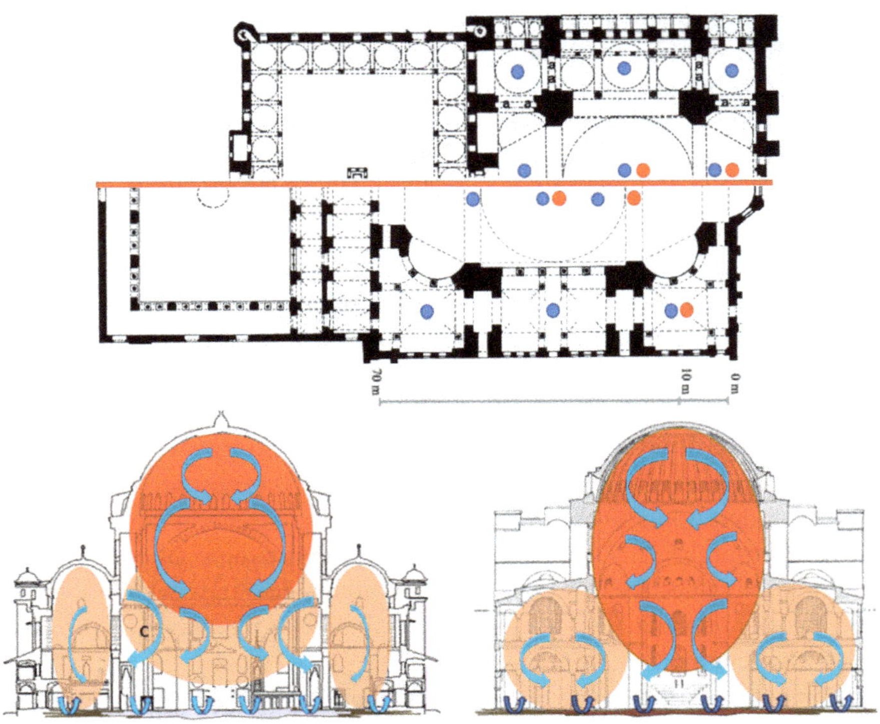

Figure 16. Above: partial plans of Süleymaniye Mosque (**above**) and Hagia Sophia (**below**); below: conceptual section view of sound energy flows of Süleymaniye Mosque (**on the left**) and Hagia Sophia (**on the right**).

6. Conclusions

Within the context of this study, sound fields of two sacred multi-domed historical structures, namely Hagia Sophia and Süleymaniye Mosque, are analyzed with a focus on their architectonic and material attributes, and applied alterations in basic restoration works. The historical significance of the monuments attracted the attention of different research groups, ending up with extensive data in the literature both in acoustics and also in many other fields such as structure, material science, restoration, or art history. A comprehensive study on the acoustics of these sacred spaces is necessitated, by a comparative analysis over acoustical field tests held in different years and over wide-ranging literature on their material and architectural features. The methodology included the field tests carried both within the scope of this research as well as the previously published test results by other researchers. Acoustical simulations were utilized for comparison of unoccupied versus occupied conditions and also for discussion on original materials.

The major difference in between the two monuments is that Hagia Sophia was originally designed to function as a church, while Süleymaniye is a mosque. The basillican layout of Hagia Sophia in comparison to the square and central plan of Süleymaniye Mosque resulted in volumetric differences. Both structures have a highly reflective upper structure, with stone and plaster claddings, while also have many sound scattering elements as of muqarnases or stalactites and space dividing arches. Another major difference in effect of acoustics is that Süleymaniye Mosque has a carpet floor finish whereas Hagia Sophia has a marble floor cladding.

The major acoustical outcomes of this study can be summarized as below:

- Having a larger volume (almost double) and reflective floor finish, Hagia Sophia resulted in higher reverberation times in comparison to Süleymaniye Mosque in mid to high frequencies. This result is appropriate for the original function of Hagia Sophia considering the liturgical music held in churches and cathedrals, in comparison to mosques where liturgical music is seldom.
- In lower octave bands, Süleymaniye Mosque has greater reverberation values in comparison to Hagia Sophia. The excessive reverberation in low frequencies of Süleymaniye Mosque is discussed for its original state by a trial on applying historical lime-based plasters through simulations. Inherently good sound absorption characteristics of historical multi-layered lime-based plasters have resulted in much proper acoustical parameter results within Süleymaniye Mosque.
- Due to the effective surface area, the slight difference on the plasters' sound absorption performance significantly changes the overall acoustical character of sacred spaces. Thus, the physical and chemical characteristics of plasters used in restorations and renovations are highly important, not only for other building physics aspects, but as well for acoustics science.
- The high reverberance of such mega-structures may negatively affect speech intelligibility due to its masking effect. However, a very short reverberation rating would cause a dry acoustical environment and reduce the envelopment and spaciousness in a religious space. Thus, the optimization studies of reverberation times in sacred spaces should not disregard the spiritual aspects.
- The recent restoration work on Süleymaniye Mosque when compared to previous years' field test results has ended up in lower and much more optimal reverberation times in mid to high frequencies, on average. This is most probably due to the removal of cement-based plasters, but is still not totally effective. However, the low frequencies are even higher. One reason for lower decay rates of previous field tests in low frequencies is thought to be due to the insufficient impulse response length.
- A concave dome is a representation of the wholly world, while at the same time this geometry is the primary reason for sound focusing and all relevant acoustical defects as of echoesBurnt-clay pots (Sebu's) as discussed in Süleymaniye Mosque, might had cured the low frequency sound energy built-up within the Mosque, as well as provided acoustical asymmetry in overcoming acoustical defects as of sound foci or echo formation. This approach should be a motivation for today's scientific research with much developed acoustical technology.

- On the other hand, the particular geometrical configuration of multi-domed case structures and interrupted plan-layout with arches, and elements such as mahfili's, elephant feet's, columns, and piers, help to overcome fluttering echoes in between parallel wall surfaces and also help an even distribution of sound.
- All of the field tests in Hagia Sophia were held after the major architectural revisions in relation to the function of this sacred space. The most recent field tests held within the scope of this research indicates that the reverberation times in overall frequency spectrum is lowered by 1–2 s.
- Another discussion point, briefly mentioned in this study, is that the sacred monuments provide not a single decay with a single reverberation time but multiple sound energy decays, which is rarely observed in modest sized single-space structures. The significance of multiple-decay, in form of early and late energy, is that the early decay enhances clarity or definition of sound, while the late decay contributes to the reverberance that complements spiritual needs.

To conclude, the comparative analyses as done in this research are beneficial for acquiring insight to the acoustical outcomes of repairs and restorations. However, the discussion on the basic acoustical metrics that have been studied for years on historical sacred mega-structure is a challenge, especially for the cases where there are multiple-sound energy decays. It should be noted that the unusual sensation (hearing) of sound in such monumental sacred spaces may be explained by overlapping decay curves and their acoustical outcomes, which also might have contributed to them being called out as the world's significant architectural master-pieces, for both the selected case studies as well as the other sacred edifices of the world.

Acknowledgments: The author would like to express her gratitude for the given permissions by General Directorate of Pious Foundations for Süleymaniye Mosque field tests, and to Turkish Ministry of Culture and Tourism, and General Directorate of Turkish Cultural Heritage and Museums for Hagia Sophia field tests. Mehmet Çalışkan, Ayşe Tavukçuoğlu and Ning Xiang are gratefully acknowledged for their guiding and support during different phases of this research.

Conflicts of Interest: The authors declare no conflict of interest.

References

1. Abdelazeez, M.K.; Hammad, R.N.; Mustafa, A.A. Acoustics of King Abdullah Mosque. *J. Acoust. Soc. Am.* **1991**, *90*, 1441–1445. [CrossRef]
2. Suárez, R.; Alonso, A.; Sendra, J.J. Virtual acoustic environment reconstruction of the hypostyle mosque of Cordoba. *App. Acoust.* **2018**, *140*, 214–224. [CrossRef]
3. *CAHRISMA, Conservation of the Acoustical Heritage by the Revival and Identification of Sinan's Mosques' Acoustics*; Project report No: ICA3-CT-1999-00007, Work package 2, Deliverables; Yıldız Technical University: Istanbul, Turkey.
4. Abdou, A.A. Measurement of acoustical characteristics of mosques in Saudi Arabia. *J. Acoust. Soc. Am.* **2003**, *113*, 1505–1517. [CrossRef] [PubMed]
5. Sü, Z.; Yılmazer, S. The acoustical characteristics of the Kocatepe Mosque in Ankara, Turkey. *Arch. Sci. Rev.* **2008**, *51*, 21–30. [CrossRef]
6. Gül, Z.S.; Çalışkan, M. Impact of design decisions on acoustical comfort parameters: case study of Doğramacızade Ali Paşa Mosque. Technical Note. *App. Acoust.* **2013**, *74*, 834–844. [CrossRef]
7. Magrini, A.; Magnani, L. Models of the influence of coupled spaces in Christian churches. *Build. Acoust.* **2005**, *12*, 115–139. [CrossRef]
8. Luigi, E.; Martellotta, F. Acoustics as a cultural heritage: the case of Orthodox churches and of the "Russian church" in Bari. *J. Cult. Heritage* **2015**, *16*, 912–917.
9. Martellotta, F. Identifying acoustical coupling by measurements and prediction-models for St. Peter's Basilica in Rome. *J. Acoust. Soc. Am.* **2009**, *126*, 1175–1186. [CrossRef]
10. Anderson, J.S.; Anderson, M.B. Acoustic coupling effects in St Paul's Cathedral, London. *J. Sound Vib.* **2000**, *236*, 209–225. [CrossRef]
11. Alonso, A.; Suárez, R.; Sendra, J.J. On the assessment of the multiplicity of spaces in the acoustic environment of cathedrals: The case of the cathedral of Seville. *App. Acoust.* **2018**, *141*, 54–63. [CrossRef]

12. Suárez, R.; Alonso, A.; Sendra, J.J. Intangible cultural heritage: the sound of the Romanesque Cathedral of Santiago de Compostela. *J. Cult. Heritage* **2015**, *16*, 239–243. [CrossRef]
13. Pedrero, A.; Ruiz, R.; Díaz-Chyla, A.; Díaz, C. Acoustical study of Toledo Cathedral according to its liturgical uses. *App. Acoust.* **2014**, *85*, 3–33. [CrossRef]
14. Kleiner, M.; Klepper, D.L.; Torres, R.R. *Worship Space Acoustics*, 1st ed.; J. Ross Publishing: Plantation, FL, USA, 2010; pp. 227–271.
15. Sü Gül, Z.; Xiang, N.; Çalışkan, M. Investigations on sound energy decays and flows in a monumental mosque. *J. Acoust. Soc. Am.* **2016**, *140*, 344–355. [CrossRef] [PubMed]
16. SüGül, Z. Assessment of Non-Exponential Sound Energy Decays within Multi-Domed Monuments by Numerical and Experimental Methods. Ph.D. Thesis, Middle East Technical University (METU), Ankara, Turkey, January 2015.
17. Sü Gül, Z.; Çalışkan, M.; Tavukcuoglu, A.; Xiang, N. Assessment of acoustical indicators in multi-domed historic structures by non-exponential energy decay analysis. *Acoust. Aust.* **2018**, *46*, 181–192. [CrossRef]
18. Klenbauer, W.E.; White, A.; Matthews, H. *Hagia Sophia*; Scala Publishers: London, UK, 2004.
19. Oyhon, E.; Etingü, B. Hagia Sophia, Church of Divine Wisdom. In *Churches in İstanbul*; YKY: Istanbul, Turkey, 1999.
20. Mark, R.; Çakmak, A.S. *Hagia Sophia from the age of Justinian to the Present*; Cambridge University Press: New York, NY, USA, 1992.
21. Kahler, H.; Mango, C. *Hagia Sophia*; Frederick A. Praeger: New York, NY, USA, 1967.
22. Mainstone, R.J. *Hagia Sophia: Architecture, structure and liturgy of Justinian's Great Church*; Thames and Hudson: London, UK, 1988.
23. Weitze, C.A.; Rindel, J.H.; Christensen, C.L.; Gade, A.C. *The Acoustical History of Hagia Sophia Revived through Computer Simulation*; Forum Acusticum: Seville, Spain, 2002.
24. Abel, J.S.; Woszczyk, W.; Ko, D.; Levine, S.; Hong, J.; Skare, T.; Wilson, M.J.; Coffin, S.; Lopez-Lezcano, F. Recreation of the Acoustics of Hagia Sophia in Stanford's Bing Concert Hall for the Concert Performance and Recording of Cappella Romana. In Proceedings of the International Symposium on Room Acoustics, Toronto, ON, Canada, 9–11 June 2013.
25. Akgündüz, A.; Öztürk, S.; Baş, Y. *Kiliseden Müzeye Ayasofya*; Osmanlı Araştırmaları Vakfı: Fatih/İstanbul, Turkey, 2006; pp. 174–180, 280–285.
26. Eyice, S. *Ayasofya*; Yapı Kredi: Istanbul, Turkey, 1984.
27. Cantay, T. *Süleymaniye Camii*; EREN Yayıncılık: Istanbul, Turkey, 1989.
28. T.R. Prime Ministry Directorate General of Foundations Archive. 2011.
29. Saatçi, S. Temelden Aleme İnşaat Süreci. In *Bir Şaheser Süleymaniye Külliyesi*; Mülayim, S., Ankara, T.C., Eds.; Kültür ve Turizm Bakanlığı: Ankara, Turkey, 2007; pp. 57–74.
30. Yılmaz, Y. *Kanuni Vakfiyesi Süleymaniye Külliyesi*; Vakıflar Genel Müdürlüğü: Ankara, Turkey, 2008; pp. 87–134.
31. Necipoğlu-Kafadar, G. The Süleymaniye Complex in İstanbul: an interpretation. *Muqarnas* **1985**, *3*, 92–117. [CrossRef]
32. Mungan, I. Strüktür Çözümü. In *Bir Şaheser Süleymaniye Külliyesi*; Mülayim, S., Ankara, T.C., Eds.; Kültür ve Turizm Bakanlığı: Ankara, Turkey, 2007.
33. Kuban, D.A. Symbol of ottoman architecture: The Süleymaniye. In *Ottoman Architecture*; Antique Collectors' Club: Suffolk, UK, 2010; pp. 277–294.
34. Kolay, I.A.; Çelik, S. Ottoman stone acquisition in the mid-sixteenth century: The Süleymaniye Complex in Istanbul. *Muqarnas* **2006**, *23*, 251–272. [CrossRef]
35. Barkan, L.O. *Süleymaniye Camii ve İmareti İnşaatı (1550-1557)*; Türk Tarih Kurumu Matbaası: Ankara, Turkey, 1972; Volume 1, pp. 14–45.
36. Eyüpgiller, K. Restitüsyon ve Renovasyon. In *Bir Şaheser Süleymaniye Külliyesi. Mülayim. Mülayim, S.*; Mülayim, S., Ankara, T.C., Eds.; Kültür ve Turizm Bakanlığı: Ankara, Turkey, 2007; pp. 193–232.
37. İrteş, S. Kalemişi, Cam ve Revzen. In *Bir Şaheser Süleymaniye Külliyesi. Mülayim. Mülayim, S.*; Mülayim, S., Ankara, T.C., Eds.; Kültür ve Turizm Bakanlığı: Ankara, Turkey, 2007; pp. 293–328.
38. Ersen, A.; Nilgün, O.; Akbulut, S.S.; Yıldırım, B.S. Süleymaniye Camii 2007–2010 yılları restorasyonu ve restorasyon kararları. *Restorasyon* **2011**, *3*, 6–27.
39. Acar, S. Süleymaniye'nin düşündürdükleri. *Tasarım* **2000**, *102*, 108–117.

40. A Discussion on the Acoustics of Süleymaniye Mosque for its Original State. Available online: http://www.radikal.com.tr/turkiye/256_bos_kupun_sirri-1028387 (accessed on 17 July 2013).
41. ISO 3382-1. *Acoustics-Measurement of Reverberation Time of Rooms with Reference to Other Acoustical Parameters*. ISO: Geneva, Switzerland, 2009.
42. Sü Gül, Z.; Çalışkan, M.; Tavukcuoglu, A. On the acoustics of Süleymaniye Mosque from past to present. *Megaron* **2014**, *9*, 201–216. [CrossRef]
43. Kayılı, M. Mimar Sinan'ın Camilerindeki Akustik Verilerin Değerlendirilmesi. *Mimarbaşı Koca Sinan: Yaşadığı Çağ ve Eserleri.* **1988**, *1*, 545–555.
44. Topaktaş, L. Acoustical properties of classical ottoman mosques, simulation and measurements. Master's Thesis, Middle East Technical University (METU), Ankara, Turkey, July 2003.
45. Naylor, G.M. ODEON—another hybrid room acoustical model. *Appl. Acoust.* **1993**, *38*, 131–143. [CrossRef]
46. Sü Gül, Z.; Xiang, N.; Çalışkan, M. Diffusion equation based finite element modeling of a monumental worship space. *J. Comput. Acoust.* **2017**, *25*, 1–16.
47. Sü Gül, Z.; Odabaş, E.; Xiang, N.; Çalışkan, M. Diffusion equation modeling for sound energy flow analysis in multi domain structures. *J. Acoust. Soc. Am.* **2019**, *145*, 2703–2717. [CrossRef] [PubMed]
48. Ahnert, W.; Feistal, S.; Behrens, T. Speech Intelligibility Prediction in very Large Sacral Venues. In Proceedings of the Meetings on Acoustics ICA2013, Montreal, QC, Canada, 2–7 June 2013.
49. Tavukçuoğlu, A.; Aydın, A.; Çalışkan, M. Tarihi Türk Hamamlarının Akustik Nitelikleri: Özgün Hali ve Bugünkü Durumu. In Proceedings of the TAKDER 9th National Congress, Ankara, Turkey, 26–27 May 2011.
50. SüGül, Z.; Çalışkan, M. Acoustical Design of Turkish Religious Affairs Mosque. In Proceedings of the 21st International Congress on Acoustics ICA2013, Montreal, QC, Canada, 2–7 June 2013.
51. *ASHRAE Handbook HVAC Applications*, SI ed.; American Society of Heating, Refrigerating, and Air Conditioning Engineers: Atlanta, GA, USA, 2011.
52. SüGül, Z.; Tavukcuoglu, A.; Çalışkan, M.A. Discussion on the Acoustics of Süleymaniye Mosque for its Original State. In Proceedings of the 9th International Symposium on the Conservation of Monuments in the Mediterranean Basin (MONUBASIN), Ankara, Turkey, 3–5 June 2014.
53. Bos, H.L.; Van Den Oever, M.J.A.; Peters, O.C.J.J. Tensile and compressive properties of flax fibres for natural fibre reinforced composites. *J. Mater. Sci.* **2002**, *37*, 1683–1692. [CrossRef]
54. Dalmay, P.; Smith, A.; Chotard, T.; Sahay-Turner, P.; Gloaguen, V.; Krausz, P. Properties of cellulosic fibre reinforced plaster: Influence of hemp or flax fibres on the properties of set gypsum. *J. Mater. Sci.* **2010**, *45*, 793–803. [CrossRef]
55. Orfali, W.A. Sound Parameters in Mosques. In Proceedings of the 153rd Meeting of ASA, Salt Lake City, UT, USA, 4–8 June 2007.

3

Acoustic Black Holes in Structural Design for Vibration and Noise Control

Chenhui Zhao and Marehalli G. Prasad *

Department of Mechanical Engineering, Stevens Institute of Technology, Hoboken, NJ 07030, USA; czhao1@stevens.edu

* Correspondence: mprasad@stevens.edu

Abstract: It is known that in the design of quieter mechanical systems, vibration and noise control play important roles. Recently, acoustic black holes have been effectively used for structural design in controlling vibration and noise. An acoustic black hole is a power-law tapered profile to reduce phase and group velocities of wave propagation to zero. Additionally, the vibration energy at the location of acoustic black hole increases due to the gradual reduction of its thickness. The vibration damping, sound reduction, and vibration energy harvesting are the major applications in structural design with acoustic black holes. In this paper, a review of basic theoretical, numerical, and experimental studies on the applications of acoustic black holes is presented. In addition, the influences of the various geometrical parameters and the configuration of acoustic black holes are presented. The studies show that the use of acoustic black holes results in an effective control of vibration and noise. It is seen that the acoustic black holes have a great potential for quiet design of complex structures.

Keywords: acoustic black hole; structure design; noise and vibration control

---

1. Introduction

It is known that with the development of high-speed machinery, the control of unwanted vibration and noise are very important for their stability and reliability, as well as the environmental noise impact [1]. The two well-known methods for passive control of structural vibrations which also results in a reduction of noise are constrained layer damping and tuned dynamic absorbers [2]. The first method is based on using a viscoelastic layer attached to the structure and the second method needs an attachment of additional weight [3] to the target structure. Additionally, the active vibration control devices are also used for vibration damping [4]. However, these active methods require consistent input energy and more complex electro-mechanical design. Thus, for the reasons of limitation of size, budget, or weight, sometimes it is not possible, and also is not desirable, to use these above methods. There is always a need for an effective design of structures for vibration and noise control [5].

In recent times micro-devices, such as portable electronics and wireless remote sensors, are developed and widely used [6]. Most of these low-power electronics are powered by battery. However, even for the long-lasting batteries, they still need to be replaced because of their limited lifecycle. For some applications, such as sensors deployed in remote locations or inside the human body [7], it is challenging and costly, or even impractical. Energy harvesting is the process of capturing and converting ambient energy in the environment into usable electrical energy to extend the life of batteries, which makes the devices self-sustainable and environmental-friendly. Piezoelectric vibration energy harvesting (PVEH) is one of the typical energy harvesting methods. In the design of portable micro devices, the challenge is to reduce the weight and size of the host structure. Thus, approaches to increase the energy harvested from the vibrations of the host structures are desirable [8].

Recently, an approach for passive vibration control, acoustic black holes (ABH), has been developed. An ABH is usually a power-law tapered profile built on structures, such as beams, plates, and shells, where the vibration energy is concentrated due to the reduction of wave speed [9] (as shown in Figure 1). Therefore, due to this concentration effect of the ABH, the vibration energy can be absorbed by attaching small amount of damping material at the ABH location, which also results in reduced sound radiation. Additionally, the performance of energy harvesting is enhanced by attaching piezoelectric material at the ABH location [10]. An ABH is a tailing method which cuts material away from the host structure, and it also decreases the usage of damping layer, so it decreases the weight of host structures. Therefore, it is a good option for vibration and noise control of lightweight structures.

The ABH effect was first discovered by Pekeris in 1946 [11]. He exploited the central physical principle of ABH, namely the phase velocity of sound waves that propagate in a stratified fluid are progressively decreased to zero with increasing depth. In 1988 Mironov determined that a flexural wave propagates in a thin plate slows down and needs infinite time to reach a tapered edge [12]. Later, Krylov first used the name "acoustic black hole" to this effect [13], and applied ABH on beams and plates, also indicating that the ABH approach results in an increased amount of energy to be absorbed by adding a small amount of material attenuation near the ABH locations [14–17]. Then Conlon developed further numerical and experimental work to analyze the ABH effect on vibration and sound radiation of thin plates [18–20]. Later, researchers from different countries around the world worked on the ABH effect on structural vibration control, sound radiation, and vibration energy harvesting. Recently, a review on mechanics problem of the ABH structure was presented by Ji et al. [21]. This study systematically introduced the theoretical study on mechanics for 1D and 2D structures and a summary of applications of ABH. Another review on the applications of ABH on vibration damping and sound radiation was conducted by Chong et al. [22]. Due to the increase of complexity of structure design with ABH and the limitation of traditional manufacturing methods, such as milling [23], 3D printing technology is applied. In the study of Chong et al. [22], a numerical and experimental study on vibration response of the 3D-printed ABH beams was also developed. Furthermore, a series of studies on dynamic and static properties [24] and applications in vibration damping [25] and energy harvesting [8,26] of 3D-printed structures embedded with ABH was also investigated by other researchers.

This paper presents a review of recent studies on the use of ABH in structural design. The paper presents studies on the applications of ABH in structural vibration control, noise reduction, and vibration energy harvesting. In addition, the review particularly focusses on the influence of geometrical parameters of 1D ABH and the layout of the 2D ABH on the structural response in order to make the ABH features more efficient in structural design.

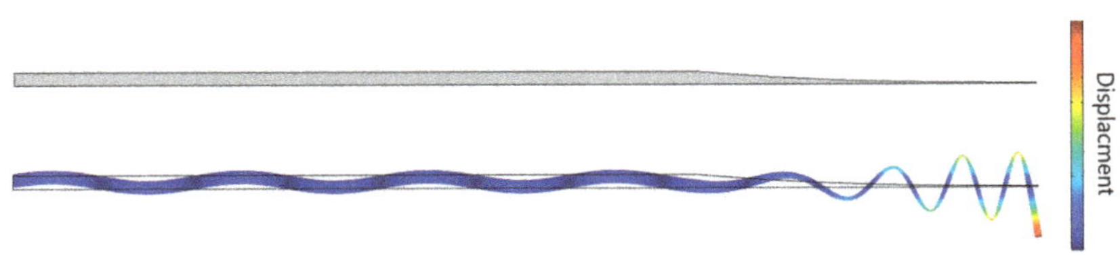

Figure 1. ABH concentration effect, the amplitude of the incident wave increases significantly when propagating to the end of the ABH wedge.

2. Theoretical Analysis

Mironov [12] showed that the bending wave speed goes to zero for beams and plates whose thickness decreases according to:

$$h(X) = aX^m, \tag{1}$$

where h is the thickness, a is constant, m is exponent of the power-law curve, and X is the distance from the tip of the ideal power-law curve.

However, in reality, due to the limitation of manufacturing technics, it is impossible to build a zero thickness, so there will be a residual thickness h_1 at the free end. Then the equation of the power-law curve (1D ABH) becomes:

$$h(x) = \varepsilon x^m + h_1, \tag{2}$$

where the exponent m is a positive rational number and $m \geq 2$, parameter ε is a constant, x is the distance from the tip of the power-law curve with residual thickness, and the scheme is shown in Figure 2.

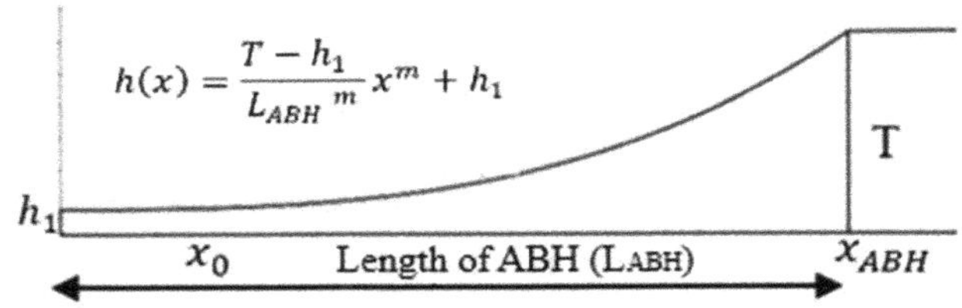

Figure 2. Schematic of a 1D ABH [27].

For the specific thickness of the beam T, Parameter ε affect the length of ABH:

$$L_{ABH} = \sqrt[m]{\frac{T - h_1}{\varepsilon}}, \tag{3}$$

thus:

$$\varepsilon = \frac{T - h_1}{L_{ABH}{}^m}, \tag{4}$$

then the ABH power-law curve is:

$$h(x) = \frac{T - h_1}{L_{ABH}{}^m} x^m + h_1, \tag{5}$$

where T is the thickness of the beam, h_1 is the residual thickness of the ABH part, L_{ABH} is the length of the ABH, and m is the exponent of the power-law profile [27].

The phase velocity C_p and group velocity C_g are given by:

$$C_p = \sqrt[4]{\frac{E}{3\rho(1 - \nu^2)}} \sqrt{\omega h(x)}, \tag{6}$$

$$C_g = \sqrt[4]{\frac{4E}{3\rho(1 - \nu^2)}} \sqrt{\omega h(x)}, \tag{7}$$

where E is the Young's modulus, ν is the Poisson's ratio, ρ is the mass density, $h(x)$ is the varying thickness at ABH of the beams, and ω is the angular frequency of the flexural wave. When $h_1 = 0$ and $x \to 0$, the phase velocity and group velocity tend to zero [9,15].

Propagation time T_0 from x_{ABH} to x_0 is shown in Figure 2:

$$T_0 = \int_{L_{ABH}}^{x_0} \frac{1}{c_g} dx = \sqrt[4]{\frac{12\rho(1 - \nu^2)}{E\omega^2}} \frac{1}{2 - m} \left(x_0{}^{1-m/2} - L_{ABH}^{1-m/2} \right), \tag{8}$$

when x_0 tends to 0, $h_1 = 0$ then T_0 tends to infinity, only if $m \geq 2$ [12].

Equations (6)–(8) indicate that the ABH can alter wave speed to decrease and that also result in the concentration of vibration energy at the ABH location when $m \geq 2$.

If the host structure has a non-zero-loss factor, the reflection coefficient and wave number can be presented as below:

$$R = e^{\left(-2\int_{x_{ABH}}^{x} Im\, k(x)dx\right)}, \tag{9}$$

$$k(x) = \sqrt[4]{12}\sqrt{\frac{k_l}{h(x)}}, \tag{10}$$

$$k_l^2 = \frac{\rho(1-\nu^2)\omega^2}{E}, \tag{11}$$

where E is the Young's modulus, ν is the Poisson's ratio, ρ is the mass density, h is the thickness of the plate, and ω is the angular frequency of flexural wave [28,29].

Equations (9)–(11) indicate that when x tends to zero, if the residual thickness h_1 equal to zero, wave numbers $k(x)$ tend to infinity at ABH locations, and the reflection coefficient tends to zero, which indicates that no wave can escape from the ABH location [30].

When the ABH is partially covered with a damping layer, the reflection coefficient at any point in the damping area can be expressed as:

$$R_0(x) = \left(\sqrt{\frac{\varepsilon x^m}{h_1}} + \sqrt{\frac{\varepsilon x^m}{h_1} + 1}\right)^{-K} \times exp\left\{-K_2 \frac{\sqrt{\frac{\varepsilon x^m}{h_1}}}{\sqrt{\frac{\varepsilon x^m}{h_1} + 1}}\right\} \tag{12}$$

$$K_2 = \frac{3 \times 12^{1/4} {k_p}^{1/2} v}{4\varepsilon^{1/2}} \frac{E_2}{E_1} \frac{\delta}{h_0} \tag{13}$$

$$K = \frac{12^{1/4} {k_p}^{1/2} \eta}{2\varepsilon^{1/2}}, \tag{14}$$

where v is the loss factor of the material of the absorbing layer, η is the loss factor of the wedge material. δ is the layer thickness and E_2/E_1 is the ratio of Young's moduli of the absorbing layer and the plate, respectively [31,32].

A 2D ABH can be seen as a rotation of the 1D ABH by 360 degrees. When a wave propagates in the 2D ABH, it deviates into the center of the ABH [17,32]. O'Boy et al. developed the theoretical analysis for the thin plate [33]. A systematic summary of theoretical analysis containing a 2D ABH is studied by Ji et al. [21]. When attaching the damping material on the plate, the structure can be seen as a composite structure. The loss factor can be expressed as:

$$\eta_{comp}(r) = \frac{\eta_D \frac{E_D h_D}{E_w h(r)}\left[4\left(\frac{h_D}{h(r)}\right)^2 + 6\left(\frac{h_D}{h(r)}\right) + 3\right]}{1 + \frac{E_D h_D}{E_w h(r)}\left[4\left(\frac{h_D}{h(r)}\right)^2 + 6\left(\frac{h_D}{h(r)}\right) + 3\right]} \tag{15}$$

where E_D is the Young's modulus of damping, η_D is the loss factor of the damping material. E_w is the elasticity modulus of the plate in plural form. $h(r)$ is the thickness of the plate and h_D is the thickness of damping layer. r is the distance to the center of ABH [33]. When r decreases, the loss factor increases which means more vibration energy is absorbed.

The reflection coefficient of ABH becomes:

$$R_0 = exp\left\{-2\frac{3 \times 12^{1/4}\omega^{1/2}\left[\rho(1-\nu^2)\right]^{1/4}}{4\varepsilon^{1/2}E^{1/4}} \int_{R_t}^{R_i} \eta_{comp}(r)\frac{1}{r}dr\right\} \tag{16}$$

where R_i is the radius of whole ABH and R_t is the truncation length which means the radius inner hole [33]. The theoretical analysis of the flexural wave propagation in ABH with damping being developed and more new mathematical models are currently under investigation. For example, a semi-analytical model to analyse an Euler-Bernoulli beam with ABH and its full coupling with the damping layers coated over its surface is presented by Tang et al. [34].

3. Applications to Structural Design with Acoustic Black Holes

3.1. *Application of ABHs to Vibration Control*

3.1.1. Vibration Control of Beams with ABHs

In 2003, Krylov first used the name "acoustic black hole" [13]. The theoretical and numerical work [29] on beams with a one-dimensional acoustic black hole (1D ABH), as shown in Figure 3, was developed. Figures 4 and 5 show the effect of various truncation length x_0 and thickness of absorbing film δ on the reflection coefficient of the ABH. First, it can be observed from the behavior of the reflection coefficient for the uncovered wedge (solid curve) that a small truncation can result in a large increase of the refection coefficient. Second, it can be observed that the reflection coefficient increases with the increase of truncation length. Third, damping layers with higher relative stiffness can decrease the reflection coefficient further. Finally, the reflection coefficient decreases with the increase of excitation frequency. These two figures indicate that the presence of thin absorbing layers on the surfaces of ABH wedges can result in very low reflection coefficients of flexural waves from their edges.

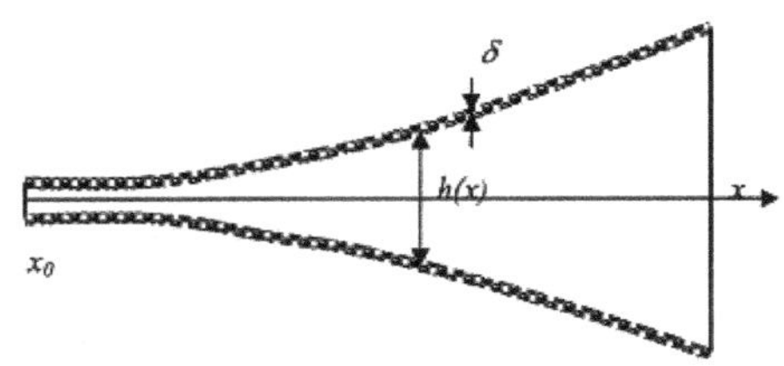

Figure 3. Truncated quadratic wedges covered by thin damping layers. x_0 is the truncation length. Reprinted with permission from [29]. Copyright Elsevier, 2004.

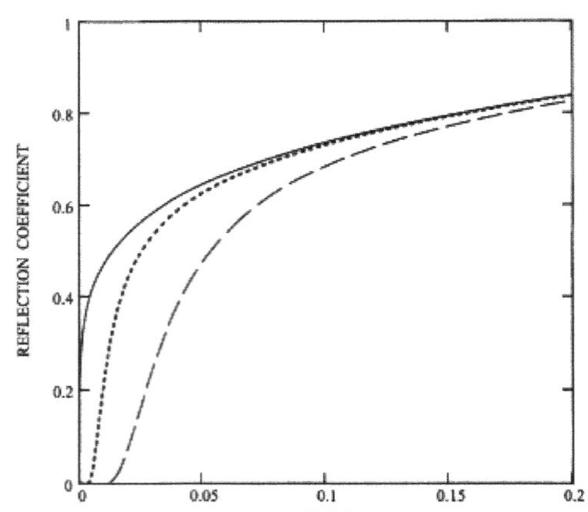

Figure 4. Reflection coefficient for the wedge covered by a thick absorbing film. The solid curve corresponds to an uncovered wedge, and dotted and dashed curves correspond to wedges covered by thin absorbing films with the values of relative stiffness $E_2/E_1 = 2/30$ and $E_2/E_1 = 2/3$, respectively; the film material loss factor ν is 0.2, and the film thickness δ is 5 μm. Reprinted with permission from [29]. Copyright Elsevier, 2004.

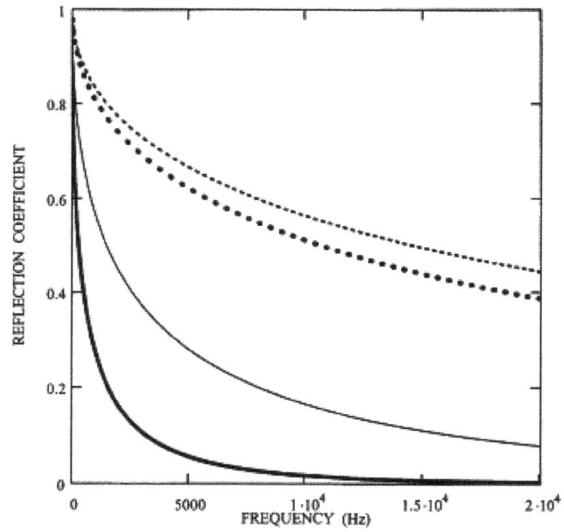

Figure 5. Frequency dependence of the reflection coefficient R: x_0 = 1.5 and 2.5 cm (thicker and thinner curves, respectively); solid and dotted curves correspond to wedges with absorbing films and to uncovered wedges; the film material loss factor ν is 0.2, and the film thickness δ is 5 μm. Reprinted with permission from [29]. Copyright Elsevier, 2004.

A numerical study of beams with a spiral ABH was developed by Jeon et al. [35,36]. The spiral ABH is a compact and curvilinear shape by using an Archimedean spiral with a uniform gap-distance between adjacent baselines of the spiral as shown in Figure 6. Figure 7 shows the driving point mobility of the beam with a 720 mm spiral ABH compared with reference uniform beam (black line). The beam with the ABH is 10% lighter than the reference beam, but it reduces the resonant peak levels to 90% without additional damping. This indicates that the spiral ABH has great potential for vibration damping.

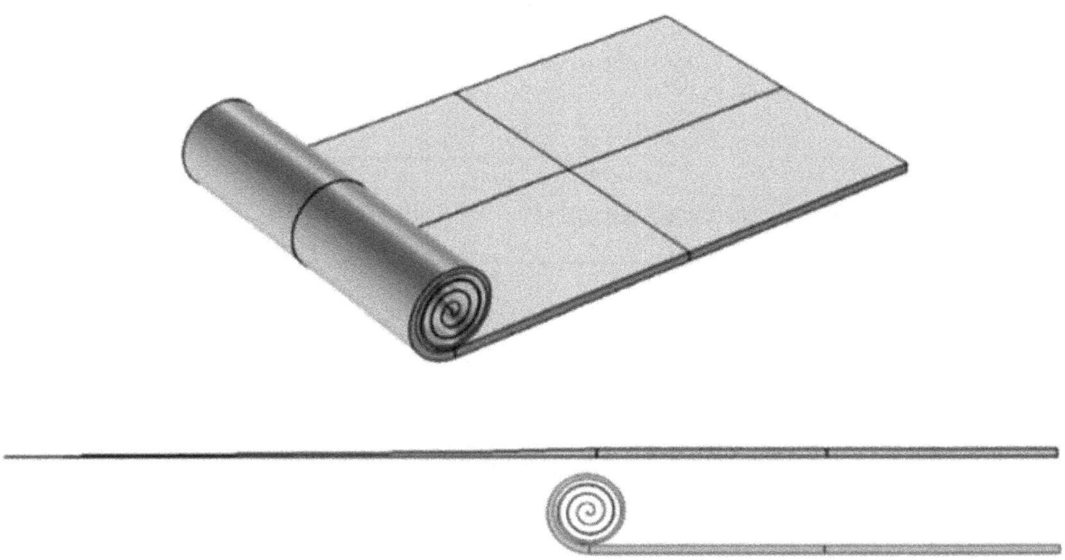

Figure 6. Shape of the beam with a spiral ABH. The length of the ABH is 720 mm. Reprinted with permission from [36]. Copyright Acoustic Society of America, 2017.

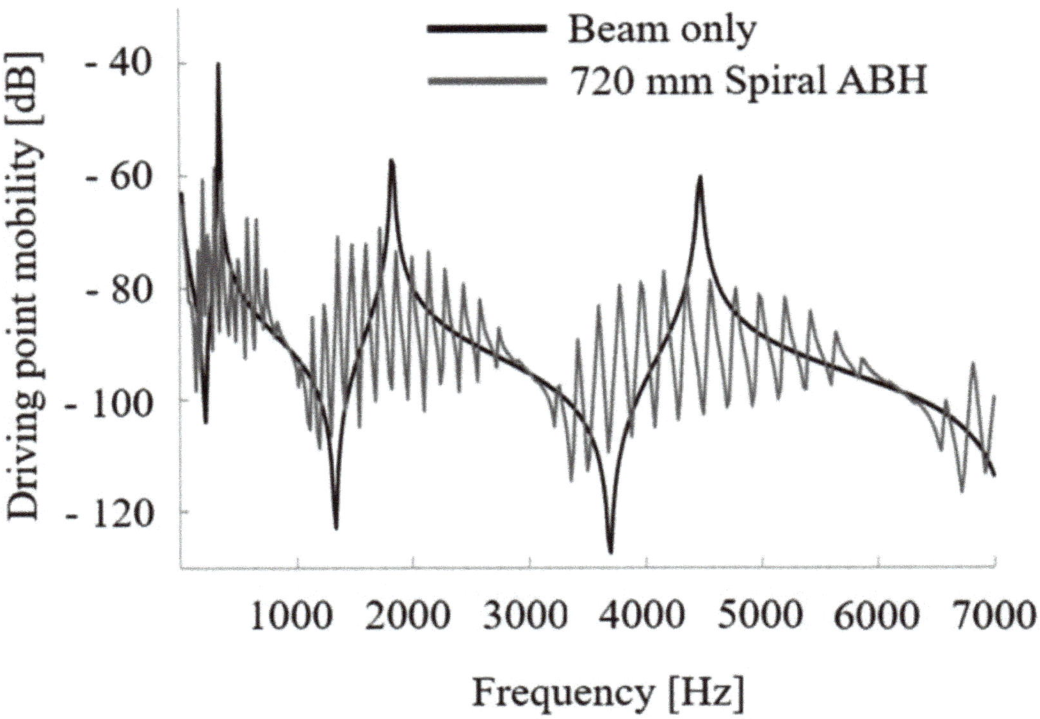

Figure 7. Driving point mobility of the beam with a 720 mm spiral ABH (grey line) and the reference uniform beam (black line). Reprinted with permission from [36]. Copyright Acoustic Society of America, 2017.

Recently, Zhou and Cheng proposed a numerical and experimental work to develop an ABH that featured a resonant beam damper, as shown in Figure 8 (ABH-RBD) [37]. Figure 9 shows the measured driving and cross point mobility of the host beam with and without ABH-RBD. It can be observed that a significant vibration reduction is obtained with mounting the ABH-RBD. This study shows a great damping treatment of ABH-RBD to control the vibration of the host structure.

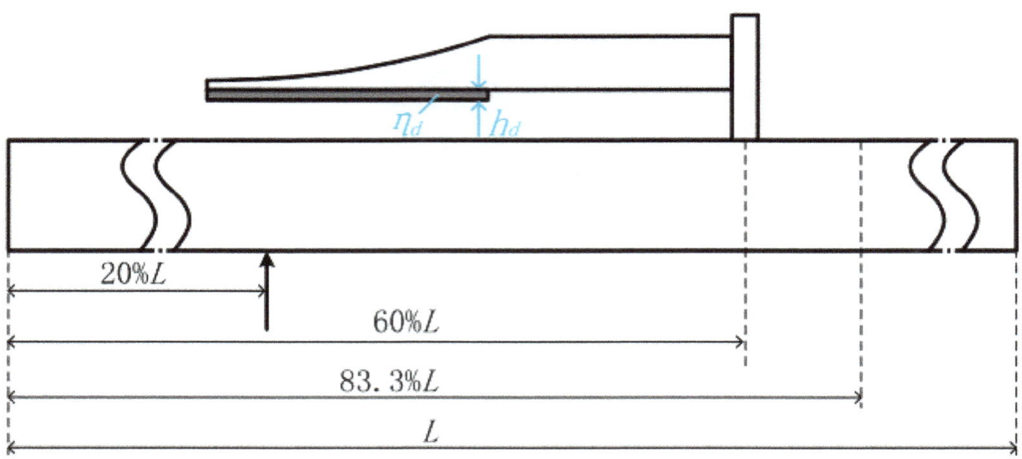

Figure 8. Schematics of the host beam with an ABH-RBD. Reprinted with permission from [37]. Copyright Elsevier, 2018.

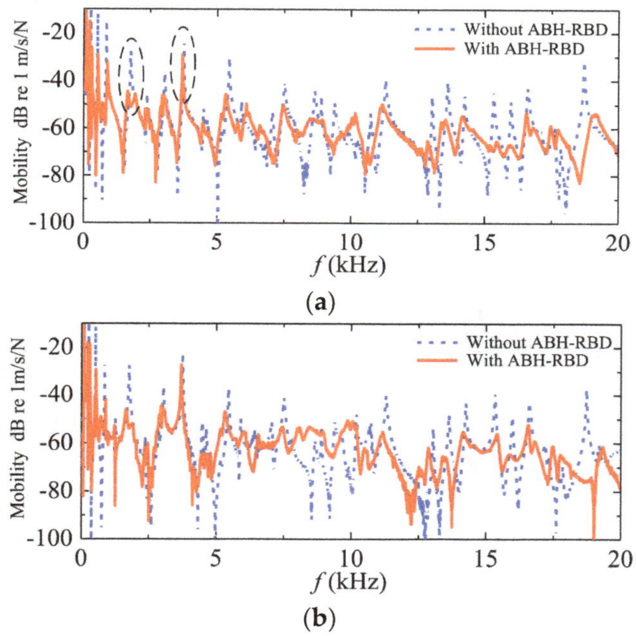

Figure 9. Measured (**a**) driving and (**b**) cross-point mobility of the primary host beam with and without ABH-RBD. Reprinted with permission from [37]. Copyright Elsevier, 2018.

3.1.2. Vibration Damping of Plates with ABH

The vibration performance of the plates with various 1D ABH slots was studied by Bowyer [38]. One of the samples is shown in Figure 10a. From the experimental result of this sample, it can be observed that a substantial reduction of acceleration is obtained in comparison with the reference plate. The experimental results of other steel plates and composite plates with various slots also show acceptable damping performance. The composite plates have good damping performance even without the damping layer attached due to the large material loss factor.

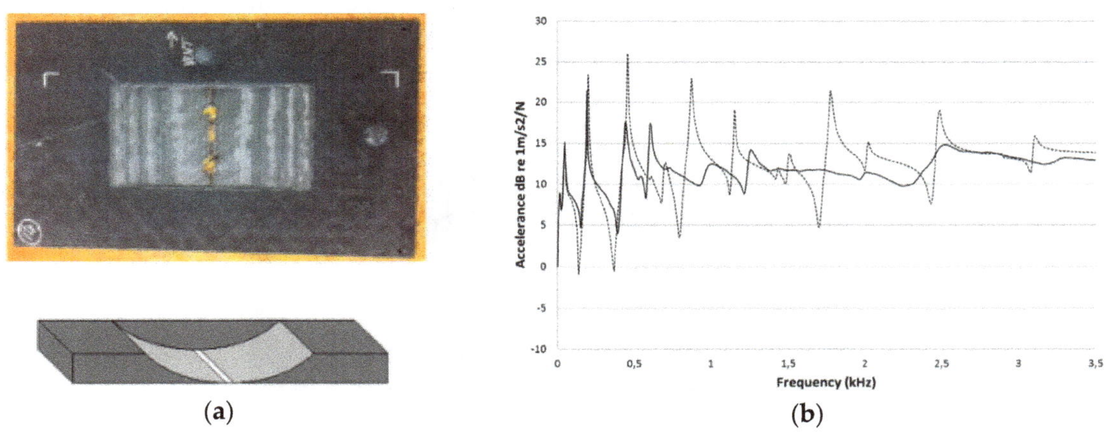

Figure 10. (**a**) Carbon fiber composite sample and longitudinal cross-section of the sample; and (**b**) Acceleration for the plate with ABH slot (solid line) and reference plate (dashed line). Reprinted from [38] under a CC BY 4.0 license. Copyright Bowyer, E.P. and Krylov, V.V., 2016.

Experimental investigation on damping flexural vibrations using two-dimensional acoustic black holes (2D ABHs) was developed by Bowyer et al. [3]. Figure 11 shows the schematics of the experimental samples. The forced excitation was applied to the center of the plate via shaker over a frequency range of 0–9 kHz. The experimental results show that, in comparison with the reference plate, the plate embedded with single 2D ABH with a damping layer provides little damping below 3 kHz. In the region of 3.8–9 kHz, damping varies between 3–8 dB, and maximum damping occurs at 6.6 kHz. Due to the long wave length, the damping performance of the ABH is not evident at low frequency, but it has better damping performance for higher frequencies. It is also indicated by this study that the plates with multiple holes in the current random layout does not significantly improve the damping performance of the 2D ABHs in comparison with the plate with a single 2D ABH.

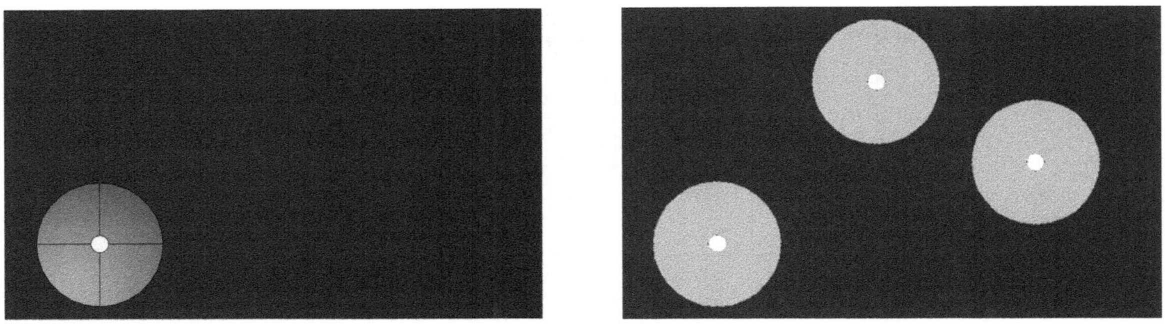

(a) (b)

Figure 11. (**a**) A singular 2D ABH with a center hole, and (**b**) three 2D ABH with central holes. Adapted from [3] under a CC BY 3.0 license. Copyright Bowyer, E.P. and Krylov, V.V., 2010.

Later, another experimental investigation of damping flexural vibrations in plates was also developed by Bowyer et al. [39]. Figure 12 shows the steel plate containing an array of six 2D ABH. Figure 13 clearly shows that, compared with the reference plate without ABH, the acceleration of the plate with six ABHs attaching damping layers sharply decreases. Additionally, it shows that the minimum frequency of effective damping performance is about 1.5 kHz. These two experimental investigations indicate that ABHs with damping layers can decrease the vibration of plates and the number and configuration of ABHs affect the performance of vibration damping. The increase in the number of ABHs can expand the effective frequency of damping performance of the plate with a 2D ABH and damping layer.

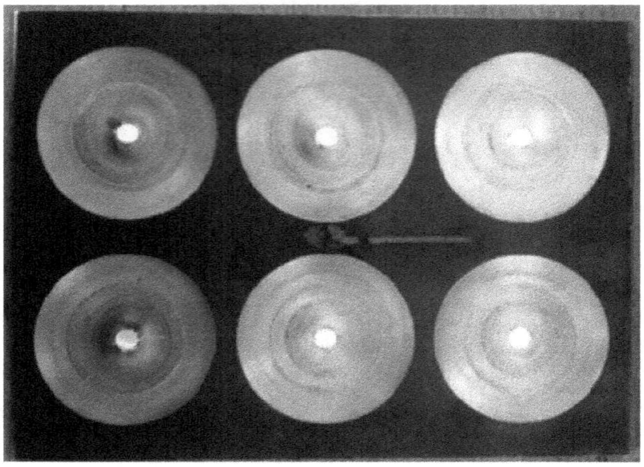

Figure 12. Manufactured steel plate containing an array of six 2D ABHs. Reprinted with permission from [39]. Copyright Elsevier, 2013.

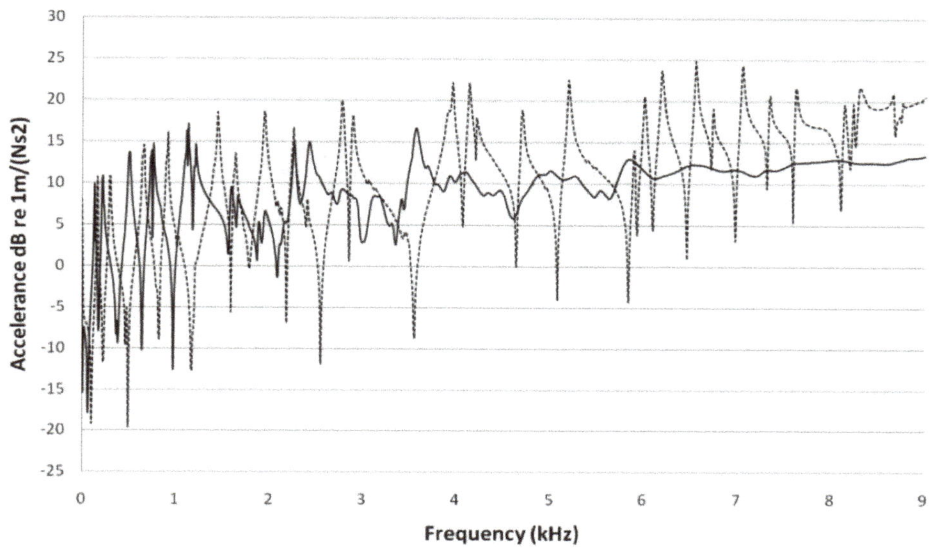

Figure 13. Measured acceleration for a plate containing six 2D ABHs with 14 mm central holes and additional damping layers (solid line) and a reference plate (dashed line). Reprinted with permission from [39]. Copyright Elsevier, 2013.

Other shapes of plates with ABH are also studied. Mobilities for a circular plate with a central ABH with a thin damping layer and a constrained layer are studied by O'Boy and Krylov [40]. The point- and cross-mobilities show a suppression of resonant peaks which is up to 17 dB compared with the reference plate. A numerical and experimental study of the acoustic black hole effect for vibration damping in elliptical plates has been studied by Georgiev et al. [41]. Elliptical plates with ABH and without ABH were tested. An elliptical plate with disks of resin placed at the location of the ABHs and an elliptical plate completely covered by resin were also tested. Figure 14 shows velocity fields of plates with ABH (b) and without ABH (a). The excitation force was applied to the left focus whereas the ABH is in the right one. It can be observed that the plate with ABH (b) has a lower amplitude of vibration. Figure 15 shows the point mobility measured at the left focus of the plate. It can be observed that the point mobility of the plate with ABH with the damping material at its location was reduced over 2 kHz.

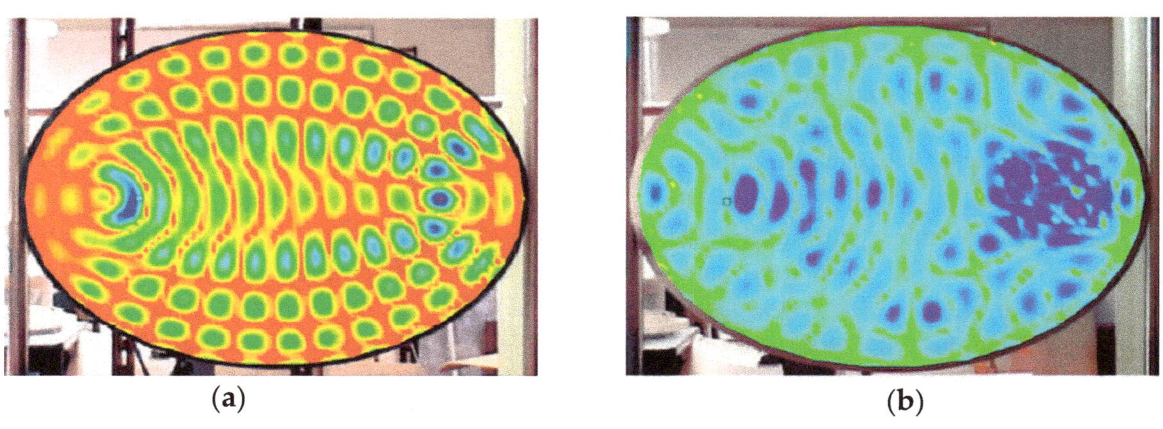

Figure 14. Velocity fields of an elliptical plate (**a**) without ABH at 8671 Hz, and (**b**) with ABH at 8117 Hz. Reprinted with permission from [41]. Copyright Elsevier, 2011.

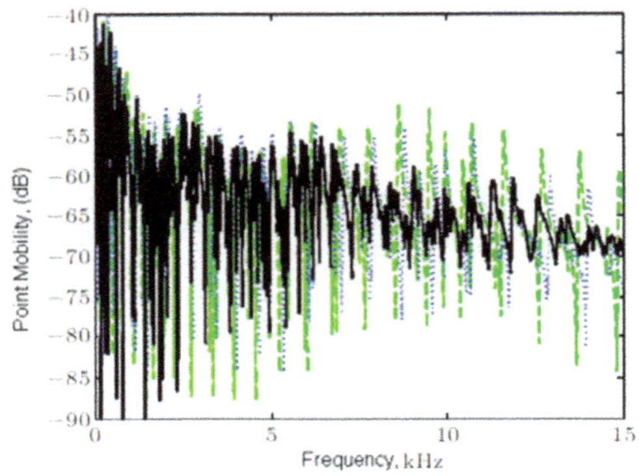

Figure 15. Measured point mobilities of the elliptical plate. The black line shows the plate with ABH by attaching damping material at its location, the dashed line shows the plate without ABH by attaching damping at the same location, and the green line shows the plate without ABH by covering the damping material over the whole plate. Reprinted with permission from [41]. Copyright Elsevier, 2011.

Not only were plates with ABH with damping material, but also ABH with dynamic vibration absorbers (DVA) were studied by Jia et al., as shown in Figure 16 [42]. It can be observed from the simulation results that there is a reduction of over 10 dB at major response peaks over 1 kHz for the plate with ABH with DVA. This result shows great potential of combining ABHs and DVAs for vibration control.

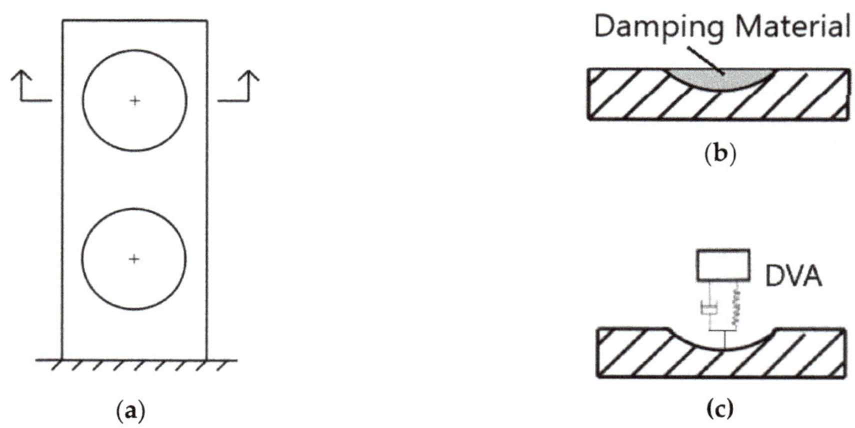

Figure 16. Cross-section of the plate structure considered in the FE model and the experiments: (**a**) a top-view of plate structure with two ABHs, (**b**) a plate embedded with ABH and damping layer, and (**c**) a plate embedded with ABH and DVA. Adapted with permission from [42]. Copyright American Society of Mechanical Engineers, 2015.

Since the structures with multiple ABHs have great damping effect for vibration and sound, such structures can be considered as metamaterials [43]. Semperlotti and Zhu developed a meta-structure based on the concept of ABH [44]. This load-bearing thin-wall structure element enables propagation characteristics comparable with resonant metamaterials without the fabrication complexity. The experimental work shows that the ABH treatments can significantly improve the damping effect.

A waveguide is designed to observe travelling waves by Foucaud et al. [45], which is inspired by artificial cochlea. The experimental study (shown in Figure 17) uses a varying width plate immersed in fluid and terminated with an ABH. It shows that an ABH used as an anechoic end improves the quality of measurements and the accuracy of tonotopic maps due to the attenuation of reflected waves.

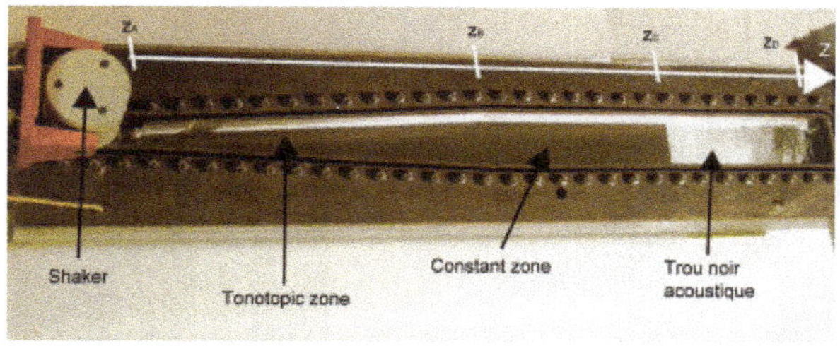

Figure 17. The photo for the assembled experimental setup. Reprinted with permission from [45]. Copyright Elsevier, 2014.

3.1.3. Vibration Damping of Turbofan Blades with ABH

An experimental work of damping of flexural vibrations in turbofan blades using ABH was studied by Bowyer and Krylov [46]. Figure 18a shows the fan blade profile with a tapered ABH geometry. The experimental setup with four experimental samples are shown in Figure 19. Figure 20 shows the measurement of acceleration for a twisted reference blade (dashed line) compared to a blade with a 1D ABH and damping layer (solid line). It can be observed that the acceleration has about 50% reduction at 60 Hz and 360 Hz. This indicates that the trailing edges of the 1D ABH with appropriate damping layers are efficient in the reduction of airflow-excited vibrations of the fan blades. This study shows the great potential of the ABH in jet engine design to reduce flexural vibration in the blades, thus reducing internal stresses in the blades and increasing their fatigue life cycle.

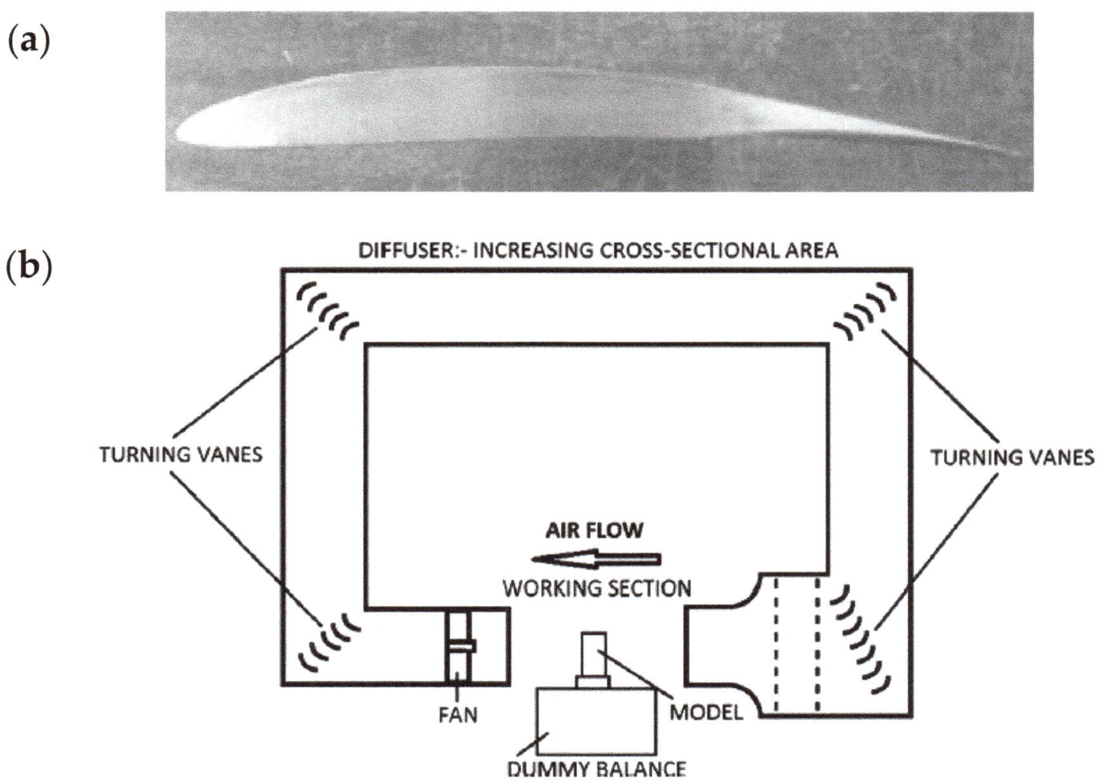

Figure 18. (a) Fan blade profile with tapered ABH geometry. (b) Experimental setup. Reprinted from [46] under a CC BY 3.0 license. Copyright Bowyer, E.P. and Krylov, V.V., 2014.

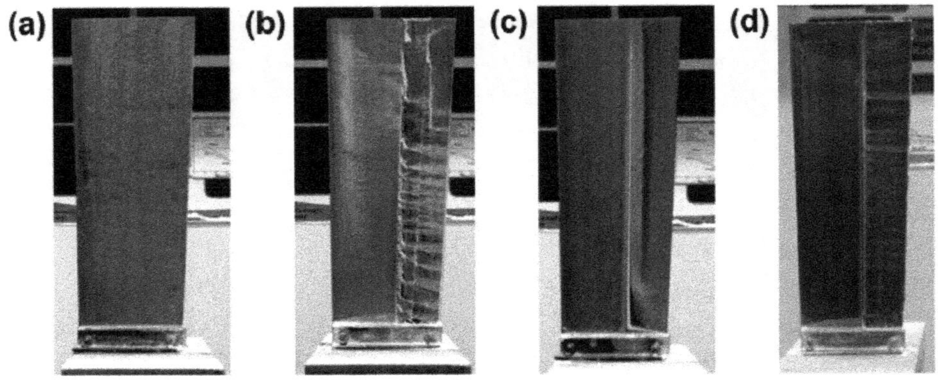

Figure 19. Experimental setup of closed-circuit wind tunnel and samples of the test. Experimental samples: Flow visualization diagram for: (**a**) reference fan blade, (**b**) fan blade with a power-law wedge, (**c**) fan blade having a power-law wedge with a single damping layer, and (**d**) fan blade having a power-law wedge with a shaped damping layer. Reprinted from [46] under a CC BY 3.0 license. Copyright Bowyer, E.P. and Krylov, V.V., 2014.

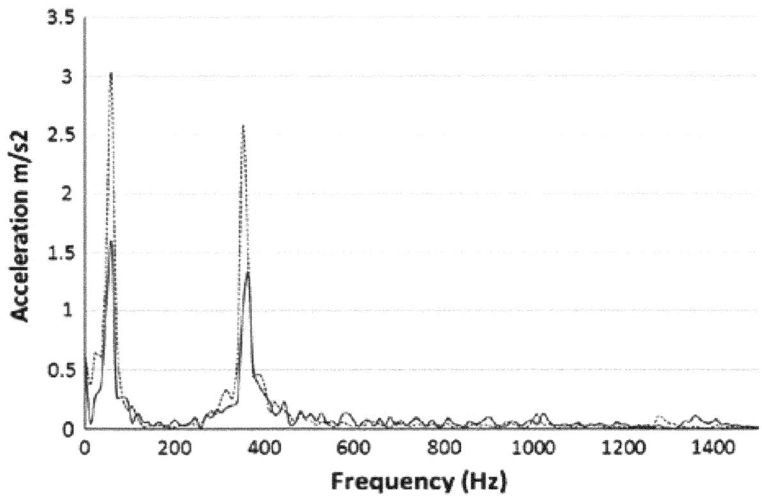

Figure 20. Acceleration for a twisted reference blade (dashed line) compared to a twisted blade with a 1D ABH and damping layer (solid line). Reprinted from [46] under a CC BY 3.0 license. Copyright Bowyer, E.P. and Krylov, V.V., 2014.

3.2. Application of ABH to Sound Reduction

An experimental work was developed by Bowyer and Krylov [47]. A 300 × 400 mm × 5-mm thick plate embedded with six 2D ABHs (Figure 21) with damping at the center of each ABH was suspended vertically. The excitation force was applied centrally on the plate. The results compare the sound radiation power level of a plate containing six 2D ABHs with a damping layer with the sound radiation power level of the reference plate, which are shown in Figure 22. Below 1 kHz there is little reduction in the sound power level. Between 1 and 3 kHz, the sound power level is reduced by 10–18 dB in comparison with the reference plate, and the maximum reduction in the sound radiation occurs at 1.6 kHz. Above 3 kHz, almost all responses in sound radiation are absorbed. This indicates that the plate with ABHs having damping layers effectively reduce the sound radiation of the steel plate.

Figure 21. Sample of the plate with six 2D ABHs. Reprinted from [47] under a CC BY license. Copyright Bowyer E.P. and Krylov, V.V., 2012.

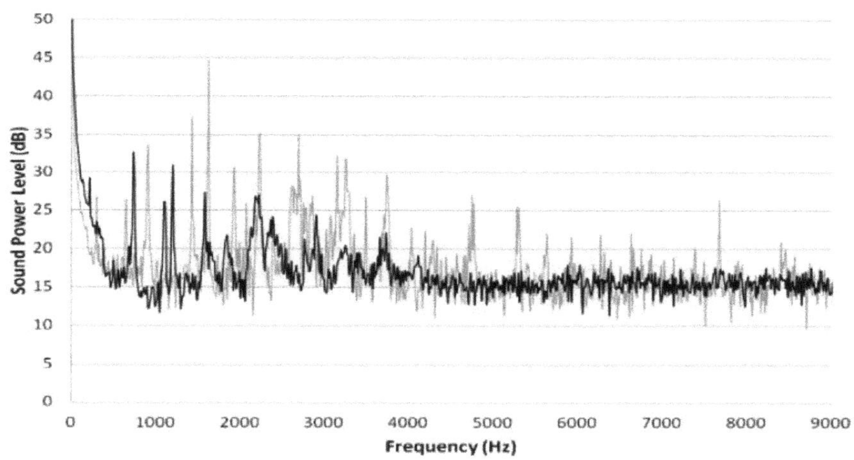

Figure 22. Sound power level comparison for a plate containing six indentations of power-law profile with a damping layer (black line) compared to a reference plate (grey line). Reprinted from [47] under a CC BY license. Copyright Bowyer E.P. and Krylov, V.V., 2012.

Feurtado and Conlon developed a numerical and experimental investigation of sound power of a plate with an array of ABH [48]. A 4 × 5 array of 10-cm diameter 2D ABHs, which is shown in Figure 23, was machined into a 6.35-mm thick, 61 cm × 91 cm aluminum plate. The ABH with various diameters of damping layers was tested to assess the effects of the amount of damping layer on ABH performance. The experimental setup is shown in Figure 24. The plate was mounted on a frame and excited with band-limited white noise. Figure 25 shows one-third octave band radiated sound power tested by an intensity probe for a uniform plate and ABH plate with various diameters of damping material. It can be observed that an ABH with a damping layer effectively reduced the radiated sound power over 1.5 kHz compared with the reference plate (blue line). It can also be observed that sound radiation of the plate with the diameters of the damping layer of 6.75 cm and 10 cm have almost the same performance, namely higher radiated sound power. The 3.5 cm diameter damping layer shows comparable performance to the larger damping diameters.

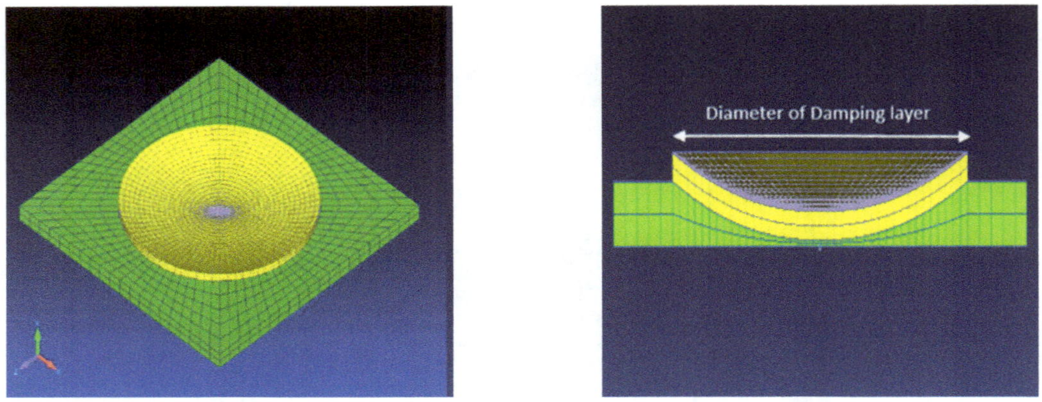

(a) (b)

Figure 23. (**a**) FEA model and (**b**) cross-section of an aluminum plate (green) with a 2D ABH attached damping layer (yellow). Reprinted with permission from [48]. Copyright American Society of Mechanical Engineers, 2015.

Figure 24. Aluminum plate with a 4 × 5 array of embedded ABHs with full-diameter damping layers in a frame with a mechanical point drive. Reprinted with permission from [48]. Copyright American Society of Mechanical Engineers, 2015.

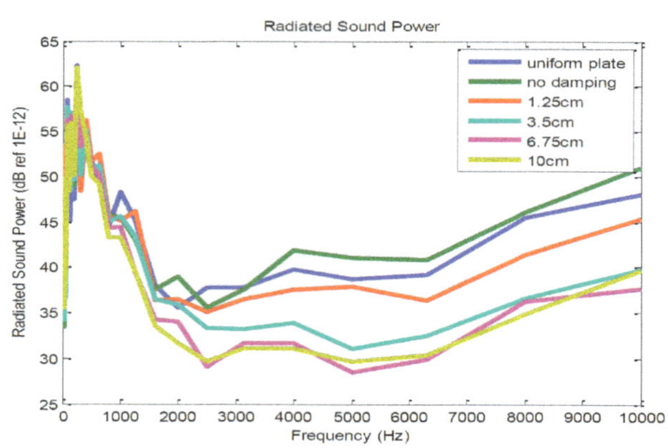

Figure 25. One-third octave band radiated sound power for a uniform plate and ABH plate with varying diameters of damping material. Reprinted with permission from [48]. Copyright American Society of Mechanical Engineers, 2015.

A practical experiment is studied by Bowyer and Krylov to investigate the effect of ABH on sound radiation of engine cover [49].The ABH were machined on two plates and then bonded into the engine cover with glue as shown in Figure 26. Figure 27 shows the sound radiation from a reference engine cover (dashed line) in comparison with the engine cover attaching plates with ABH (black line) at 2100 rpm with the bonnet closed. A total average reduction from the reference specimen of 6.5 dB was recorded. This indicates that engine covers with ABHs can decrease the sound radiation from the vehicle engine.

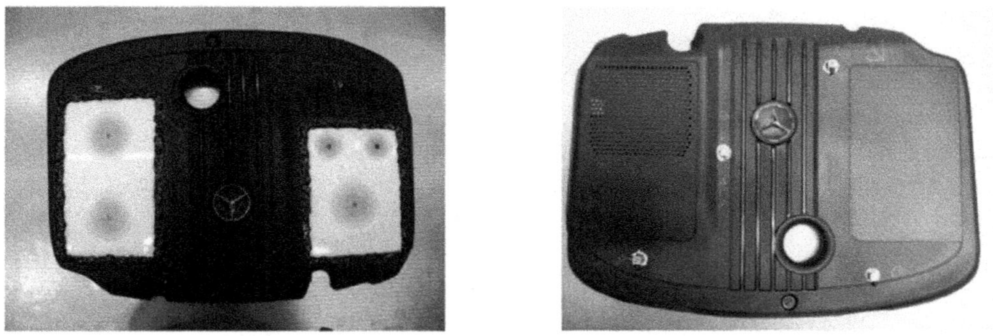

(a) (b)

Figure 26. Engine cover attaching two plates with ABHs (**a**) and reference cover (**b**). Reprinted from [49] under a CC BY-NC-ND 4.0 license. Copyright Bowyer E.P. and Krylov, V.V., 2015.

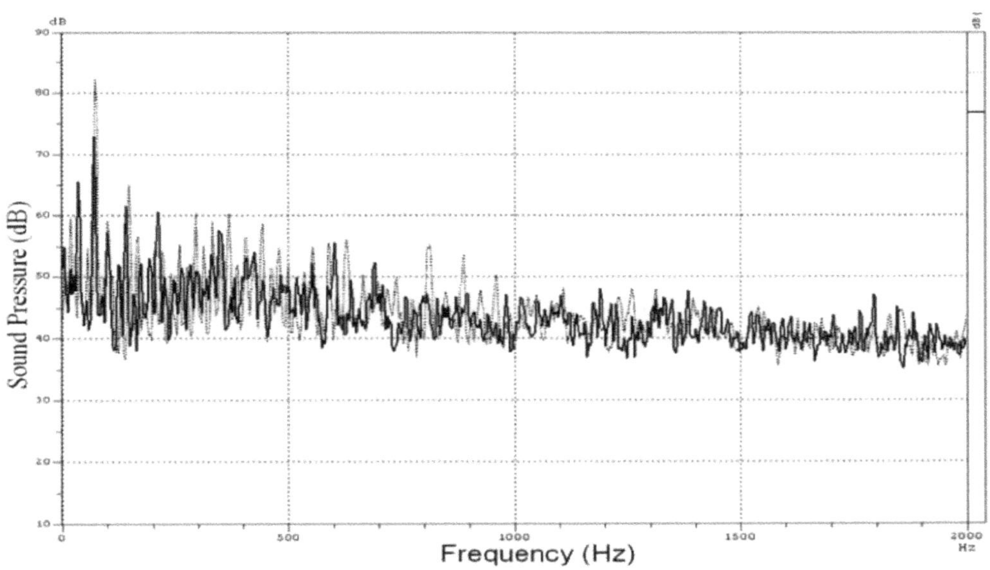

Figure 27. Sound radiation from a reference engine cover (dashed line) compared with the engine cover attaching plates with ABH (black line) at 2100 rpm with the bonnet closed. Reprinted from [49] under a CC BY-NC-ND 4.0 license. Copyright Bowyer E.P. and Krylov, V.V., 2015.

An experimental study on sound absorption in air of ABH based inhomogeneous acoustic waveguides [50] was developed by Azbaid et al. [51,52]. The inner radius of the structures as shown in Figure 28 are built following the linear function (exponent m = 1) and power-law function (exponent m = 2), respectively. Using two microphone transfer function methods, the experiment results show a substantial reduction in the reflection coefficient. The adding of absorbing porous materials results in a further reduction of the reflection coefficient.

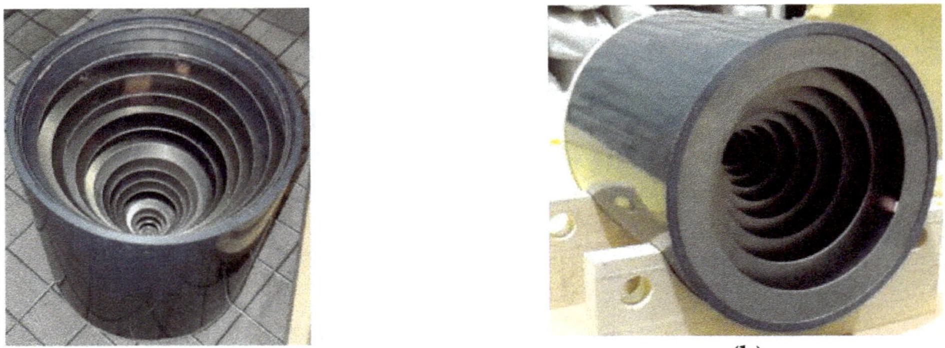
(a) (b)

Figure 28. Photo of (**a**) a linear ABH (exponent $m = 1$) and (**b**) a quadratic ABH (exponent $m = 2$). Reprinted from [51] under a CC BY-NC-ND 4.0 license. Copyright Azbaid El Ouahabi, A., Krylov, V.V. and O'Boy, D.J., 2015.

A numerical study on Helmholtz resonators (HR) using ABH to enhance the energy absorption performance was developed by Zhou and Semperlotti [53]. The HR cavity shown in Figure 29 is partitioned by a flexible plate-embedded ABH and the numerical results show a great increase in the energy absorption of the HR system.

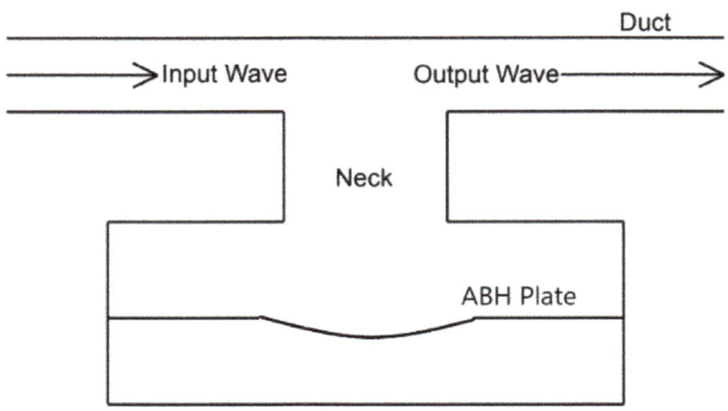

Figure 29. Schematic of the HR system with ABH. Adapted from [53]. Copyright Noise Control Foundation, 2016.

3.3. Application of ABH to Vibration Energy Harvesting

Zhao and Conlon developed a numerical study to investigate the structures tailored with ABHs to enhance vibration energy harvesting under both steady state and transient excitation [10]. Figure 30 shows the schematic of the plate with five equally spaced 1D ABH grooves attached with transducers. The plate is excited on the right side by a 400 N force sweeping from 0 to 10 kHz. The external resistance is 1 Ω. The numerical results of this study show the performance of the energy harvesting under steady state excitation and transient excitation. It can be observed that, by comparing with the flat plate, the normalized energy ratio of the plate with ABH increases drastically up to 80% in the 5–10 kHz frequency band at all the five ABH locations, and it increases most at Location 2. This study shows that the structure with ABH can drastically increase the efficiency of the energy harvesting. Later, another numerical and experimental investigation of a plate with three 2D ABHs (as shown in Figure 31) was also studied by Zhao and Conlon [54]. The ABHs effectively focus broadband energy to the center of the ABH. This also indicates that the focusing ability of the ABH is independent of the spectral and spatial characteristics of the external mechanical load.

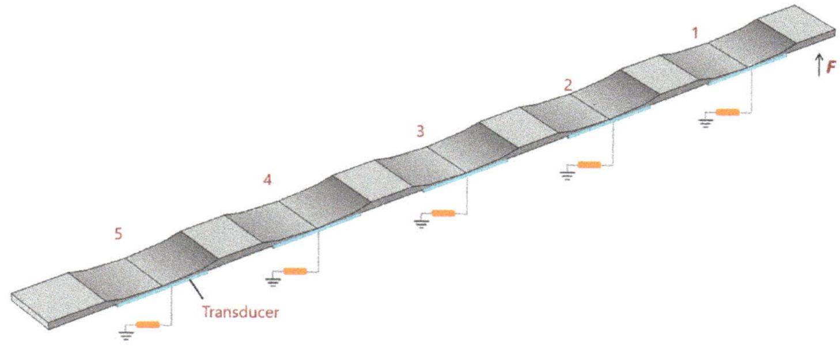

Figure 30. A schematic of the plate with five equally spaced 1D ABH grooves, and one of the ABH with the surface mounted piezo-transducer. Adapted from [10]. Copyright IOP Publishing, 2014.

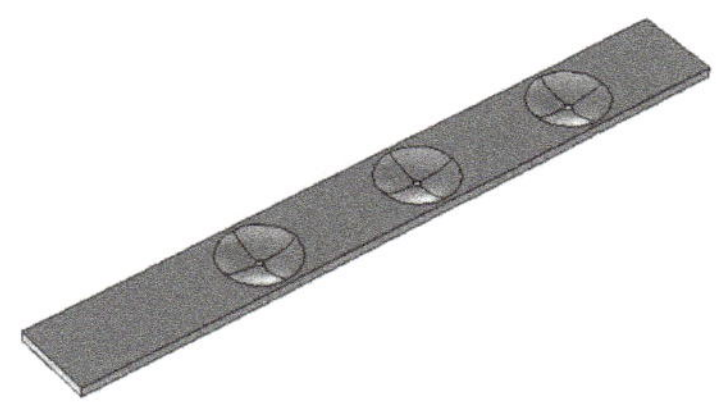

Figure 31. A schematic of the plate with three equally-spaced 2D ABHs. Adapted from [54]. Copyright IOP Publishing, 2015.

An experimental study on vibration energy harvesting using a cantilever beam with a modified ABH cavity was developed by Zhao and Prasad [8]. The cantilever beam is designed with an ABH cavity near the fixed end due to the presence of higher strain energy. The experimental setup of a cantilever beam with modified ABH cavity attaching a piezo sensor is shown in Figure 32. Figure 33 shows the energy concentration effect of the ABH cavity at 2000 Hz. Figure 34 shows the experimental results of the voltage power spectra. The vertical axis presents the decibel value referring to 1 volt. It can be observed that the beam with ABH cavity (red line) has a higher voltage output level within the frequency range from 900–500 Hz, the voltage output is 10 dB higher than the beam without the ABH (blue line). The increases due to the ABH cavity are substantial, even without considering the neutralization of the piezo material. Embedding a modified ABH cavity uses less piezoelectrical material and decreases the weight of the host structure. Tunability can be achieved by adjusting the length of the ABH cavity. Thus, ABH cavity design into structures has good potential in increasing the energy harvested.

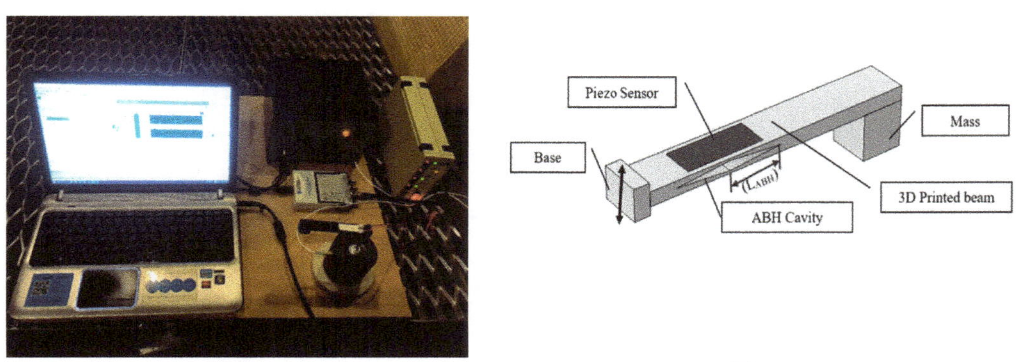

(a) (b)

Figure 32. (**a**) The experimental setup of a cantilever beam with modified ABH cavity with an attached piezo sensor. (**b**) Schematic of a 3D-printed beam with an ABH cavity [8].

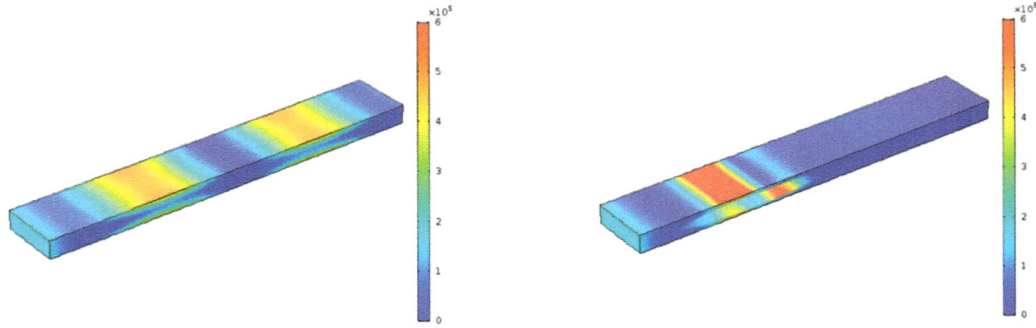

(a) (b)

Figure 33. Numerical results for the energy of (**a**) the cantilever beam without an ABH cavity and (**b**) with an ABH cavity at 2000 Hz [8].

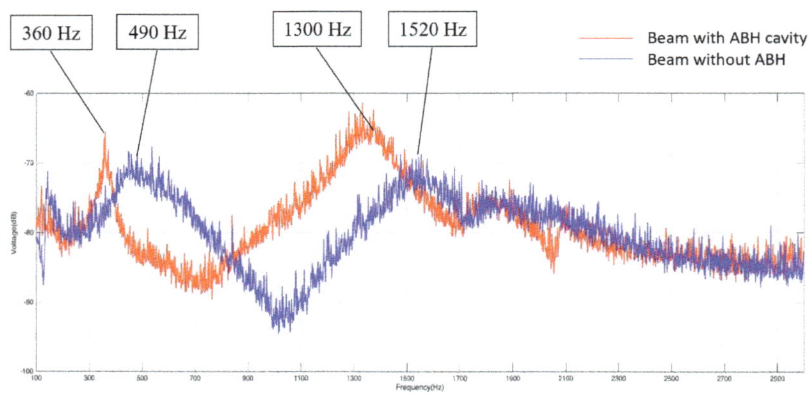

Figure 34. Experimental result for the voltage power spectrum. The blue line is for the beam without the ABH cavity and red line is for the beam with the ABH cavity [8].

A numerical study on energy harvesting using multiple ABH cavities as shown in Figure 35 was studied by Liang et al. [26]. The size of piezoelectric patches is designed to be relatively smaller than the wavelengths of ABH features, avoiding neutralization of the electric charge. Figure 36 shows the numerical result of power harvested. The results show that the beam with ABH cavities are more effective for vibration energy harvesting than uniform structures.

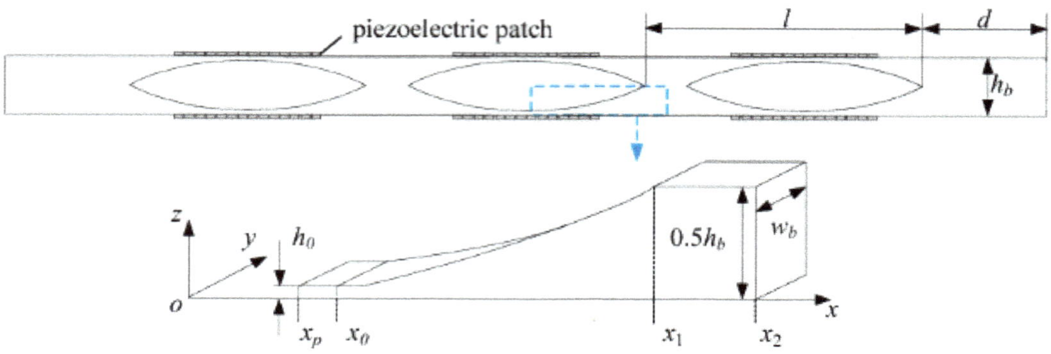

Figure 35. Experimental result for the voltage power spectrum. The blue line is for the beam without the ABH cavity and red line is for the beam with the ABH cavity. Reprinted with permission from [26]. Copyright IEEE, 2018.

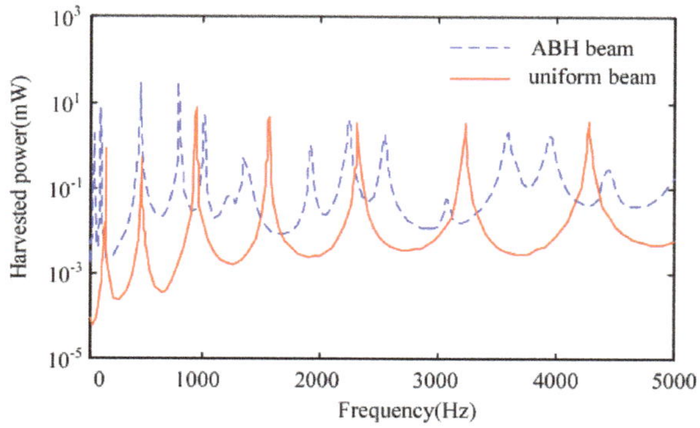

Figure 36. Harvested power spectra from the beam with and without ABHs, when R = 1 Kω. Reprinted with permission from [26]. Copyright IEEE, 2018.

3.4. Discussion

Applications of ABH for vibration control, sound radiation and vibration energy harvesting are presented. Generally, 1D ABH edges or grooves on beams and multiple 2D ABHs on plates with damping materials or piezo transducers are the major methods for the design of structures to enhance the vibration damping, sound reduction, and energy harvesting. Some specific structures embedding ABHs for practical applications, such as turbo fan blades, engine covers, etc., are also reviewed.

4. Design of Structures Using Acoustic Black Holes

4.1. Design of 1D ABH

The length of the ABH (L_{ABH}), the exponent m, and the residual thickness h_1 are the three geometrical parameters that affect the shape of 1D ABHs. Exponent m affects the depth of the ABH curve when the length of the ABH and residual thickness are fixed. Figure 37a shows the various shapes of power-law profiles with different values of exponent m. $m = 0$ means the beam is of uniform thickness (no ABH). L_{ABH} affects the length of ABH part when m is fixed as shown in Figure 37b. For 2D ABH, L_{ABH} represents the radius of outer circle abstracting the radius of inner hole.

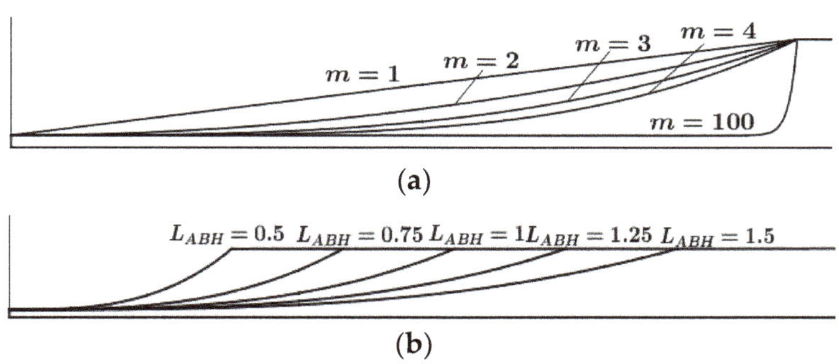

Figure 37. (**a**) Shapes of power law profiles with various m, and (**b**) shapes of power law profiles with various L_{ABH}(in).

In the study by Krylov, which is mentioned in Section 3.1.1 [29], the concept of "truncation" is applied. For a given value of exponent m, the truncation length is controlled by two variables, the length of the ABH L_{ABH} and the residual thickness h_1. In this numerical study, it can be observed that when the truncation length is larger than about 0.01 m the reflection coefficient of the beam with ABH (the solid line in Figure 4) increases sharply. This indicates that even a small truncation length results in a large increase in the reflection coefficient, which weakens the ABH effect. Thus, a sharp end of an ABH is critical for structural design. It is noted that a small amount of damping material can effectively restrain the increase of reflection coefficient.

A numerical and experimental study on sound radiation of a beam with a 1D ABH is developed by Li and Ding [55]. In Figure 38, it can be seen that the increase of the truncation thickness results in decrease of radiated sound power in the frequency from 43 to 160 Hz and increase from 626 Hz to 6 kHz. The reason for this phenomenon was discussed and explained. The increase of truncation thickness causes added mass and equivalent stiffness effects, which makes an enhancement of the low frequency bandwidth. On the other hand, an increase of truncation thickness causes an increase of the reflection coefficient and a decrease of the degree of wave concentration in the ABH, therefore weakening noise suppression at high frequencies. This indicates that an optimization of the truncation thickness is required for the best performance of noise reduction.

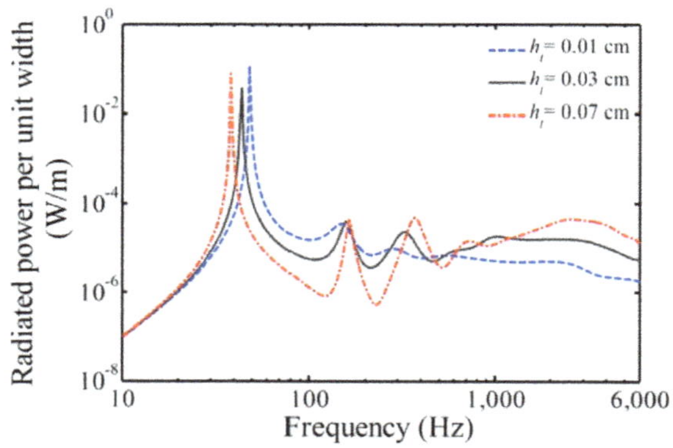

Figure 38. Sound radiation from the beams with various truncation thicknesses. Reprinted with permission from [55]. Copyright Elsevier, 2019.

A wavelet-decomposed energy model is developed and investigated by Tang and Cheng [56]. A new type of beam structure to achieve broad attenuation bands in relatively low frequencies is proposed. The model can be used to predict the frequency bounds of the band gaps. A parametric analysis is also illustrated. The increase of m and decrease of the truncation thickness h_0 would decrease the lower band gaps so as to enhance the ABH effect.

A numerical study on the influence of geometrical parameters of ABHs on vibration of cantilever beams was studied by Zhao and Prasad [57]. Figure 39 shows the 3rd mode shape simulation results of the beam without ABH [58] and beams with 1D ABH with various m values. It can be observed that the amplitude of displacement increases at the ABH location (with the increase of m value) and decreases at other locations of nodes.

The numerical study on the sound radiation from vibrating cantilever aluminum beams with various exponent m, which are 10 inches long, 1 inch wide, and 1/8 inch thick, with various shapes of ABHs at the free end, and the residual thickness h_1 equal to 1/64 inch is developed by Zhao and Prasad [27]. The results show that, in this group of beams with given residual thickness and length of the ABH, the concentration effect performs better when m is less than 5. However, when m is larger than 5, the ABH loses its concentration effect. Figure 40b shows how the logarithmic ratio of near field sound radiation at the free end and the other three anti-node locations changes with various m values, which essentially shows how the m values affect the concentration effect. This indicates that the m value has a maximum limitation. Another theoretical and numerical study on vibration energy concentration of the 1D ABH was developed by Li and Ding [59]. The length of the edge part with comparatively large deflection deformation L_e, as show in Figure 41, was investigated. With the given parameters of the ABH, the results show that more than 80% strain energy and 96% kinetic energy are trapped within a small area (L_e) from the wedge tip, which shows the energy concentration effect. For a fixed value of L_e, the peak ratio of energy occurs in the range of m from 2.5–3.0 for this case. This study also indicates that the exponent m has an optimum value to provide the best concentration effect.

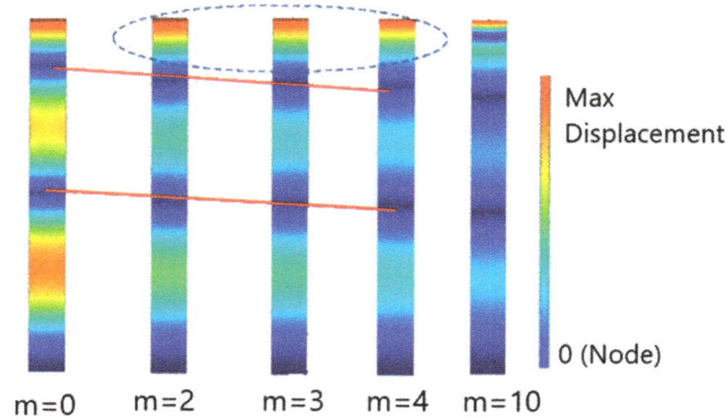

Figure 39. Third model shape simulation results. The bottom is the fixed end, and the top is the free end. Blue color shows the locations of nodes and red color shows the peak value locations [57].

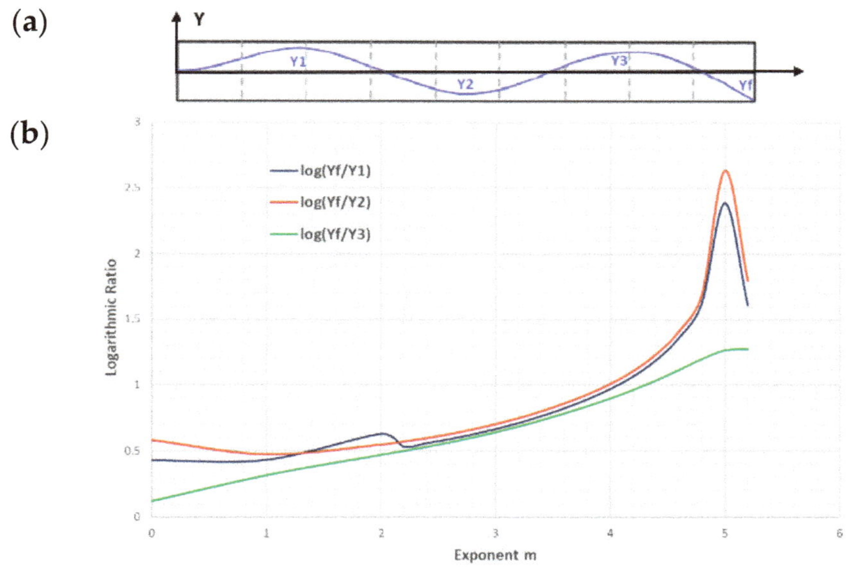

Figure 40. (**a**) The fourth mode of cantilever beam [58]. (**b**) The logarithmic ratio of displacement at free end and other three anti-nodes along with the various m values in Group 1 with excitation of 1600 Hz [57].

Figure 41. Schematic in 1D shows the ABH length of the edge part with a comparatively large deflection deformation L_e. Adapted from [59]. Copyright SAGE Publications, 2018.

One possible explanation of the phenomenon is that when the m value is larger, the ABH part becomes so thin that the uniform part of the beam and ABH part becomes two isolated vibration systems and it loses the pattern of the response of the beam vibration, the ABH part performs as a untuned dynamic absorber of the uniform part, which can be seen in Figure 39, the beam with m = 10 generated an extra node at the ABH location. A previous study developed by Feurtado and Conlon [9] applied another assumption. In this study, it is shown that the violation of smoothness criterion is a significant design problem for ABHs. Smoothness criterion is an assumption that the change of the local wave number is small enough over distances to make the ABH perform better, which is stated by Mironov [12]. Wave number variation (NWV) is investigated to assess the validity of the smoothness criterion. Equation (17) shows that NWV is a function of frequency and position and need to be much less than 1.

$$\frac{dk(x)}{dx}\frac{1}{k^2} = \left(\frac{1}{2}\right)\left(\frac{E}{\rho\omega^2 12(1-v^2)}\right)^{1/4}\frac{1}{h^{1/2}}\frac{dh}{dx} \ll 1, \tag{17}$$

where E is the Young's modulus, v is the Poisson's ratio, ρ is the density, h is the varying thickness at the ABH of the beams, and ω is the angular frequency of the flexural wave. According to this theory, an increased m value increases the NWV and the smoothness criterion is violated for a greater range of frequencies. For $m = 10$, the NWV is above one for the frequency range of study, which is shown in Figure 42. Further work investigates that NWV < 0.4 is a good rule for designing ABHs [60].

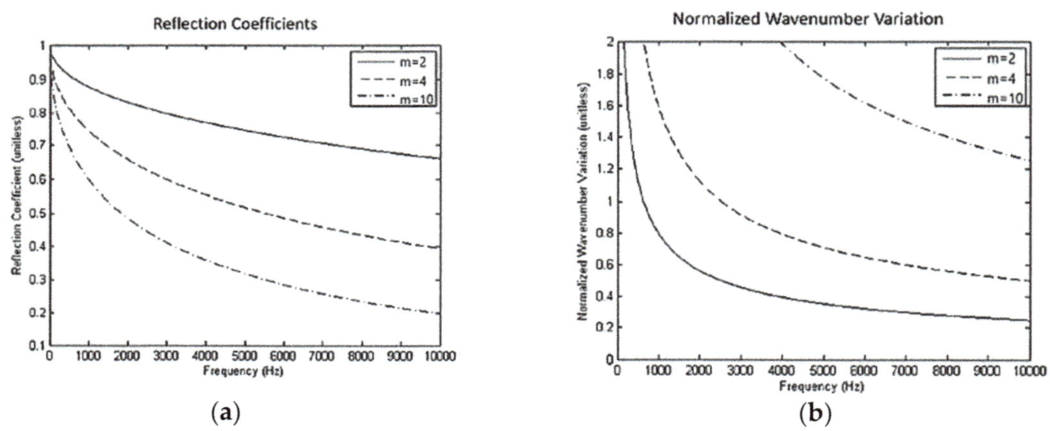

Figure 42. (**a**) Reflection coefficients and NWV for 5-cm long ABHs with m = 2, 4, 10. (**b**) Increased m increases the NWV and violates the smoothness criterion. Reprinted with permission from [9]. Copyright Acoustic Society of America, 2014.

In the study of Denis et al. which investigated ABH as a vector for energy transfer from the low frequencies to high frequencies in the framework of nonlinear vibration, a parametric study on a number of beams shown in Figure 43 was developed [61]. Longer additional termination, which means longer L_{ABH}, can improve the efficiency of energy transfer. It is also noted that a 2D ABH can present larger L_{ABH} with minimal thickness and obtains larger modal density in the low-frequency range.

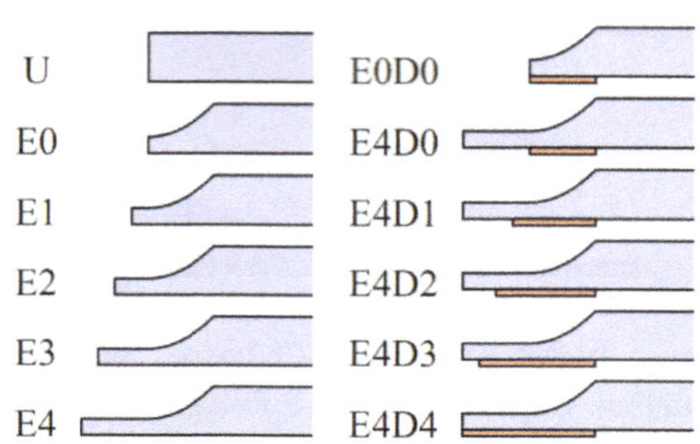

Figure 43. Beams with various lengths of ABHs and various lengths of damping. Reprinted with permission from [61]. Copyright Elsevier, 2017.

The effect of the length of ABH is also investigated by Zhao and Prasad [8,27]. For a given beam at a given excitation frequency there are specific optimum values of L_{ABH}. The simulation results of total energy at the ABH location of beams with the ABH cavity (shown in Figure 32b) is shown in Figure 44. It is seen that the beam with L_{ABH} of 19 mm has maximum energy output. This indicates that the ABH is tunable by changing L_{ABH} to obtain a higher concentration of vibration energy [8].

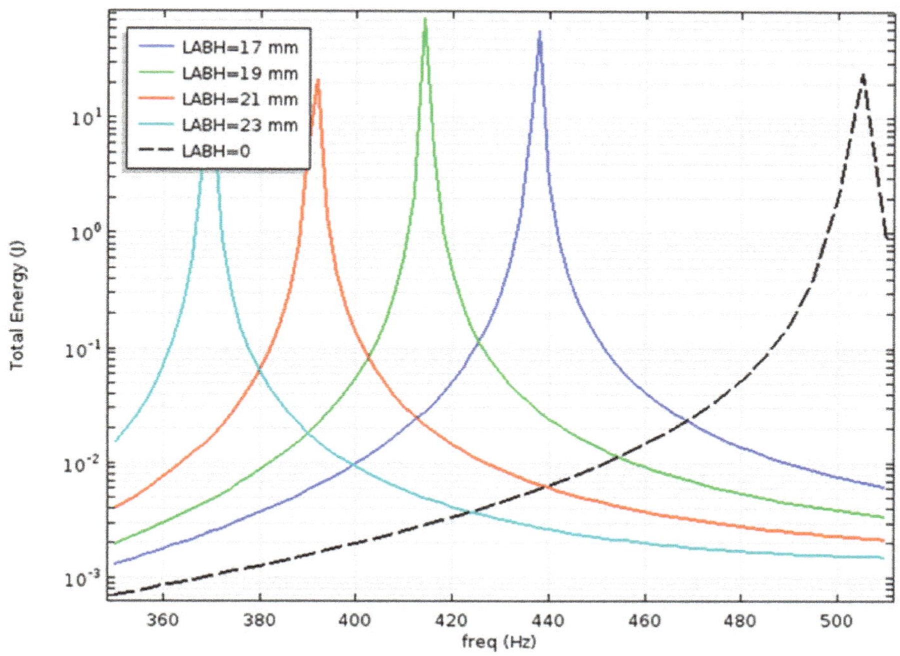

Figure 44. Simulation results of total energy output of 3D-printed beams with ABH cavities with various L_{ABH} [8].

From previous studies in this section, it can be seen that, all the three geometrical parameters need an optimization design to get a best performance on vibration damping and sound reduction. An investigation of the optimization design and position of an embedded 1D ABH is studied by McCormick [62]. This study is on a thin simply-supported beam with a 1D ABH at some position along the beam. The relevant design variables are shown in Figure 45, which are the length of the ABH L_{ABH}, the portion of the damped taper L_d, the thickness of the damping layer h_d, the drive point location D, and the offset between the center of the beam and the center of the ABH B. The optimization result is shown in Figure 46. Comparing with the uniform beam, the simple damped beam can decrease the surface-averaged velocity, but increases the mass by 5%. The optimized design can decrease the total mass of the beam by 15% and decreases the total surface-averaged velocity response by 12 dB, which has more reduction than a simple damped beam. The aim of the study is to figure out the ABH design that minimizes the total mass of the beam and, at the same time, minimizes the total surface-averaged velocity response. This benefits the design of ABHs for vibration reduction without adding mass.

As shown in the studies on vibration of beams and plates with ABH in Section 3.1, there is a minimum effective frequency, below which the ABH damping performance is not evident. A numerical and experimental study on the investigation of disappearance of the ABH effect is presented by Tang and Cheng [63]. It shows that cut-on frequency bands close to the low-order local resonant frequencies of the beam exist and the ABH effect failure in these frequency bands. The failure frequencies of the beam are delimited by the excitation point in order to avoid the phenomenon in the structural design.

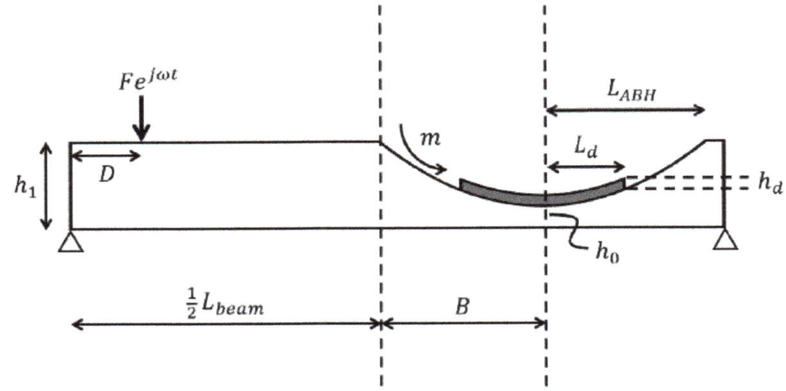

Figure 45. Schematic of the simple supported beams with the relevant design variables: Length of ABH L_{ABH}, the portion of the damped taper L_d, the thickness of the damping layer h_d, the drive point location D, and the offset between the center of the beam and the center of the ABH B. Reprinted with permission from [62]. Copyright McCormick, C.A.; Shepherd, M.R., 2018.

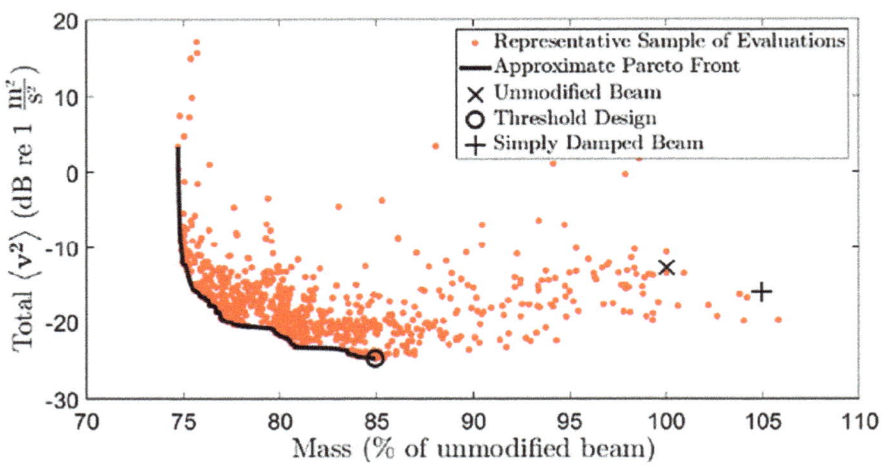

Figure 46. Approximate Pareto front (black curve) after 10,000 evaluations. Objective function evaluations for a sample of designs are denoted by red dots. The black × shows the function evaluation for an unmodified beam. The black circle shows the optimization. The black + shows the evaluation of a uniform beam with the same damping. Reprinted with permission from [62]. Copyright McCormick, C.A.; Shepherd, M.R., 2018.

4.2. Design of a 2D ABH

For a 2D ABH, it is to be noted that, besides geometrical parameters, the spatial layout of the ABH, the material of the host structure and damping material, the diameter of center hole, and the number of ABH also affect the performance of the ABH.

A numerical analysis of flexural ray trajectories in 2D ABH is developed by Huang et al. [32]. Figure 47 shows that for a given value of exponent m, the 2D ABH with thinner residual thickness (h_1 = 0.0002 m) absorbs more propagation energy than the one with residual thickness of h_1 = 0.001 m. In this study, the results also show the ratio of the rays converging to the center is 99.69% in the ideal ABH. The wave focalization can be enhanced with a larger m and a small h_1.

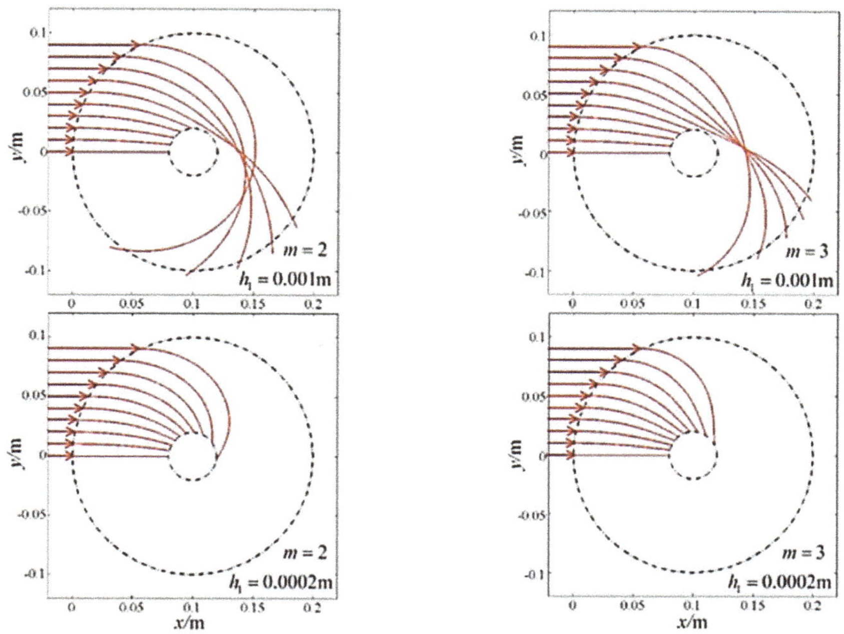

(a) (b)

Figure 47. Numerical analysis of flexural ray trajectories in 2D ABH with various geometrical analysis (**a**) m = 2; and (**b**) m = 3. Reprinted with permission from [32]. Copyright Elsevier, 2018.

A study on various layout of ABH on glass fiber composite plates with 1D and 2D ABH was studied by Bowyer and Krylov [16]. In this study, the glass fiber composite honeycomb sandwich panels with two 2D ABHs with various configurations (shown in Figure 48a) are investigated and the experimental results of acceleration are compared with a reference sandwich panel without ABH plates. The results show that theoretically the structure in Figure 48b should be the optimum layout to perform with the best vibration damping.

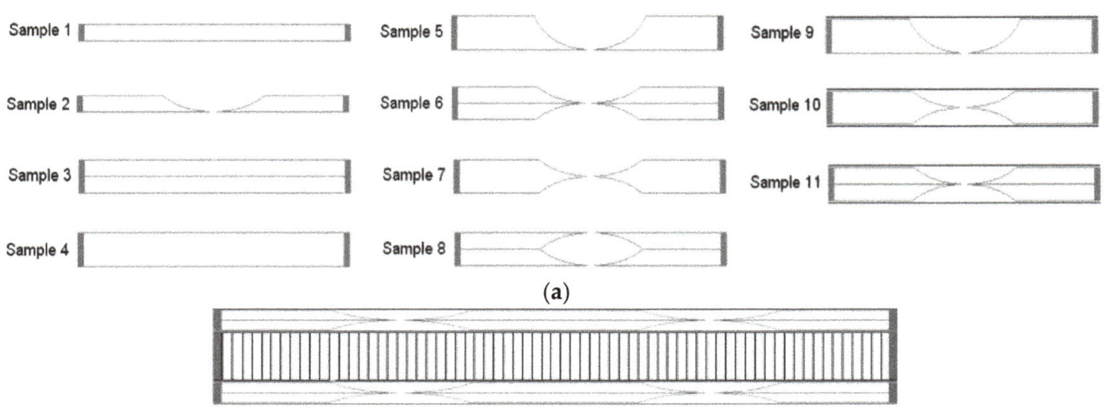

(a)

(b)

Figure 48. Cross-section of (**a**) various configurations of ABHs and (**b**) optimum configurations of ABHs. Reprinted from [16] under a CC BY 3.0 license. Copyright Bowyer E.P. and Krylov, V.V., 2014.

Due to the limitation of manufacturing, the ABH may not be perfect. A study on the wave energy focalization in a plate with imperfect 2D ABH is developed by Huang et al. [64]. The imperfect 2D ABH uses a polynomial profile instead of a power-law profile. The results still show drastic increases in energy density around the tapered area. Chong also indicates that the smooth and stepped ABH beams are slightly different [22]. Denis et al. also find that the controlled imperfection of the tip of the 1D ABH causes a decrease of the reflection coefficient, which indicates that imperfect extremities are not detrimental to the performance of the ABH effect [65]. These two studies lower the stringent requirement of the ideal power-law thickness variation of ABH.

A numerical study on the influence of number of ABHs on a plate in vibration is carried out by Conlon et al. [66,67]. The space averaged acceleration and total radiated sound power results for the 5 × 5 ABH (shown in Figure 49a) and uniform plates are compared in Figure 50. This shows that the plate with the ABH array has very effective results on vibration and noise reduction. Another comparison of total radiated sound power response of ABH panel between the panel with a 5 × 5 ABH array (Figure 49a) and the panel with 13 ABHs (Figure 49b) is shown in Figure 51. It can be seen that for higher frequencies (over 5 kHz) increasing of the number of ABHs results in more sound reduction, but in lower frequency this effect can only be observed in a few areas. This type of ABH array was also applied on the phononic crystals (PC) by Zhu and Semperlotti [68] to investigate peculiar dispersion characteristics.

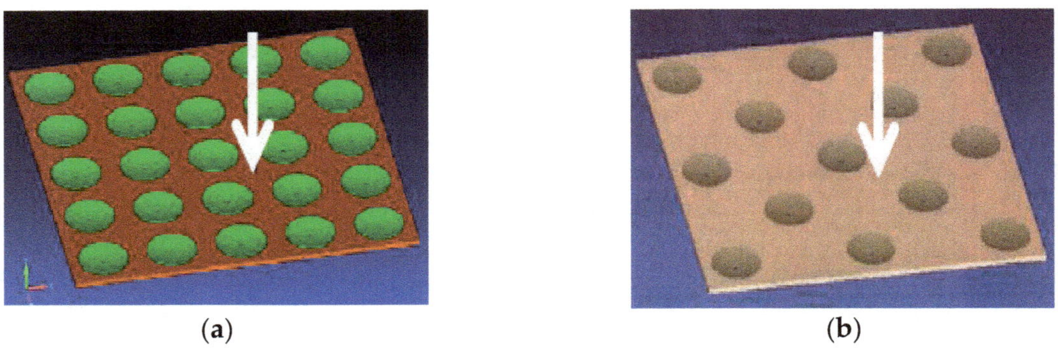

Figure 49. (**a**) Schematic of a panel with 5 × 5 ABH and (**b**) a panel with 13 ABHs. Reprinted with permission from [66]. Copyright Conlon, S.C.; Fahnline, J.B.; Shepherd, M.R.; Feurtado, P.A., 2015.

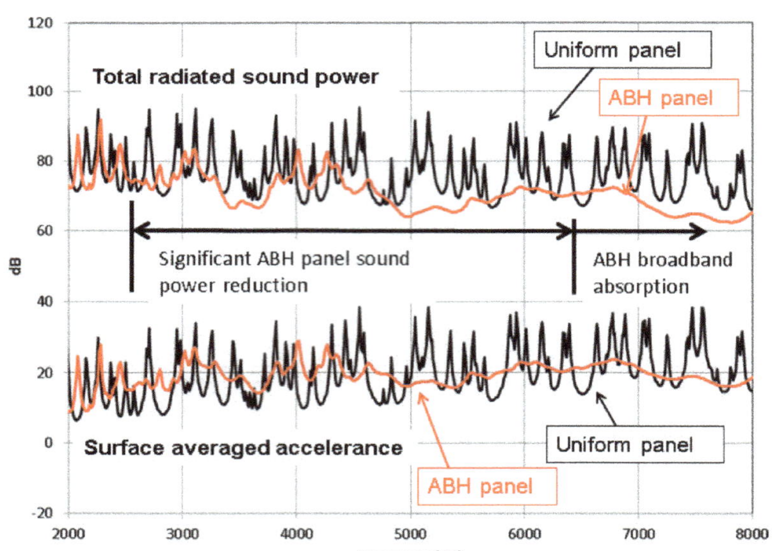

Figure 50. The space averaged acceleration (**bottom**) and total radiated sound power (**top**) results for the 5 × 5 ABHs (read line) and uniform plates (black line). Reprinted with permission from [66]. Copyright Conlon, S.C.; Fahnline, J.B.; Shepherd, M.R.; Feurtado, P.A., 2015.

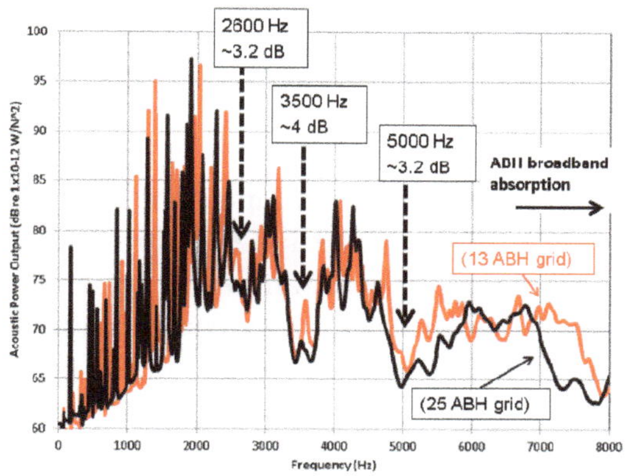

Figure 51. Total radiated sound power results for the panel with 5 × 5 ABHs and panel with 13 ABHs. Reprinted with permission from [66]. Copyright Conlon, S.C.; Fahnline, J.B.; Shepherd, M.R.; Feurtado, P.A., 2015.

4.3. Design of Damping Layer

The study Feurtado and Conlon [48] mentioned in Section 3.2 indicates that the radius of the damping layer has an optimum value. When the diameter is larger than the optimum value, the damping effect increases slightly. A time domain experimental study based on a laser visualization system is developed by Ji et al. [31]. The wave propagation and attenuation in a plate-embedded 1D ABH is investigated. It is also noted that there exists an optimal thickness for the damping layer. Krylov indicate that the damping material with higher material loss factor makes the damping effect of the ABH have better performance [29]. Denis et al. indicate that a good compromise has been found by using a long additional termination with a moderate length of the damping layer in the framework of nonlinear vibration, which can develop into a turbulent regime [61]. An optimization study was developed by Lee et al. [69] and indicates that the damping performance can be enhanced by treating the tip with an appropriate size of damping layer. Another study developed by Tang et al. [34] shows that the stiffness of the damping layer plays is more critical than mass for a better damping effect but, meanwhile, the effect of the added mass still needs to be accurately designed especially when the thickness of damping is considerable to the tip of the ABH wedge.

4.4. Studies on 3D-Printed Structures with ABH

Due to the limitation of traditional manufacturing methods, a power-law curve is difficult to manufacture. Bowyer and Krylov indicate that milling is not a practical way to machine structures with 2D ABH and suggest that 3D printing technology can be applied to produce these structures [23]. Due to the well-known advantages of 3D printing [70], vibration and sound radiation of 3D-printed structures are investigated by researchers. Zhou et al. developed a numerical study to investigate the dynamic and static properties of 3D-printed double-layered compound structures with ABHs [24]. Due to small residual thickness of the ABH profile, the structure with ABHs has low local stiffness and high stress concentration. The double-layered compound structure shows good damping performance and also significantly improves the static properties in structural stiffness and strength. Zhao and Prasad developed an experimental case study of vibration energy harvesting of a 3D-printed beam with a modified ABH cavity [8]. Liang et al. developed a numerical simulation for vibration energy harvesting of a 3D-printed beam with multiple ABH cavities [26]. Chong et al. numerically and experimentally studied the dynamic responses of 3D-printed beam with damped ABH grooves [22]. Rothe et al. investigated the dynamic behavior of the 3D-printed beams using a 3D- hexahedron finite element and an isotropic linear elastic homogenized material model [25]. A cantilever beam is embedded with the ABH at its free ends and fully fills the ABH area with flexible thermoplastic polyurethane (TPU) to make the overall thickness uniform. In comparison with the uniform beam, the beam-embedded ABH with TPU shows good damping performance.

5. Concluding Remarks

This review has presented the recent theoretical and numerical studies on ABH. Applications of ABH on beam- and plate-type structures have been demonstrated. It is shown that the use of ABH in structural design is effective in controlling vibration and noise without adding additional mass. This is particularly important for the design of lightweight structures, such as aircraft panels. ABH has also shown good promise in vibration energy harvesting, however, its practical application is still under research.

The current studies show that for 1D ABH, the geometric parameters are critical for the ABH effect. For 2D ABH, besides geometrical parameters, additional variables, such as the diameter of the center hole, the spatial layout of ABH array, and the number of ABHs, influence the performance of the ABH effect. Additionally, the material of the host structure and damping material are important for both 1D and 2D ABH. Optimization of the geometric parameters can significantly improve the damping effect of ABH. The number of ABHs and their spatial layout can expand the effective frequency range of ABH. A higher loss factor for both the host structure and damping layer can further improve the damping effect of ABH using less additive damping. However, there is a need for research efforts to further understand the interdependence of various geometrical parameters so that structural optimization studies can be carried out in designing structures with ABHs for better performance in vibration and noise control.

3D printing technology makes the manufacturing of more complex structures possible, and it is being applied in structures with ABH features. Further studies are required to apply ABH widely to real-life structures and carry out the application for more complex structures. Thus, it is observed from this review of various studies that the use of ABH in structural design for vibration and noise control is significantly effective and has great potential for research and industrial applications.

Funding: This research received no external funding.

Acknowledgments: The first author thanks the support from Department of Mechanical Engineering of Stevens Institute of Technology.

Conflicts of Interest: The authors declare no conflict of interest.

References

1. Rao, M.D. Recent applications of viscoelastic damping for noise control in automobiles and commercial airplanes. *J. Sound Vib.* **2003**, *262*, 457–474. [CrossRef]
2. Denis, V. Vibration Damping in Beams Using the Acoustic Black Hole Effect. Ph.D. Thesis, Universit´e du Maine, Le Mans, France, 2014.
3. Bowyer, E.P.; O'Boy, D.J.; Krylov, V.V.; Gautier, F. Experimental investigation of damping flexural vibrations using two-dimensional Acoustic 'Black Holes'. In Proceedings of the International Conference on Noise and Vibration Engineering (ISMA 2010), Leuven, Belgium, 20–22 September 2010.
4. Nishimura, I.; Kobori, T.; Sakamoto, M.; Koshika, N.; Sasaki, K.; Ohrui, S. Active tuned mass damper. *Smart Mater. Struct.* **1992**, *1*, 306. [CrossRef]
5. Mead, D.J. *Passive Vibration Control*; Wiley: Chichester, UK, 1999.
6. Li, H.; Tian, C.; Deng, Z. Energy harvesting from low frequency applications using piezoelectric materials. *Appl. Phys. Rev.* **2014**, *1*, 041301. [CrossRef]
7. Sojan, S.; Kulkarni, R. A Comprehensive Review of energy harvesting techniques and its potential applications. *Int. J. Comput. Appl.* **2016**, *139*, 14–19. [CrossRef]
8. Zhao, C.; Prasad, M.G. Studies on vibration energy harvesting using a cantilever beam with a modified Acoustic Black Hole cavity. In Proceedings of the Inter-Noise 2018, Chicago, IL, USA, 26–29 August 2018.
9. Feurtado, P.A.; Conlon, S.C.; Semperlotti, F.A. Normalized wave number variation parameter for Acoustic Black Hole design. *J. Acoust. Soc. Am.* **2014**, *136*, 148–152. [CrossRef] [PubMed]

10. Zhao, L.; Conlon, S.C.; Semperlotti, F. Broadband energy harvesting using Acoustic Black Hole structural tailoring. *Smart Mater. Struct.* **2014**, *23*, 065021. [CrossRef]
11. Pekeris, C.L. Theory of propagation of sound in a half-space of variable sound velocity under conditions of formation of a shadow zone. *J. Acoust. Soc. Am.* **1946**, *18*, 295–315. [CrossRef]
12. Mironov, M.A. Propagation of a flexural wave in a plate whose thickness decreases smoothly to zero in a finite interval. *Sov. Phys. Acoust.* **1988**, *34*, 318–319.
13. Krylov, V.V.; Bowyer, E.P. Acoustic Black Holes: A new approach to vibration damping in light-weight structures. *Proc. Inst. Acoust.* **2003**, *35*, 184–191.
14. Krylov, V.V.; Shuvalov, A.L. Propagation of localised flexural vibrations along plate edges described by a power law. *Proc. Inst. Acoust.* **2000**, *22*, 263–270.
15. Krylov, V.V. Acoustic Black Holes and their applications for vibration damping and sound absorption. In Proceedings of the International Conference on Noise and Vibration Engineering (ISMA 2012), Leuven, Belgium, 17–19 September 2012; pp. 933–944.
16. Bowyer, E.P.; Krylov, V.V. Experimental investigation of damping flexural vibrations in glass fiber composite plates containing one- and two-dimensional Acoustic Black Holes. *Compos. Struct.* **2014**, *107*, 406–415. [CrossRef]
17. Krylov, V.V. Acoustic Black Holes: Recent developments in the theory and applications. *IEEE Trans. Ultrason. Ferroelectr. Freq. Control* **2014**, *61*, 1296–1306. [CrossRef] [PubMed]
18. Conlon, S.C.; Fahnline, J.; Semperlotti, F. Numerical analysis of the vibroacoustic properties of plates with embedded grids of Acoustic Black Holes. *J. Acoust. Soc. Am.* **2015**, *137*, 447. [CrossRef] [PubMed]
19. Conlon, S.C.; Semperlotti, F. Passive control of vibration and sound transmission for vehicle structures via embedded Acoustic Black Holes. In Proceedings of the Noise-Con 2013, Denver, CO, USA, 26–28 August 2013.
20. Conlon, S.C.; Feurtado, P.A. Progressive phase trends in plates with embedded Acoustic Black Holes. *J. Acoust. Soc. Am.* **2018**, *143*, 921. [CrossRef] [PubMed]
21. Ji, H.; Huang, W.; Qiu, J.; Cheng, L. Mechanics problems in application of Acoustic Black Hole structures. *Adv. Mech.* **2017**, *47*, 201710.
22. Chong, B.M.P.; Tan, L.B.; Lim, K.M.; Lee, H.P. A review on Acoustic Black-Holes (ABH) and the experimental and numerical study of ABH-featured 3D printed beams. *Int. J. Appl. Mech.* **2017**, *9*, 1750078. [CrossRef]
23. Bowyer, E.P.; Krylov, V.V. Acoustic Black Hole manufacturing for practical applications and the effect of geomatical and material imperfections. In Proceedings of the Inter-Noise 2016, Hamburg, Germany, 21–24 August 2016.
24. Zhou, T.; Tang, L.; Ji, H.; Qiu, J.; Cheng, L. Dynamic and static properties of double-layered compound Acoustic Black Hole structures. *Int. J. Appl. Mech.* **2017**, *9*, 1750074. [CrossRef]
25. Rothe, S.; Blech, C.; Watschke, H.; Vietor, T.; Langer, S.C. Numerical modelling of additively manufactured Acoustic Black Holes. In Proceedings of the Inter-Noise 2018, Chicago, IL, USA, 26–29 August 2018.
26. Liang, Y.; Ji, H.; Qiu, J.; Cheng, L.; Wu, Y.; Zhang, C. Enhancement of piezoelectric energy harvesting using ABH structural tailoring. In Proceedings of the 2018 IEEE/ASME International Conference on Advanced Intelligent Mechatronics (AIM), Auckland, New Zealand, 9–12 July 2018.
27. Zhao, C.; Prasad, M.G. Studies on influence of geometrical parameters of an Acoustic Black Hole. In Proceedings of the Noise-Con 2017, Grand Rapids, MI, USA, 12–14 June 2017.
28. O'Boy, D.J.; Krylov, V.V.; Kralovic, V. Damping of flexural vibrations in rectangular plates using the Acoustic Black Hole effect. *J. Sound Vib.* **2010**, *329*, 4672–4688. [CrossRef]
29. Krylov, V.V.; Tilman, F.J.B.S. Acoustic 'Black Holes' for flexural waves as effective vibration dampers. *J. Sound Vib.* **2004**, *274*, 605–619. [CrossRef]
30. Zhao, L. Passive vibration control based on embedded Acoustic Black Holes. *J. Vib. Acoust.* **2016**, *138*, 041002. [CrossRef]
31. Ji, H.; Luo, J.; Qiu, J.; Cheng, L. Investigations on flexural wave propagation and attenuation in a modified one-dimensional Acoustic Black Hole using a laser excitation technique. *Mech. Syst. Signal Process.* **2018**, *104*, 19–35. [CrossRef]
32. Huang, W.; Ji, H.; Qiu, J.; Cheng, L. Analysis of ray trajectories of flexural waves propagation over generalized Acoustic Black Hole indentations. *J. Sound Vib.* **2018**, *417*, 216–226. [CrossRef]

33. O'Boy, D.J.; Bowyer, E.P.; Krylov, V.V. Damping of flexural vibrations in thin plates using one and two dimensional Acoustic Black Hole Effect. In Proceedings of the 10th International Conference on Recent Advances in Structural Dynamics (RASD 2010), Southampton, UK, 12–14 July 2010.
34. Tang, L.; Cheng, L.; Ji, H.; Qiu, J. Characterization of Acoustic Black Hole effect using a one-dimensional fully-coupled and wavelet-decomposed semi-analytical model. *J. Sound Vib.* **2016**, *374*, 172–184. [CrossRef]
35. Jeon, W. Vibration damping using a spiral Acoustic Black Hole. In Proceedings of the Inter-Noise 2016, Hamburg, Germany, 21–24 August 2016.
36. Lee, J.Y.; Jeon, W. Vibration damping using a spiral Acoustic Black Hole. *J. Acoust. Soc. Am.* **2017**, *141*, 1437. [CrossRef] [PubMed]
37. Zhou, T.; Cheng, L. A resonant beam damper tailored with Acoustic Black Hole features for broadband vibration reduction. *J. Sound Vib.* **2018**, *430*, 174–184. [CrossRef]
38. Bowyer, E.P.; Krylov, V.V. Slots of power-law profile as Acoustic Black Holes for flexural waves in metallic and composite plates. *Structures* **2016**, *6*, 48–58. [CrossRef]
39. Bowyer, E.P.; O'Boy, D.J.; Krylov, V.V.; Gautier, F. Experimental investigation of damping flexural vibrations in plates containing tapered indentations of power-law profile. *Appli. Acoust.* **2013**, *74*, 553–560. [CrossRef]
40. O'Boy, D.J.; Krylov, V.V. Damping of flexural vibrations in circular plates with tapered central holes. *J. Sound Vib.* **2011**, *330*, 2220–2236. [CrossRef]
41. Georgiev, V.B.; Cuenca, J.; Gautier, F.; Simon, L.; Krylov, V.V. Damping of structural vibrations in beams and elliptical plates using the Acoustic Black Hole effect. *J. Sound Vib.* **2011**, *330*, 2497–2508. [CrossRef]
42. Jia, X.; Du, Y.; Zhao, K. Vibration Control of variable thickness plates with embedded Acoustic Black Holes and dynamic vibration absorbers. In Proceedings of the ASME 2015 Noise Control and Acoustics Division Conference, San Francisco, CA, USA, 9–12 August 2015.
43. Krylov, V.V. Acoustic black holes as elements of metamaterials. In Proceedings of the IOP Physical Acoustics Group Tutorial Day, Loughborough, UK, 17 September 2015.
44. Semperlotti, F.; Zhu, H. Acoustic meta-structures based on periodic Acoustic Black Holes. *J. Acoust. Soc. Am.* **2015**, *137*, 2265. [CrossRef]
45. Foucaud, S.; Michon, G.; Gourinat, Y.; Pelat, A.; Gautier, F. Artificial Cochlea and Acoustic Black Hole travelling waves observation: Model and experimental results. *J. Sound Vib.* **2014**, *333*, 3428–3439. [CrossRef]
46. Bowyer, E.P.; Krylov, V.V. Damping of flexural vibrations in turbofan blades using the Acoustic Black Hole effect. *Appl. Acoust.* **2014**, *76*, 359–365. [CrossRef]
47. Bowyer, E.P.; Krylov, V.V. Sound Radiation of rectangular plates containing tapered indentations of power-law profile. In Proceedings of the 164th Meeting of the Acoustical Society of America, Kansas City, MO, USA, 22–26 October 2012.
48. Feurtado, P.A.; Conlon, S.C. Experimental analysis of vibration and radiated sound power reduction using an array of Acoustic Black Holes. In Proceedings of the Inter-Noise 2015, San Francisco, CA, USA, 9–12 August 2015.
49. Bowyer, E.P.; Krylov, V.V. A review of experimental investigations into the Acoustic Black Hole effect and its applications for reduction of flexural vibrations and structure-borne sound. In Proceedings of the Inter-Noise 2015, The 44th International Congress and Exposition on Noise Control Engineering, San Francisco, CA, USA, 9–12 August 2015.
50. Mironov, M.A.; Pislyakov, V.V. One-dimensional acoustic waves in retarding structures with propagation velocity tending to zero. *Acoust. Phys.* **2002**, *48*, 347–352. [CrossRef]
51. Azbaid, E.; Krylov, V.V.; O'Boy, D.J. Experimental investigation of the Acoustic Black Hole for sound absorption in air. In Proceedings of the 22nd International Congress on Sound and Vibration, Florence, Italy, 12–16 July 2015.
52. Azbaid, E.; Krylov, V.; O'Boy, D.J. Investigation of the Acoustic Black Hole termination for sound waves propagating in cylindrical waveguides. In Proceedings of the Inter-Noise 2015, San Francisco, CA, USA, 9–12 August 2015.
53. Zhou, X.; Semperlotti, F. The energy absorption properties of Helmholtz resonators enhanced by Acoustic Black Holes. In Proceedings of the Inter-Noise 2016, Hamburg, Germany, 21–24 August 2016.
54. Zhao, L.; Conlon, S.C.; Semperlotti, F. An experimental study of vibration based energy harvesting in dynamically tailored structures with embedded Acoustic Black Holes. *Smart Mater. Struct.* **2015**, *24*, 065039. [CrossRef]

55. Li, X.; Ding, Q. Sound radiation of a beam with a wedge-shaped edge embedding Acoustic Black Hole feature. *J. Sound Vib.* **2019**, *439*, 287–299. [CrossRef]
56. Tang, L.; Cheng, L. Broadband locally resonant band gaps in periodic beam structures with embedded Acoustic Black Holes. *J. Appl. Phys.* **2017**, *121*, 194901. [CrossRef]
57. Zhao, C.; Prasad, M.G. Studies on sound radiation from beam with Acoustic Black Hole. In Proceedings of the Noise-Con 2016, Providence, RI, USA, 13–15 June 2016.
58. Mustakerov, I.; Borissova, D. Multi-criteria model for optimal number and placement of sensors for structural health monitoring: lexicographic method implementation. *Adv. Model. Optim.* **2014**, *16*, 103–112.
59. Li, X.; Ding, Q. Analysis on vibration energy concentration of the one-dimensional wedges-shaped Acoustic Black Hole structure. *J. Intell. Mater. Syst. Struct.* **2018**, *29*, 2137–2148. [CrossRef]
60. Feurtado, P.A.; Conlon, S.C. Investigation of boundary-taper reflection for Acoustic Black Hole design. *Noise Control Eng. J.* **2015**, *63*, 460–466. [CrossRef]
61. Denis, V.; Pelat, A.; Touzé, C.; Gautier, F. Improvement of the Acoustic Black Hole effect by using energy transfer due to geometric nonlinearity. *Int. J. Non-Linear Mech.* **2017**, *94*, 134–145. [CrossRef]
62. McCormick, C.A.; Shepherd, M.R. Optimal design and position of an embedded one-dimensional Acoustic Black Hole. In Proceedings of the Inter-Noise 2018, Chicago, IL, USA, 26–29 August 2018.
63. Tang, L.; Cheng, L. Loss of Acoustic Black Hole effect in a structure of finite size. *Appl. Phys. Lett.* **2016**, *109*, 014102. [CrossRef]
64. Huang, W.; Ji, H.; Qiu, J.; Cheng, L. Wave energy focalization in a plate with imperfect two-dimensional Acoustic Black Hole indentation. *J. Vib. Acoust.* **2016**, *138*, 061004. [CrossRef]
65. Denis, V.; Pelat, A.; Gautier, F. Scattering effects induced by imperfections on an Acoustic Black Hole placed at a structural waveguide termination. *J. Sound Vib.* **2016**, *362*, 56–71. [CrossRef]
66. Conlon, S.C.; Fahnline, J.B.; Shepherd, M.R.; Feurtado, P.A. Vibration control using grids of Acoustic Black Holes: How many is enough? In Proceedings of the Inter-Noise 2015, San Francisco, CA, USA, 9–12 August 2015.
67. Conlon, S.C.; Fahnline, J.B.; Semperlotti, F.; Feurtado, P.A. Enhancing the low frequency vibration reduction performance of plates with embedded Acoustic Black Holes. In Proceedings of the Inter-Noise 2014, Melbourne, Australia, 16–19 November 2014.
68. Zhu, H.; Semperlotti, F. Phononic thin plates with embedded Acoustic Black Holes. *Phys. Rev. B* **2015**, *91*, 104304. [CrossRef]
69. Lee, J.; Jeon, W. An optimal design of 1D Acoustic Black Hole with damping layer for plate geometry. In Proceedings of the 23rd International Congress of Sound and Vibration, Athens, Greece, 10–14 July 2016; pp. 99–104.
70. Kamran, M.; Saxena, A.A. Comprehensive study on 3D printing technology. *MIT Int. J. Mech. Eng.* **2016**, *6*, 63–69.

4

Heat Transfer Enhancement due to Acoustic Fields: A Methodological Analysis

Alessandro Franco * and Carlo Bartoli

Department of Energy, Systems, Territory and Constructions Engineering (DESTEC), University of Pisa, 56126 Pisa, Italy; carlo.bartoli@unipi.it

* Correspondence: alessandro.franco@unipi.it

Abstract: The aim of this paper is to expose the main involved physical phenomena underlying the alteration of convective heat transfer in a heat exchanger subjected to imposed vibrations. This technique seems to have interesting features and industrial applications, such as for efficiency increases, heat transfer rate control and cleanliness action. However, a clear description and comprehension of how vibrations may alter the convective heat transfer coefficient in a heat exchanger has still not been reached due to the complexity of the involved physical mechanisms. For this reason, after a presentation and a schematization of the analyzed thermodynamic system, the fundamental alterations of the thermo-fluid dynamics fields are described. Then, the main involved physical phenomena are exposed for the three cases of gaseous, monophasic liquid and boiling liquid mediums. Finally, on the basis of the characteristics of these described phenomena, some considerations and indications of general validity are presented.

Keywords: heat exchangers; heat transfer enhancement; acoustic fields; vibrations; thermo-fluid dynamics; fouling mitigation

1. Introduction

The influence that the acoustic field has on convective heat transfer has attracted the interest of the scientific and technical community since the 1930s, with the pioneering works of important authors such as Martinelli and Boelter [1]. Starting from the 1960s, intense research activity, mainly experimental, has been carried out.

The aforementioned research activity led to contributions that can be subdivided in two different categories of works.

The first category deals with the study (almost completely experimental) of the physics of phenomena in very simplified configurations, with the primary purpose of clarifying which are the mechanisms underlying the convective heat transfer alteration. A lot of cases are available that involve different fluids in different regimes of flow, for example: gas in natural convection [2–8], gas in forced convection [9–12], liquid in natural convection [13–18], liquid in forced convection [19–23] and liquid in pool-boiling [24–28]. The results obtained are very interesting.

Considering the results exposed in the different papers it is possible to observe that the increase of convective heat transfer coefficient can reach maximum values close to 1000%, but this value is largely dependent on the experimental conditions (for example frequency and amplitude of oscillations, temperature and pressure of work, speed of the fluid etc.) and can reach minimum values under 10%.

From a general prospective, it is possible to observe that the liquid, in particular in subcooled conditions, is the medium for which the maximum increments are reached, though not negligible results have been gained for air and boiling; the lower the speed of the fluid, the higher the measured increments (so that the best conditions are for, in the order, natural convection, laminar regime and turbulent regime); the highest increments are reached giving vibrational energy directly to the heating surface rather than to the fluid in convection. Even though this achieves good results, some problems are still present: there is a great variability in results among the different articles considered in the paper [2–28], indicating that similar experimental conditions may lead to very different convective heat transfer enhancements; there is not agreement on what is the fundamental physical mechanism of alteration, while each author tries to give their own explanation of involved physics; no mathematical model able to quantitatively predict the convective heat transfer has been developed yet. A possible practical application of the enhancement of heat transfer using ultrasounds is observed in the cooling of electronic components.

The second group of papers deals instead with the experimental study of the effects that voluntary vibrational exiting of a heat exchanger can have on its global operation. In these studies, a preexisting heat exchanger is subject to excitation forces giving energy to its structure, and its behavior is then compared with that in absence of vibrations. There are two main lines of investigation. First of all, the increase of the global efficiency of the heat exchanger is measured in some papers by Legay and coworkers [29–32]: the results are very interesting, because in some configurations an increase close to 300% of the heat transfer coefficient is observed in typical configurations of heat exchangers. However, there is still no a clear understanding of the fundamental physical mechanism of alteration in the heat exchanger, and therefore which are the most favorable working and transducing conditions necessary for gaining a good enhancement.

The second line of investigation is concentrated instead on the possibility of cleaning up a fouled liquid to liquid heat exchanger employing the effects of acoustic cavitation [33–35].

Once again, the results are very interesting: almost a complete cleaning of the heat exchanger is possible without stopping and opening the device. Summarizing, the use of vibrational excitation in the heat exchangers seems to be an interesting engineering application. It could be used for:

(1) increasing the global efficiency of the heat exchanger;
(2) controlling the heat exchanging rate (by varying the power of transduction);
(3) keeping clean the heat exchanger without the need of frequent stops.

Nevertheless, it's still not clear how to choice position, frequency and energy of vibration to get a good and efficient action.

With the purpose of giving a clear background of the physics of the phenomenon, in this article the problem is approached by using the fundamental point of view of applied thermodynamics. After initial definitions of the thermodynamic system in analysis, the generic modifications of the different thermo-fluid dynamics variables are presented and discussed both for laminar and for turbulent flows. Then, these generic modifications are used for describing and discussing the main involved phenomena of convective heat transfer alteration for the three mediums of gas, monophasic liquid and boiling liquid. Finally, based on the presented physical phenomena, some general considerations and indications are given for each of the three fluid mediums.

2. Heat transfer Enhancements by Means of Acoustic Fields and Ultrasounds: An Analysis of the Available Experimental Results

In order to quantify the effect of ultrasound on the heat transfer coefficient, an enhancement factor *EF* is usually defined. This can be considered as the ratio of the value observed with ultrasound divided by the value under silent conditions. In the case of a thermal control system, it corresponds to the ratio of the overall heat transfer coefficient in the presence of ultrasonic waves divided by the overall heat transfer coefficient without ultrasound for the same hydrodynamic configuration:

$$EF = \frac{h_g \ with \ ultrasounds}{h_g \ without \ ultrasounds} \tag{1}$$

As discussed in the introductive section, a theoretical prediction of the effects of ultrasound on heat transfer is not fully available and a lot of analysis in the literature is based on experimental results.

In general, it can be observed that ultrasound is an efficient way to enhance heat transfer performances of liquid flow in laminar regime up to the level usually observed for turbulent flow conditions. On the other hand, a theoretical analysis of the problem and the definition of the heat enhancement effect of ultrasounds appear to be really complex: the extremely fast and small-scale phenomenon have interaction with turbulence inside the fluid.

Moreover, the problem can be complicated by the fact that the acoustic field propagations could be sensitive to fluid property variations, as well as by the complex geometry involved. In any case, the heat transfer enhancement factor surely depends on the following elements:

- the fluid under analysis;
- the phase of the fluid under analysis;
- the operating conditions of the fluid;
- the ultrasonic frequency of the generator, f;
- the ultrasonic generator power, P_{gen};
- the heat transfer surface characteristics;
- the geometry of the system;
- the material of the surface determining the possible formation of chemical substances.

Several authors have investigated the heat transfer enhancement due to ultrasonic waves using different fluids and in different operating conditions. The studies on the heat transfer enhancement by means of ultrasounds mainly focused on measuring the heat transfer rate change with such experimental parameters as the vibration frequency, the distance between the vibration transducer and heat source. Moreover, the heat transfer enhancement depends on a great number of conditions, such as ultrasonic wave power and frequency, dimension of heated wire, chamber dimension and liquid properties, affect the heat transfer enhancement degree.

In each experimental analysis, exposed in the literature since the 1960s, considerable efforts to investigate the effects of individual parameter changes, like the fluid have been made [2–28]. Furthermore, without generalization of those observations into reduced or dimensionless parameters, the collection of experimental data showing the influence of the factors above is severely limited unless the same conditions are realized in applications. Obviously they affect flow in a variety of ways depending on the conditions of the fluid. In principle, pressure waves caused by ultrasound transmitted in a medium cause an oscillation of the molecules around their mean position determining alternating compression and decompression of the medium.

The collected data, although in general obtained in experimental apparatuses, appear to confirm, in situations of practical interest, that the enhancement effect of the convective heat transfer coefficient can be remarkable in different specific situations.

In case of gas and liquids the enhancement factor can be very high (average values more than 200%), while being less remarkable in the case of two-phase fluid (average value of about 50%).

High enhancement (average value more than 400%) can be obtained in the case of natural convection, while smaller values (average value of about 120%) can be observed in the case of forced convection. Concerning the type of generation, better results can be observed with mechanical excitation rather than with acoustical excitation.

In Table 1 some details about different combinations of fluid and operating conditions are discussed regarding the various papers considered in the introduction, taking into account only the papers that analyze physical phenomena.

Table 1. An overview of the results obtained in references [2–28].

Reference	Phase	Type of Convection	Type of Generations	Approximate EF Value
[2]	Gas	Natural	Acoustic	75%
[3]	Gas	Natural	Acoustic	100%
[4]	Gas	Natural	Acoustic	1200%
[5]	Gas	Natural	Acoustic	
[6]	Gas	Natural	Acoustic	
[7]	Gas	Natural	Mechanical	400%
[8]	Gas	Natural	Mechanical	
[9]	Gas	Forced	Acoustic	80%
[10]	Gas	Forced	Acoustic	Laminar flow (50%)
[10]	Gas	Forced	Acoustic	Turbulent flow (25%)
[11]	Gas	Forced	Acoustic	25%
[12]	Gas	Forced	Mechanical	130%
[13]	Liquid	Natural	Acoustic	400%
[14]	Liquid	Natural	Acoustic	
[15]	Liquid	Natural	Acoustic	
[16]	Liquid	Natural	Mechanical	
[17]	Liquid	Natural	Mechanical	800%
[18]	Liquid	Natural	Mechanical	30%
[19]	Liquid	Forced	Acoustic	70%
[20]	Liquid	Forced	Acoustic	100%
[21]	Liquid	Forced	Acoustic	90%
[22]	Liquid	Forced	Mechanical	280%
[23]	Liquid	Forced	Mechanical	Laminar flow (450%)
[23]	Liquid	Forced	Mechanical	Turbulent flow (10%)
[24]	Pool-boiling		Acoustic	60%
[25]	Pool-boiling		Acoustic	10%
[26]	Pool-boiling		Mechanical	62%
[27]	Pool-boiling		Mechanical	24%
[28]	Pool-boiling		Mechanical	100%

3. Fundamental Modifications of Thermo-Fluid Dynamics by the Actions of an Acoustic Field

The acoustic field can be imposed with an acoustical transducer, that is a mechanic object which gives vibrational energy to the fluid medium in the control region. In this first case, we would define this ensemble of elements as an acoustically excited system.

The other possibility is the presence of a mechanical transducer, that is a mechanical device which gives vibrational energy to a wall of the control region (and then, from this wall, to the fluid medium). In this case, the ensemble of elements can be defined as a mechanically excited system.

In both the cases the fundamental open question is to understand how the vibrational energy may alter the standard convective heat transfer mechanisms. It's important to observe that, both in the cases of acoustically and mechanically excited systems, there could be more than one inlet or outlet sections, as there could be more than a portion of wall subjected to convective heat transfer.

In the specific literature on the subject, three main kinds of fluids have been taken into consideration:

1. gas or gas mixture, for example oxygen, nitrogen, air etc.;
2. single phase fluid (liquid), for example water in standard conditions;
3. two phase fluid, in particular a liquid and its vapor (e.g., boiling and condensing regimes).

In the cases (2) and (3) it is also important to take into consideration the presence of dissolved gases in the liquid (in particular air): those dissolved gases may cause the insurgence of a specific convective transfer alteration mechanism.

The periodic excitation action of the transducer is assumed to be sinusoidal in time. This action is completely determined if the following two parameters have been specified:

1. the frequency f, i.e., the number of oscillations of the transducer per unit of time;
2. the amplitude δ, i.e., the maximum displacement of the transducer from its relaxed position.

Depending on the frequency of the excitation, the transducer actions may be classified as:

1. infrasounds, if it is $f < 20$ Hz;
2. sounds, if it is 20 Hz $< f <$ 20 kHz;
3. high power ultrasounds, if it is 20 kHz $< f <$ 1 MHz;
4. low power ultrasounds, if it is $f >$ 1 MHz.

Infrasounds and sounds are known for generating dangerous acoustic pollution, so it's quite evident that they could not be employed in the majority of applications and that the application of ultrasounds has to be considered [35].

3.1. *Laminar Flows*

Let's suppose that one of the systems has reached a working condition characterized by a stationary laminar flow. Indicating with ψ the generic scalar thermo-fluid dynamics variable in a generic point $\vec{x}$ of the fluid domain (e.g., ψ could be the temperature, the pressure, a cartesian component of velocity vector etc.), we may imagine the trend of $\psi(t)$ as the one on the left. In this condition, we are saying that the system has reached a laminar equilibrium state.

Let's suppose now that the transducer has been activated for a relatively long time and let's imagine and plot the new trend of $\psi(t)$. In general, the typical trend is represented in Figure 1.

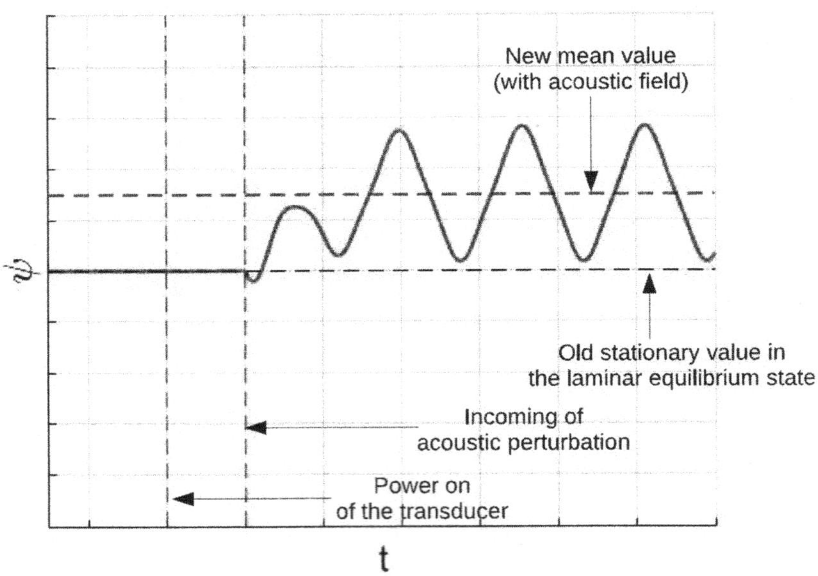

Figure 1. Two different aspects of alteration of the trend of y (laminar flow).

As it's possible to see, there are two different fundamental modifications in the trend of $\psi(t)$:

1. a new mean value of $\psi(t)$, in general different from the old one in the previous equilibrium state can be identified;
2. there is a periodic oscillation of $\psi(t)$, in general (but not always) close to sinusoidal shape.

Each physical phenomenon involved in the alteration of convective heat transfer in a laminar flow may be associated with one of the two above fundamental modifications.

3.2. *Turbulent Flows*

If the fluid has reached a working condition characterized by a stationary turbulent flow, indicating again with ψ the generic scalar thermo-fluid dynamics variable in a generic point $\vec{x}$

of the fluid domain, it is possible to recall the statistical theory of turbulence of Kolmogorov. According to this theory, the trend of each $\psi(t)$ can be written as the sum of a stationary mean value $\overline{\psi}$ and a randomly fluctuating value $\widetilde{\psi}(t)$:

$$\psi(t) = \overline{\psi} + \widetilde{\psi}(t) \tag{2}$$

so that the trend of $\psi(t)$ is like the one in the Figure 2.

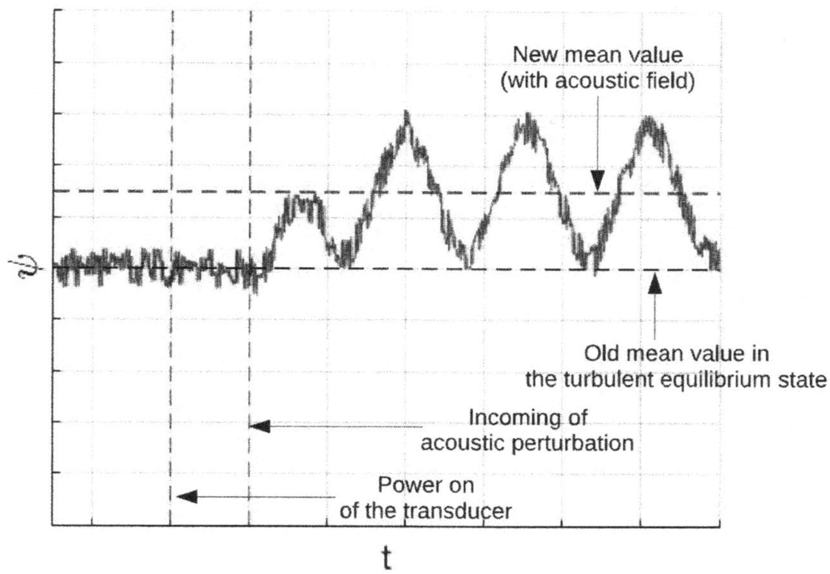

Figure 2. Two different aspects of alteration of the trend of y (turbulent flow).

In this condition, the system has reached a turbulent equilibrium state.

Let's suppose again that the transducer has been activated for a relatively long time and let's imagine and plot the new trend of $\psi(t)$. In this case, there are three different fundamental modifications in the trend of $\psi(t)$:

1. a new mean value of $\psi(t)$ that is in general different from the old one in the previous turbulent equilibrium state;
2. a periodic oscillation of $\psi(t)$, in general (but not always) very close to the sinusoidal shape;
3. a change in the characteristics of random fluctuations, which is a change in the statistical distribution of the frequencies and sizes of vortexes.

Again, it is possible to state that each physical phenomenon involved in the alteration of convective heat transfer in a turbulent flow may be associated to one of the first two fundamental modifications, while the third modification mechanism is not completely understood and, in any case, does not seem to play a key role.

4. Main Involved Phenomena of Convective Heat Transfer Alteration in Presence of an Acoustic Field: The Case of Gaseous Medium

In the case of gaseous medium, the scientific literature reports three main phenomena of convective heat transfer alteration caused by an acoustic field: thermo-viscous dissipation of acoustic field, acoustic streaming and acoustic alteration of thermal boundary layer. The illustration of the different phenomena can be exemplified for an acoustically excited system, but it can be extended without conceptual modifications to the mechanically excited systems.

4.1. *Thermo-Viscous Dissipation of Acoustic Field*

The phenomena of thermo-viscous dissipation of acoustic field is connected with the second fundamental modification mechanism, in particular with the oscillation of T(t) and of $\vec{v}(t)$. In the first case, in fact, there is a cyclical propagation of heat from compressed zones (at higher T) to rarefied zones (at lower T), so that there is a monotonic production of entropy (and then a monotonic destruction of exergy). Instead, in the second case, the oscillation of $\vec{v}(t)$ causes a cyclical action of frictions, so that there is a monotonic transformation of mechanical energy in internal energy of the gas (and then, again, a monotonic destruction of exergy). In conclusion, this phenomenon may be thought as a perpetual diffusion and transformation of organized mechanical energy provided by the transducer into disorganized internal energy of the continuum mediums: so, a general heating of the gas in the system is expected.

4.2. *Acoustic Streaming*

The phenomena of acoustic streaming are connected with the first fundamental modification mechanism, in particular with the change of the mean value of $\vec{v}(t)$. Indeed when a gas medium is subjected to an acoustic field, it is possible to observe an alteration of the trajectories of the fluid flow. As such, a macroscopic change of the internal mixing in the fluid domain may be important in determining an increase (or, in principle, a decrease) of convective heat transfer rate.

4.3. *Acoustic Modification of the Thermal Boundary Layer*

The phenomena of acoustic alteration of thermal boundary layer is connected with the second fundamental modification mechanism, in particular with the oscillation of the component of $\vec{v}(t)$ perpendicular to the portion of wall subjected to convective heat transfer. In fact, it's known from the fundamental courses on heat transfer that the convective heat transfer rate is, strictly speaking, proportional to the inverse of thermal boundary layer thickness:

$$q_{CONV} \approx \frac{1}{\delta_{TBL}} \tag{3}$$

Therefore, it's evident that in the compression phase the convective heat transfer rate increases, while in the expansion phase it decreases. However, supposing that compression and expansion have comparable displacements, it is noteworthy that the mean value in a symmetric interval $[x_M - a, x_M + a]$ of the reciprocal function $y = 1/x$ is greater than the median value $y(x_M)$. By this argument, it is possible to conclude that the acoustic alteration of thermal boundary layer conduces to an increase of the mean convective heat transfer rate.

5. Main involved Phenomena of Convective Heat Transfer Alteration in Presence of an Acoustic Field: The Case of Single Phase Liquid Medium

In the case of monophasic liquid medium, all three phenomena described above for the gas medium are still present, with the only difference that, because of its low compressibility, the acoustic alteration of thermal boundary layer is much less important. Moreover, in this case there are three more involved phenomena which do not manifest in the case of gas. They are: vaporous acoustic cavitation, gaseous acoustic cavitation and the anti-fouling effect of acoustic cavitation.

5.1. *Vaporous Acoustic Cavitation*

The phenomena of vaporous acoustic cavitation is connected with the second fundamental modification mechanism, in particular with the oscillation of the thermodynamic state P(t) and T(t). To illustrate this phenomena, let's assume first of all that the liquid in the heat exchanger is completely degassed. Let's imagine then and plot in the P-T diagram the local trajectory of the thermodynamic state: when the amplitude is quite big, the fluid locally enters in the vapor state. So, at that point a vapor bubble originates, grows and then, when the compression phase occurs, violently implodes (Figure 3). This violent implosion causes in turns shock waves and jets that, when the implosion happens close to the thermal boundary layer, can sensibly alter the thermal boundary layer, leading to its strong internal mix movements (Figure 4).

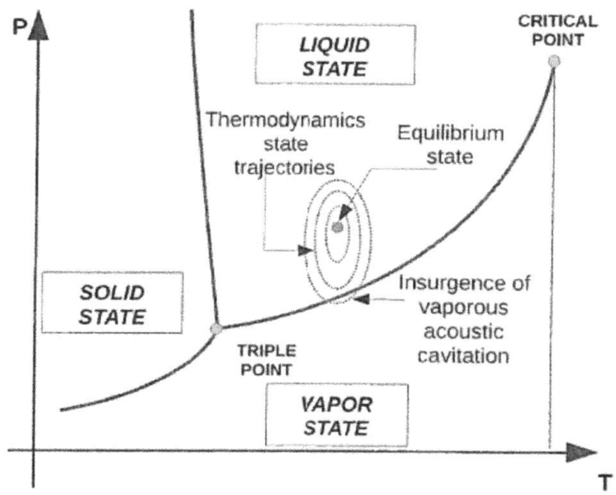

Figure 3. Schematic illustration of the phenomenon of vaporous acoustic cavitation.

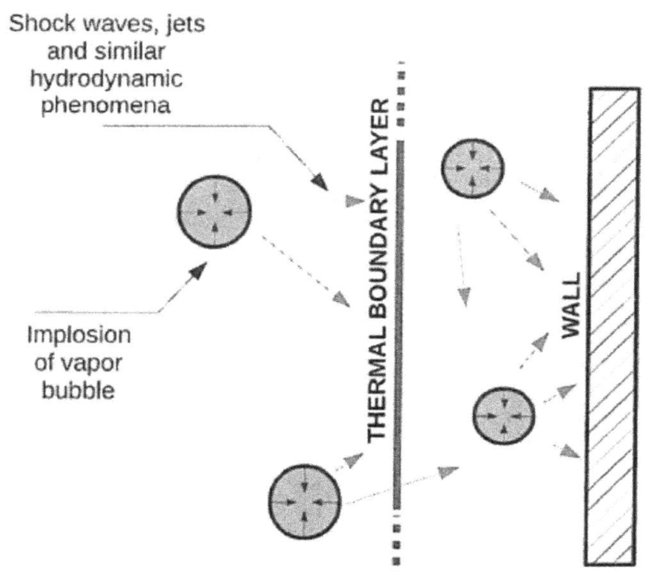

Figure 4. Schematic illustration of the phenomenon of vaporous acoustic cavitation.

5.2. Gaseous Acoustic Cavitation

Similarly to the vaporous acoustic cavitation, the phenomena of gaseous acoustic cavitation are connected with the oscillation of the thermodynamic state P(t) and T(t) too. These two phenomena are very similar to one another, but gaseous acoustic cavitation only happens when the concentration of a gas in the liquid exceeds a minimum value.

To explain this phenomena, it is necessary to recall that for every couple of liquid and gas (in particular for water and air) and for every thermodynamic condition a maximum concentration of gas exists (in terms of g/L) that could be dissolved in the liquid. So, once the concentration of the gas in the liquid is assigned, it is possible to represent a border line on the P-T plane: above this line, the gas is completely dissolved in the liquid and only a phase exists; below this line, the gas separates from the liquid and forms a new phase.

Thus, when the amplitude of the trajectory in the P-T plane is large enough, it is possible for the fluid going below the critical line: a gas bubble originates and then, when the compression phase incomes, violently implodes. In these quick movements, the hydrodynamics effects described above and illustrated in Figure 4 arise again, so that, when close to the thermal boundary layer, an increase of convective heat transfer due to internal mixing is possible. (Figure 5).

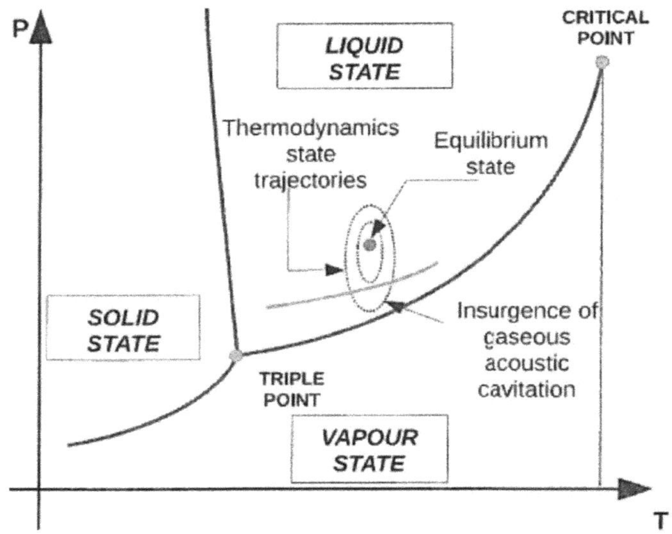

Figure 5. Illustration of the phenomena of gaseous acoustic cavitation (P-T diagram).

5.3. Anti-Fouling Effect of Acoustic Cavitation

The phenomena of anti-fouling effect of acoustic cavitation are quite particular, being indeed an indirect convective heat transfer enhancement mechanism associated with acoustic cavitation (both vaporous or gaseous). It is possible to note that in the working of a heat exchangers, a high thermal resistivity layer (the fouling layer) may originate on the walls of the machine, thus reducing its global efficiency. However, the hydrodynamic actions that originated from the bubble implosion already described below can also induce high mechanical strains in this fouling layer (Figure 6), in some cases making its disintegration and removal possible (and, therefore, a global increase of efficiency).

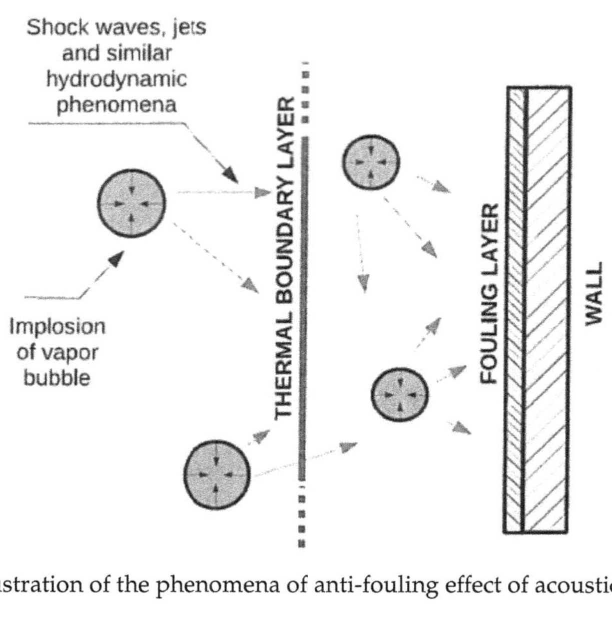

Figure 6. Illustration of the phenomena of anti-fouling effect of acoustic cavitation.

5.4. Considerations about the use of Acoustic Fields as a Way of Convective Heat Transfer Enhancement in the Case of Single-Phase Liquid

The indications given in the previous section for the gas are still valid in the case of liquid medium; moreover, for a good achievement of convective heat transfer enhancement for this kind of fluid, it is possible to

1. originate high pressure oscillations near the portion of wall subjected to convective heat transfer, so that an intense acoustic cavitation may be present there. Furthermore, from this point of view, it is quite evident that the closer the equilibrium state is to the liquid-vapor line, the easier it is to put the liquid in acoustic cavitation;
2. not degas the liquid in the heat exchanger in order to facilitate the occurrence of gaseous acoustic cavitation;
3. originate high pressure oscillation close to the zones more drastically subjected to fouling, so as to guarantee a possible removal of the fouling layer and good prevention of depositing phenomena.

Further considerations can be exposed. First of all, the difficulty of modeling the convective heat transfer process in the presence of an acoustic field is still more drastic in the case of liquid medium: now, three more basic phenomena are involved, phenomena that may in turn strongly alter the precedent thermo-fluid dynamics regime. Furthermore, the presence of two-phase separation surfaces on the bubbles of acoustic cavitation make necessary, for a detailed analysis, the relations of two-phase thermo-fluid dynamics.

Also in this case, it is highly probable that only a systematic experimental analysis could evidence the real characteristics of this enhancement technique. The use of mechanical energy for getting a heat transfer enhancement again seems to make very improbable the possibility of reaching a global higher exergy efficiency. However, in the case of liquid medium, the possibility of removing and preventing the fouling problems is very interesting and could in itself justify the utilization of acoustic fields: the heat exchanger could always work in an almost completely clean manner, with low pressure drops and without the undesirable necessity of frequent cleaning stops. By focusing the attention only on the fouling problem, it is possible to imagine a particular excited heat exchanger that periodically induces acoustic cavitation in all the parts of the walls subjected to convection: in this way, it could be possible to ensure cleaning without the constant spending of mechanical energy.

6. Main Involved Phenomena of Convective Heat Transfer with of an Acoustic Field: The Case of Boiling Liquid

The case of boiling liquid medium is, in all probability, the most complicated one. Indeed, with this kind of fluid medium all the six phenomena described above still manifest, while a new peculiar phenomenon emerges: the acoustic release of vapor bubbles. The phenomena of acoustic release of vapor bubbles is connected with the second fundamental modification mechanism, in particular with the oscillation of the strain tensor T(t) near the part of the wall subjected to convective transfer (Figure 7). In fact, the periodic action of vibrations on the vapor bubble results in an increased instability and therefore in a facilitated release of the bubble itself from the heating surface. In this way, a higher rate of latent heat departs from heating surface, causing so a global increase of convective heat transfer rate. It's important to note that the characteristics of heating surface (in particular its shape and its roughness) are very important both for boiling in presence or in absence of an acoustic field.

The indications given in the previous sections for gas and liquid medium are still valid in the case of boiling liquid; moreover, it is possible to add the further new advice for the purpose of reaching a good convective heat transfer enhancement: originate high pressure oscillations near the portion of wall subjected to convective heat transfer, so that the acoustic release of vapor bubble could be effective there.

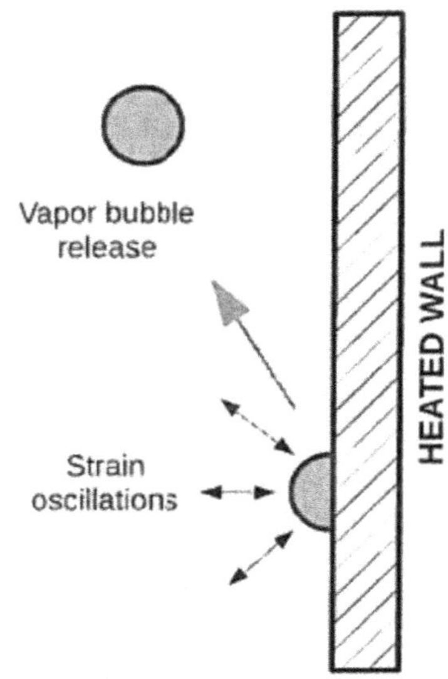

Figure 7. Illustration of the phenomenon of acoustic release of vapor bubbles.

Furthermore, it's quite evident that the complexity of a good modeling and prevision of the influence that vibrations may have in the case of boiling liquid medium reaches perhaps its maximum: not only all the six phenomena described for the other two case still manifest, but the presence of this new particular interaction between bubbles and acoustic fields seems to make farfetched the construction of an accurate forecasting theory. So, again, only an accurate experimental analysis could show the concrete effects that induced vibrations may have on boiling mechanisms.

7. Conclusions

In this article, the possibility of exiting a heat exchanger by using voluntary vibrations has been taken in consideration and analyzed using the fundamental point of view of applied thermodynamics.

The vibrational energy could be provided in the heat exchanger in two ways: directly to the fluid, that is the acoustically excited system; or to the solid structure of the heat exchanger, that is the mechanically excited system. In both cases, the proposed control volume is quite similar to the one usually presented in the standard courses of convective and mass transfer, except for the presence of a new element, which is the acoustical (or the mechanical) transducer. Therefore, with the geometry of the heat exchanger being known, as well as its working conditions and the characteristics of transduction (frequency and amplitude), the next major step is to understand how this new element can influence the convective heat transfer. So, the three fundamental modifications that the different thermo-fluid dynamics fields exhibit are:

(1) the change of the mean value;
(2) the insurgence of an oscillatory trend;
(3) only for turbulent flows, the change of aleatory oscillations characteristics.

Then, the different physical phenomena of convective heat transfer alteration individuated by the authors of bibliography are presented and explained for the three fluid medium of gas, monophasic liquid and boiling liquid. Each of these phenomena has been attributed to one of the first two fundamental modifications above.

In the case of gas, current literature has identified three phenomena of convective transfer alteration: the dissipation in heat of the introduced vibrational energy (thermo-viscous dissipation of acoustic field); the alteration of the mean flow trajectories (acoustic streaming); the intensification of conductive heat transfer rate in the thermal boundary layer (acoustic modification of thermal boundary layer).

In the case of liquid, more than the three phenomena of gas, three other phenomena have been individuated: the violent alteration and mixing of fluid flow, in particular the thermal boundary layer, through the implosion of low pressure vapor bubbles (vaporous acoustic cavitation); a similar violent alteration of vaporous acoustic cavitation, but through the implosion of a low pressure gas bubble (gaseous acoustic cavitation); the removing of eventually present fouling layers by the mechanical stress inducted by acoustic cavitation (anti-fouling effect of acoustic cavitation).

In the case of liquid boiling, all the six phenomena exposed above are still present. Furthermore, a new peculiar phenomenon has been identified: the destabilization of the vapor bubble on the heating surface and its facilitated detachment (acoustic release of the vapor bubble).

Just from this first presentation of the main involved phenomena, some qualitative considerations can be shown.

First of all, it's quite evident that, even in the simplest case of the gas medium, a lot of complicated and interrelated thermo-fluid and thermo-solid dynamics mechanisms are present: an accurate, instantaneous and punctual modeling of these mechanisms is very challenging, while some phenomenological modeling might be possible in specific working conditions for which a particular phenomenon prevails.

In addition, the use of mechanical energy for the purpose of increasing a heat transfer rate seems to determine an increase of energy use. However, in the cases of single phase and boiling liquid, the possibility of working with clean devices could be an interesting industrial application. Moreover, the possibility of a continuum controlling of the global efficiency of the heat exchanger, as well as the preventing and elimination of local hot spots, could be interesting with further features of this technique.

Finally, an extended experimental analysis of the problem is strongly recommended. Particular configurations of voluntary excited heat exchangers should be installed, and the relative increase of efficiency measured as a function of the previous equilibrium state (kind of mediums, Reynolds number, maximum ΔT, average pressure etc.) and of the transducing inputs (frequency and vibrational power).

Author Contributions: For research articles the two authors have contributed in the same way to formal analysis, investigation, data curation writing and editing.

Funding: This research received no external funding.

Acknowledgments: In this section the authors want to acknowledge the support of Francesco Niccolini, former student of the Master's Degree in Energy Engineering at the University of Pisa, that in his final dissertation developed some analytical models for studying the interaction of acoustic fields and heat transfer.

Conflicts of Interest: The authors declare no conflict of interest.

References

1. Martinelli, R.C.; Boelter, L.M.K. The Effect of Vibration upon the Free Convection from a Horizontal Tube. In Proceedings of the 5th International Congress for Applied Mechanics, Cambridge, MA, USA, 12–16 September 1938; Volume 578.
2. Kubanskii, P.N. Currents around a heated solid in a standing acoustic wave. *Trans. USSR Acad. Sci.* **1952**, *82*, 585.
3. Holman, J.P.; Mott-Smith, T.P. The Effect of High Constant Pressure Sound Fields on Free Convection Heat Transfer from a Horizontal Cylinder. *J. Aero/Space Sci.* **1959**, *20*, 188. [CrossRef]
4. Fand, R.M.; Roos, J.; Cheng, P.; Kaye, J. The local heat-transfer coefficient around a heated horizontal cylinder in an intense sound field. *J. Heat Transf.* **1962**, *84*, 245–250. [CrossRef]
5. Vainshtein, P.; Fichman, M.; Gutfinger, C. Acoustic enhancement of heat transfer between two parallel plates. *Int. J. Heat Mass Transf.* **1995**, *38*, 1893–1899. [CrossRef]
6. Hyun, S.; Lee, D.R.; Loh, B.G. Investigation of convective heat transfer augmentation using acoustic streaming generated by ultrasonic vibrations. *Int. J. Heat Mass Transf.* **2005**, *48*, 703–718. [CrossRef]
7. Lemlich, R. Effect of vibration on natural convective heat transfer. *Ind. Eng. Chem.* **1955**, *47*, 1175–1180. [CrossRef]

8. Fand, R.M.; Kaye, J. The influence of vertical vibrations on heat transfer by free convection from a horizontal cylinder. *Int. Dev. Heat Transfer Conf. Proc. Heat Tranfer. Conf. Boulder CO* **1963**, 490–498.
9. Jackson, T.W.; Harrison, W.B.; Boteler, W.C. Free convection, forced convection, and acoustic vibrations in a constant temperature vertical tube. *ASME J. Heat Transf.* **1959**, *81*, 68–74.
10. Lemlich, R.; Hwu, C.K. The effect of acoustic vibration on forced convective heat transfer. *AIChE J.* **1961**, *71*, 102–106. [CrossRef]
11. Fand, R.M.; Cheng, P. The influence of sound on heat transfer from a cylinder in crossflow. *Int. J. Heat Mass Transf.* **1963**, *6*, 571–596. [CrossRef]
12. Anantanarayanan, R.; Ramachandran, A. Effect of vibration on heat transfer from a wire to air in parallel flow. *ASME J. Heat Transf.* **1958**, *80*, 1426–1432.
13. Larson, M.B.; London, A.L. A study of the effect of ultimate vibrations on convective heat transfer to liquids. *ASME* **1962**, *62*, 62–68.
14. Fand, R.M. The influence of acoustic vibrations on heat transfer by natural convection from a horizontal cylinder to water. *J. Heat Transf.* **1965**, *87*, 309–310. [CrossRef]
15. Nakagawa, M. Analyses of acoustic streaming generated by four ultrasonic vibrators in a vessel. *Jpn. J. Appl. Phys.* **2004**, *43*, 2487. [CrossRef]
16. Deaver, F.K.; Penney, W.R.; Jefferson, T.B. Heat transfer from an oscillating horizontal wire to water. *J. Heat Transf.* **1962**, *84*, 251–254. [CrossRef]
17. Wong, S.W.; Chon, W.Y. Effects of ultrasonic vibrations on heat transfer to liquids by natural convection and by boiling. *AIChE J.* **1969**, *15*, 281–288. [CrossRef]
18. Hoshino, T.; Yukawa, H. Physical mechanism of heat transfer from heated and cooled cylinder to water in ultrasonic standing wave field. *J. Chem. Eng. Jpn.* **1979**, *12*, 347–352. [CrossRef]
19. West, F.B.; Taylor, A.T. The effect of pulsations on heat transfer-turbulent flow of water inside tubes. *Chem. Eng. Prog.* **1952**, *48*, 39–43.
20. Bergles, A.E. The influence of flow vibrations on forced-convection heat transfer. *J. Heat Transf.* **1964**, *86*, 559–560. [CrossRef]
21. Bulliard-Sauret, O.; Ferrouillat, S.; Vignal, L.; Memponteil, A.; Gondrexon, N. Heat transfer enhancement using 2 MHz ultrasound. *Ultrason. Sonochem.* **2017**, *39*, 262–271. [CrossRef] [PubMed]
22. Scanlan, J.A. Effects of normal surface vibration on laminar forced convective heat transfer. *Ind. Eng. Chem.* **1958**, *50*, 1565–1568. [CrossRef]
23. Raben, I. The use of acoustic vibrations to improve heat transfer. In Proceedings of the 1961 Heat Transfer and Fluid Mechanica Institute, Los Angeles, CA, USA, 19–21 June 1961; pp. 90–97.
24. Isakoff, S.E. *Effect of an Ultrasonic Field on Boiling Heat Transfer-Exploratory Investigation*; Heat Transfer and Fluid Mechanics Institute, Stanford University Press: Palo Alto, CA, USA, 1956; p. 15.
25. Park, K.A.; Bergles, A.E. Ultrasonic enhancement of saturated and subcooled pool boiling. *Int. J. Heat Mass Transf.* **1988**, *31*, 664–667. [CrossRef]
26. Bartoli, C.; Baffigi, F. Effects of ultrasonic waves on the heat transfer enhancement in subcooled boiling. *Exp. Therm. Fluid Sci.* **2011**, *35*, 423–432. [CrossRef]
27. Baffigi, F.; Bartoli, C. Influence of the ultrasounds on the heat transfer in single phase free convection and in saturated pool boiling. *Exp. Therm. Fluid Sci.* **2012**, *36*, 12–21. [CrossRef]
28. Nangia, K.K.; Chon, W.Y. Some observations on the effect of interfacial vibration on saturated boiling heat transfer. *AIChE J.* **1967**, *13*, 872–876. [CrossRef]
29. Monnot, A.; Boldo, P.; Gondrexon, N.; Bontemps, A. Enhancement of cooling rate by means of high frequency ultrasound. *Heat Transf. Eng.* **2007**, *28*, 3–8. [CrossRef]
30. Gondrexon, N.; Rousselet, Y.; Legay, M.; Boldo, P.; Le Person, S.; Bontemps, A. Intensification of heat transfer process: Improvement of shell-and-tube heat exchanger performances by means of ultrasound. *Chem. Eng. Process. Process Intensif.* **2010**, *49*, 936–942. [CrossRef]
31. Legay, M.; Le Person, S.; Gondrexon, N.; Boldo, P.; Bontemps, A. Performances of two heat exchangers assisted by ultrasound. *Appl. Therm. Eng.* **2012**, *37*, 60–66. [CrossRef]
32. Legay, M.; Simony, B.; Boldo, P.; Gondrexon, N.; Le Person, S.; Bontemps, A. Improvement of heat transfer by means of ultrasound: Application to a double-tube heat exchanger. *Ultrason. Sonochem.* **2012**, *19*, 1194–1200. [CrossRef] [PubMed]

33. Benzinger, W.; Schygulla, U.; Jäger, M.; Schubert, K. Anti fouling investigations with ultrasound in a microstructured heat exchanger. In Proceedings of the ECI Symposium Series, Volume RP2: Proceedings of 6th International Conference on Heat Exchanger Fouling and Cleaning—Challenges and Opportunities, Kloster Irsee, Germany, 5–10 June 2005.
34. Legay, M.; Allibert, Y.; Gondrexon, N.; Boldo, P.; Le Person, S. Experimental investigations of fouling reduction in an ultrasonically-assisted heat exchanger. *Exp. Therm. Fluid Sci.* **2013**, *46*, 111–119. [CrossRef]
35. Legay, M.; Gondrexon, N.; Le Person, S.; Boldo, P.; Bontemps, A. Enhancement of heat transfer by ultrasound: Review and recent advances. *Int. J. Chem. Eng.* **2011**. [CrossRef]

5

Archaeoacoustic Examination of Lazarica Church

Zorana Đorđević [1,*], Dragan Novković [2] and Uroš Andrić [3]

1 Institute for Multidisciplinary Research, University of Belgrade, 11000 Belgrade, Serbia
2 The School of Electrical and Computer Engineering of Applied Studies, 11000 Belgrade, Serbia; dragannovkovic72@gmail.com
3 Front Element Media, 11000 Belgrade, Serbia; uandric83@gmail.com
* Correspondence: zorana@imsi.rs; Tel.: +381-63-8995-176

Abstract: The acoustic analysis provides additional information on building tradition and related indoor practice that includes sound, thus deepening our understanding of architectural heritage. In this paper, the sound field of the Orthodox medieval church Lazarica (Kruševac city, Serbia) is examined. Lazarica is a representative of Morava architectural style, developed in the final period of the Serbian medieval state, when also the chanting art thrived, proving the importance of the aural environment in Serbian churches. The church plan is a combination of a traditional inscribed cross and a triconch. After the in situ measurement of acoustic impulse response using EASERA software, we built a computer model in the acoustic simulation software EASE and calibrated it accordingly. Following the parameters (reverberation time (T_{30}), early decay time (EDT) and speech transmission index (STI)), we examined the acoustic effect of the space occupancy, central dome and the iconostasis. In all the cases, no significant deviation between T_{30} and EDT parameter was observed, which indicates uniform sound energy decay. Closing the dome with a flat ceiling did not show any significant impact on T_{30}, but it lowered speech intelligibility. The height of iconostasis showed no significant influence on the acoustics of Lazarica church.

Keywords: archaeoacoustics; church acoustics; Lazarica church; architectural heritage; acoustic heritage; medieval building; reverberation time; speech intelligibility; acoustic simulation

---

1. Introduction

Architecture of a medieval church is a complex symbolical-metaphorical construct imbued with different references from a wide range of human experience [1]. The exclusive focus on its material reality reduces our understanding. Instead, architectural heritage requires the exploration of all the incorporated layers of meaning and their mutual relations. The multidisciplinary field of archaeoacoustics tends to introduce the balance of our comprehension of the history of a building by taking into account the aural environment influenced by various interconnected factors that create human experience in historical buildings, thus also arguing that such acoustic heritage should be considered as an intangible and inseparable layer of architectural heritage [2,3].

Christian churches have been archaeoacoustically researched for the last few decades. The comprehensive literature overview of historical church acoustics [4] considers the acoustic analysis of mostly occidental Christian churches in Europe, Asia and South America, emphasizing the need for a multidisciplinary approach that also includes perceptual and cultural aspects of the aural experience. Besides the examination of the effect of occupancy [5], speech intelligibility [6,7], and the effect of interior architectural forms and materials [8–10], it is also important to acknowledge the multidisciplinary research of acoustic vessels inbuilt in some of the historic churches [11,12], as an equally valid expression of the acoustical intentions of the builders. Archaeoacoustic research of Orthodox churches also covers a wide spectre of approaches. Among many, it includes the analysis of existing or virtually modelled church acoustics, comparison of the acoustic data from churches built in different periods [13], analysis of the live chant recorded on-site [14,15], and correlation of the development of chant and sacral architecture [16,17].

This paper contributes to the research of medieval Orthodox church heritage with its multidisciplinary approach to the investigation of mutual relatedness of architectural forms, religious practice and acoustics in the case of Serbian medieval Lazarica church. Therefore, it opens with the culture-historical context and architectural principles of building Lazarica church, leading to the acoustic requirements of the Orthodox religious indoor practice. The research methodology is based on the on-site acoustic impulse response measurement and acoustic modelling, in order to examine the acoustic influence of the presence of the faithful during the liturgy, the central dome, and the historical change in the height of the altar barrier (iconostasis). Acoustic analysis considered three parameters—reverberation time (T_{30}), early decay time (EDT) and speech transmission index (STI)—as the most relevant for two main acoustic requirements of Holy Liturgy, speech intelligibility and experience of monophonic Byzantine chant.

1.1. Culture-Historical Context of Building Lazarica Church in Kruševac, Serbia

At the end of the 14th century, the Serbian empire collapsed. After the death of the last emperor Uroš V Nemanjić in 1371, local rulers divided the land among themselves. While the Ottomans were penetrating from the south, Serbs were moving north. Among Serbian local rulers, prince Lazar Hrebeljanović who autonomously ruled the territory around the Morava rivers, most prominently tended to continue the well-established tradition of the Nemanjić dynasty [2]. However, in that restless and unstable period, new political, economic and cultural centres arose and thrived in the land of prince Lazar and his successors [4]. This inspiring zest was also reflected in church architecture, which is nowadays referred to as the Morava architectural style.

The building activity was especially induced after the reconciliation of Serbian and Byzantine church when the anathema of the Constantinople patriarch was removed and the new Serbian patriarch was elected in 1375. This was achieved thanks to, inter alia, the engagement of the monks from Mont Athos and Chilandar Monastery. The cultural relations of Serbia and Mount Athos thus grew even stronger.

Following Nemanjić's tradition of founding monasteries and church endowments, prince Lazar built the church of Ravanica monastery (1377) and then St. Stefan church in Kruševac city, also known as Lazarica [3]. The building of Lazarica (1377/81) is closely related to the work on the expansion and improvements of Kruševac city fortification [18]. Both churches—Ravanica and Lazarica—are equally important for the Morava style [19], representing the role models that achieved harmony and wholeness [18]. They marked the beginning of the most authentic building period in medieval Serbia that used the language of Byzantine constructions and shapes. It is important to note that this particular period—the end of 14th and the beginning of 15th century—was not very fruitful in Byzantine building. Due to the significant economic decay and the strengthening of the Ottoman Empire [19], the number of buildings decreased to a point of total disappearance, while in Morava Serbia construction projects flourished [20], embodying the fresh interpretation of Byzantine art [21]. The difficulty of reliably recognizing on more than one building the work of the same workshop of craftsmen leads to the conclusion that there was a strong need for the authentic art creation [20].

The development of cultural relations between Serbia and Mount Athos also affected the church architecture. The church plan in the Morava architectural style is a triconch (used in Mount Athos for monk choirs that have a significant role in the complex rites [4]) combined with an inscribed cross (traditionally used in Serbian medieval church architecture). Although applied before in Serbia, Saint Nikola Mrački church (1330), the chapel in the Rile monastery (1350) and the chapel of St. Nikola in the St. Arhangeli monastery near Prizren [7], the triconch plan became a recognizable solution for the Morava style. To designate himself as a carrier of the Nemanjić state sovereignty, it is hypothesised that prince Lazar especially looked up to the triconch church of St. Arhangeli monastery near Prizren, as an endowment of the emperor Dušan (the Mighty) Nemanjić [4]. There was also a more functional explanation as the rounded apses provided an additional chanting space on the northern and southern sides of the altar barrier, which was needed for the increased number of chanters coming to Morava Serbia from Mount Athos [8].

1.2. Architectural Design of Lazarica Church

Lazarica church was built in the Byzantine manner: alternating a row of sand stone and three rows of bricks in thick mortar couplings that emerge from the wall plane (Figure 1). It expresses a high artistic level of façade decoration, characteristic for the Morava architectural style. However, for the purpose of acoustic analysis, here we focus on the interior space organization and applied materials.

Figure 1. Lazarica church exterior view from the south-west (Photo credits: Z. Đ.).

Lazarica church has a triconch (three apses) plan combined with the compressed inscribed cross. On the intersection of cross arms rises the central dome (Figure 2). In Byzantine, and thus in Morava architecture as well, the dome had a distinctive meaning. It symbolized the heaven over the earth and complemented the image of the church as a microcosm [21] (p. 74). In Lazarica the dome is constructed on pendentives, supported by four pilasters and the attached arches (Figure 3, bottom left). The drum is circular from the inside and 8-sided from the outside. On the both ends of the transversal cross arm are chanting apses that significantly increase the space and facilitate the Holy Liturgy. The bays in naos are barrel vaulted, while the narthex is cross vaulted with a bell tower above. Since the tower is connected to the naos through one small window in the western wall of the naos, it does not affect the church acoustics.

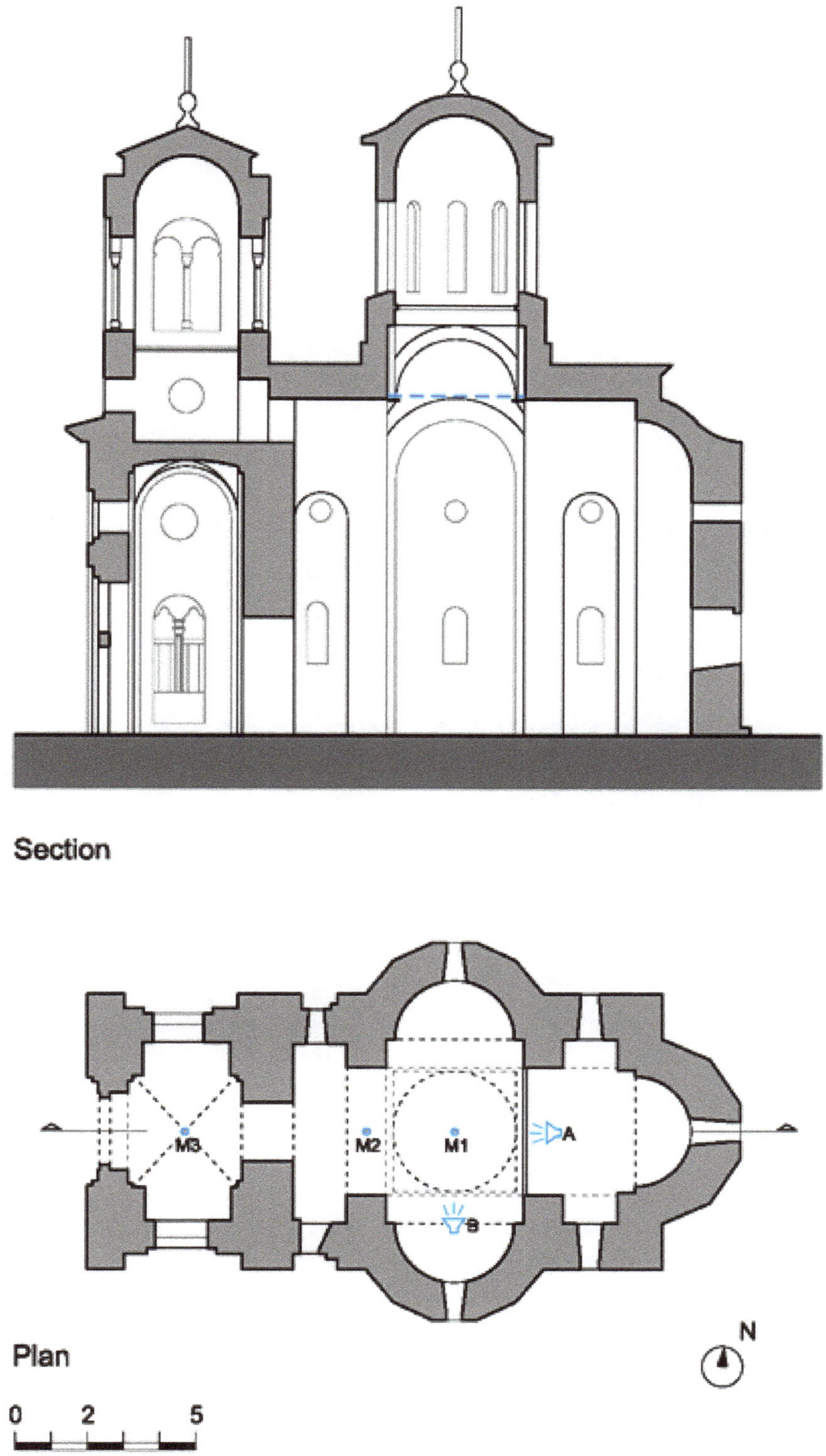

Figure 2. Plan and longitudinal section of Lazarica church with marked positions and directions of sound sources (A: Altar and B: Chanting apse) and measurement points (M1-M3). The blue dashed line in the section shows where the dome was replaced with a flat ceiling in the acoustic simulation model.

Figure 3. Lazarica church interior view from the narthex (top left), from the naos towards the altar (top right), towards the dome (bottom left), towards the southern chanting apse (bottom right) (Photo credits: Z. Đ.)

Proportional analysis of Lazarica church plan showed the combined application of two proportional methods—collapsing diagonals (based on a square) and triangulation (based on an equilateral triangle). If the length of the building brick is one module, than the side of the main proportional square is 13 modules. The dome drum circle is inscribed in this square from which all other geometric points are derived [18] (pp. 36–37).

The church interior consists of three main spaces—narthex, naos and altar—from the entrance in the west to the most sacred space on the east where the sun rises. In the altar the essence of the Orthodox Holy Liturgy takes place. Therefore, it is important to have a visual barrier which disables the faithful from approaching from the naos and to see the accomplishment of holy secrets by the priests in the altar. What we find today in Lazarica is the wooden iconostasis made in the 19th century, that is 3.7 m in width and 4.35 m in height (top of the cross is 6.5 m from the floor level, Figure 3) [18]. However, in the medieval period altar barriers were quite different. They were about 2–2.5 m high. Stone columns supported the beam, on which the curtains were hanging. Until the 14th century, the altar barrier had no icons. The central doors were beautiful gates, while the opening on the left and right had curtains. There was a short parapet wall. Comprehending that the attention of the faithful attending the liturgy is directed towards the altar barrier, over the centuries the curtains were complemented with additional rows of icons [10].

Due to the multiple architectural interventions since medieval period, in the beginning of 20th century Lazarica church was reconstructed with an intention to revive the original look from the time of prince Lazar. The main works were: taking off the façade mortar and revealing the stone-brick building play, reconstruction of the bell tower above the narthex, window details in the narthex, and rebuilding of the central dome. Nevertheless, there are no reliable data on the state of the church interior that was found. The main designer and project manager Pera Popović (1904–1908), wrote that he found the staircases leading from the narthex to the bell-tower. They do not exist anymore. There were also found segments of fresco painting from the 19th century, but it was removed during the reconstruction and the walls were mortared "straight and smooth". There is no evidence that Lazarica was fresco painted in medieval times. However, ever since Popović's reconstruction it was planned to fresco paint the church [22], but it has not been realized yet (Figure 3).

1.3. General Acoustic Requirements in Orthodox Churches in Medieaval Serbia

When we consider temple acoustics, it is important to know not only the interior shapes and finishes, but also the nature and purpose of the rites that took place there. An Orthodox temple is a place of congregation, where the Holy Liturgy is the most important act which implies communal praying, listening to the sermon, and receiving the Holy Eucharist and other sacraments. There are still in use two Liturgies written by Saint Basil the Great and Saint John Chrysostom in the second half of the 4th century. The Holy Liturgy consists of three parts. The first part is the Liturgy of Preparation or Proscomedia which is performed in silence in the northern part of the altar, where the priest prepares the bread and wine. The second part is the Liturgy of the Catechumens and the third is the Liturgy of the Faithful, both performed in a loud voice. After the Proscomedia, the church bells ringing announces the commencement of the second part of the Liturgy. The needs of the Holy Liturgy shaped the interior architecture. Thus the three parts of the Orthodox temple—narthex, naos and altar—correspond with its main acts. The narthex is for the Catechumens: those who can attend the Liturgy until a certain moment, as a part of the preparation process for receiving the Orthodox Christian Faith. The naos is the nave of the church where the faithful attend the service, standing with faces towards the altar. As the most sacred part of the temple, the altar is divided from the naos with the altar barrier, the iconostasis [23].

Byzantine chant that spread among Serbian people along with Christian faith is monophonic, with gradual melodic flow which follows rhythmically emphasized lyrics [24]. This chanting art reached its apex in the same period when the Morava architectural style developed [16]. In other words, the acoustics of a medieval Serbian church, such as Lazarica, is required to support both speech intelligibility and Byzantine chanting art. In Orthodox tradition no musical instruments are used inside a church.

2. Methodology

In accordance with ISO 3382 set of standards, monaural measurements of acoustic impulse response was carried out in the empty Lazarica church for two sound source positions (A: Altar and B: Chanting Apse) and three characteristic receiver points (M1: under the central dome, M2: in naos, and M3: in narthex), as shown on the Figure 2. As a sound source, an active loudspeaker dB Opera 110 (dBTechnologies, Valsamoggia, Italy) was used. Although the use of a directional sound source can significantly affect the obtained results of EDT and STI parameters that are analysed in this paper, the authors have decided to use such a sound source with the basic idea of establishing stronger correlation with the sound sources that emit sound energy in such spaces in real situations, which are primarily human voices. This fact should be kept in mind when it comes to the repeatability and reproducibility of measurement results. Sound field excitation was conducted from two positions that are commonly used in Orthodox religious practice, the altar and chanting apse. The excitation loudspeaker was placed at the height of 170 cm (assumed to be the average person's height) with central axe oriented towards naos.

Sine sweep signal in the range from 20 Hz to 20 kHz was used to acoustically excite the space. Acoustic impulse response measurement software EASERA (AFMG Technologies, Berlin, Germany) was used, together with Digidesign Mbox2mini audio interface and omnidirectional Behringer ECM8000 measurement microphone (Behringer, Willichm Germany) placed at assumed average height of standing listeners (170 cm).

For the purpose of sound field simulation, a detailed 3D model of the church interior was created in EASE software (AFMG Technologies, Berlin, Germany). The model was then calibrated in accordance with the results of the on-site impulse response measurements and examined for four cases: (1) empty church, (2) when faithful are present in the church, (3) when the dome is closed with a flat ceiling (Figure 2), and (4) when the iconostasis is lower, as it was in medieval times.

3. Results

3.1. Acoustic Model Calibration

The main intention of acoustic measurements conducted on-site was to investigate monophonic acoustic parameters defined by ISO 3382 set of standards, as a starting point for further analysis of basic sound field properties and comparison with values of acoustic parameters obtained by acoustic model simulation. Lazarica church interior was modelled in the EASE software and materials were assigned to all boundary surfaces. The church floor is covered with marble, which is standard material that can be found in EASE material database, but the EASE materials database does not contain a material that corresponds to Lazarica walls made of stone with a layer of mortar. Therefore, common concrete was used as the first choice for the wall material. In this respect, after comparing the first results of the simulation with the results of the on-site impulse response measurements, it was necessary to carry out calibration of the acoustic model, which was obtained by an iterative "trial and error" process. The custom values of the absorption coefficient of the church walls were defined (Table 1), which resulted in a sufficient harmonization of the results of the simulation and measurement as shown in Figure 4.

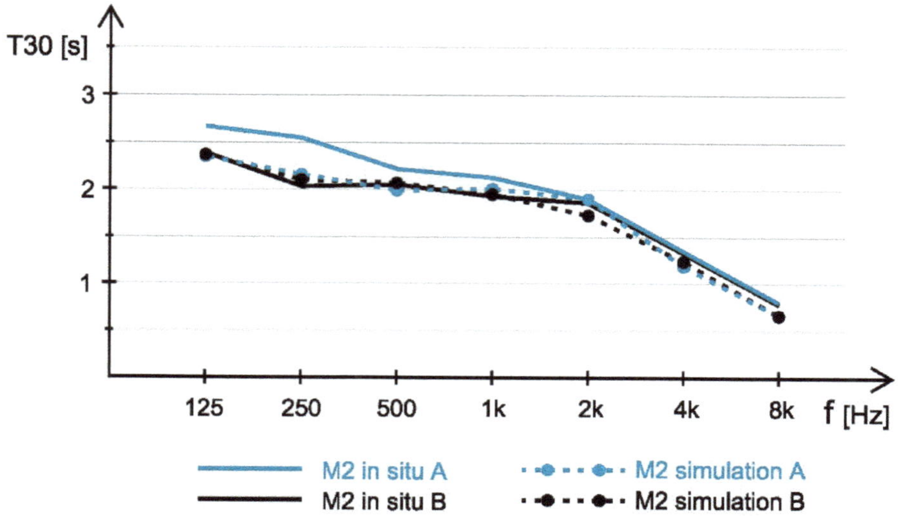

Figure 4. Measured and simulated octave values of T_{30} parameter for M2 measurement point in the case of sound excitation from the altar (A) and the chanting apse (B).

Table 1. Octave range absorption coefficient values used for wall material in the process of acoustic simulation.

Hz	125	250	500	1000	2000	4000	8000
α	0.04	0.04	0.05	0.05	0.06	0.09	0.16

3.2. Examination of the Acoustic Model

The acoustic model was used to examine the acoustic effect of (1) the space occupancy, (2) the dome and (3) the iconostasis. The first case simulates the real-life situation when faithful attend the service. The goal was to examine how the presence of people influence the church acoustic properties. An estimated 50 people are considered. The second case analysis was intended to determine the effect of the dome on the final acoustic response of the space. In this sense, a model with a flat ceiling on the place of the dome was formed (Figure 2), and its acoustic response was compared with the acoustic response of the regular model. The third case considers the altar barrier, the iconostasis, which used to be significantly lower in height in medieval period. Thus, the acoustic parameters were compared for the empty church model as it is today (with the iconostasis 4.35 m high) and as it was in medieval times (2.2 m high).

Speech transmission index (STI) is a measure of speech transmission quality, graded from 0 to 1. It is used to quantify speech intelligibility. Figure 5 visualizes the difference of the STI parameter when the church is empty (on the left) and when it is full (on the right), for the sound source positioned in the chanting apse. Table 2 shows STI values for all examined cases in Lazarica church.

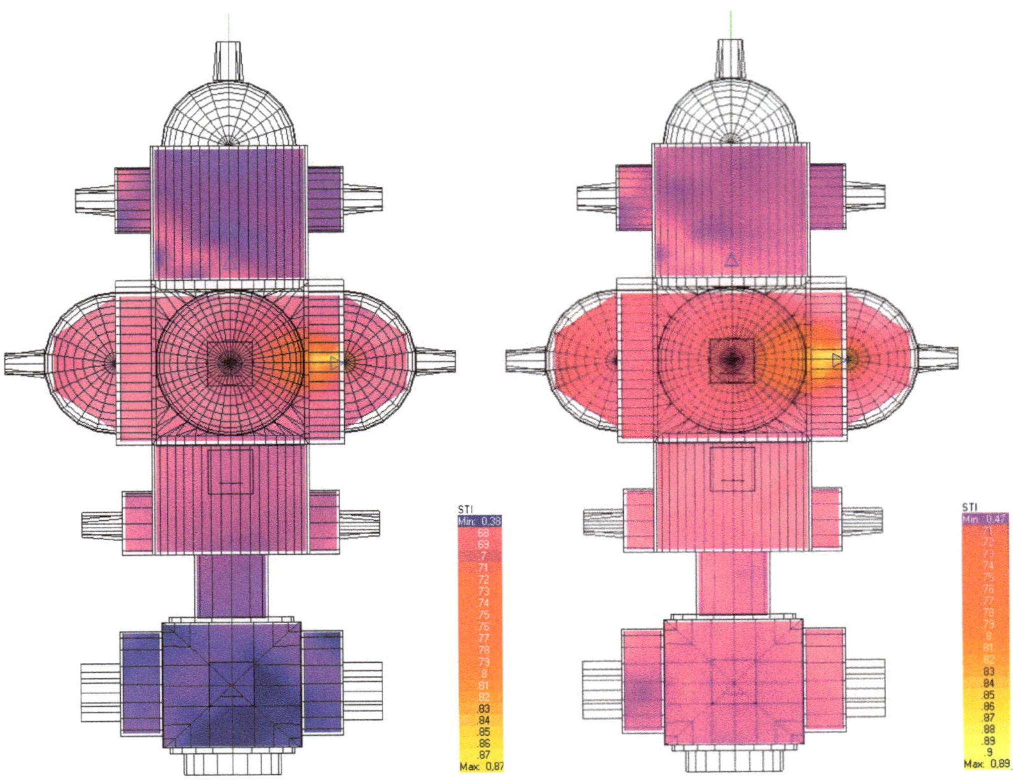

Figure 5. Mapping of speech transmission index (STI) parameter for the sound source position in the chanting apse, when the church is empty (on the left) and when the faithful are present (on the right).

Table 2. Values of speech transmission index parameter for three receiver points (M1, M2 and M3) and two sound source positions (A: Altar and B: Chanting apse) in Lazarica church.

Speech Transmission Index (STI)						
Sound Source	A: Altar			B: Chanting Apse		
Case \ Measurement Points	M1	M2	M3	M1	M2	M3
Measured in situ	0.491	0.474	0.402	0.673	0.473	0.405
Empty church	0.670	0.548	0.545	0.646	0.533	0.417
Faithful present	0.709	0.627	0.657	0.685	0.583	0.554
Without the dome	0.650	0.544	0.523	0.639	0.517	0.403
Medieval iconostasis	0.651	0.557	0.537	0.663	0.547	0.418

The octave range values for T_{30} and EDT parameters measured in situ are graphically presented in Figure 6 (the exact values are shown in the Table A1, Appendix A). The similarities of T_{30} and EDT decaying curves could be observed for naos positions (M1 and M2).

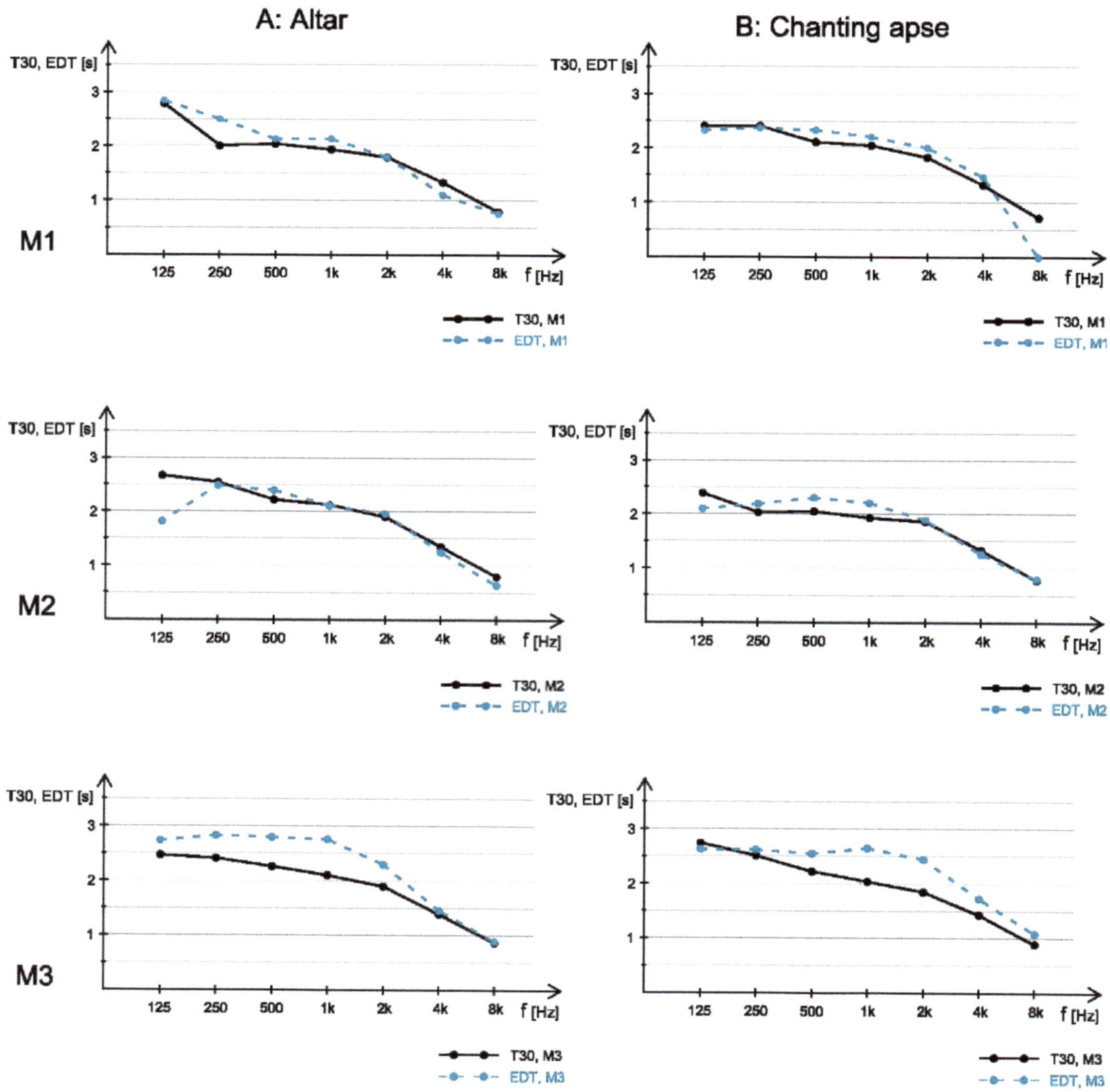

Figure 6. The octave range values for reverberation time (T_{30}) and early decay time (EDT) parameters measured in situ at M1, M2 and M3 measurement points for two sound excitation positions (altar and chanting apse).

The octave range values for T_{30} and EDT parameters for all three examined cases together with the case of the empty church obtained through the simulation process are shown in Appendix A, Tables A2–A4, respectively, for all three measurement points. As could be observed on Figure 6 there are no significant differences of T_{30} parameter values for the case without the dome, although something like this might be expected at the first glance. At the same time, the case of people present in the church derived an expected result of lowering T_{30} values (orange line on the graphs, Figure 7). The rate of this lowering is consistently around 25% at frequency of 500 Hz at all three measurement positions.

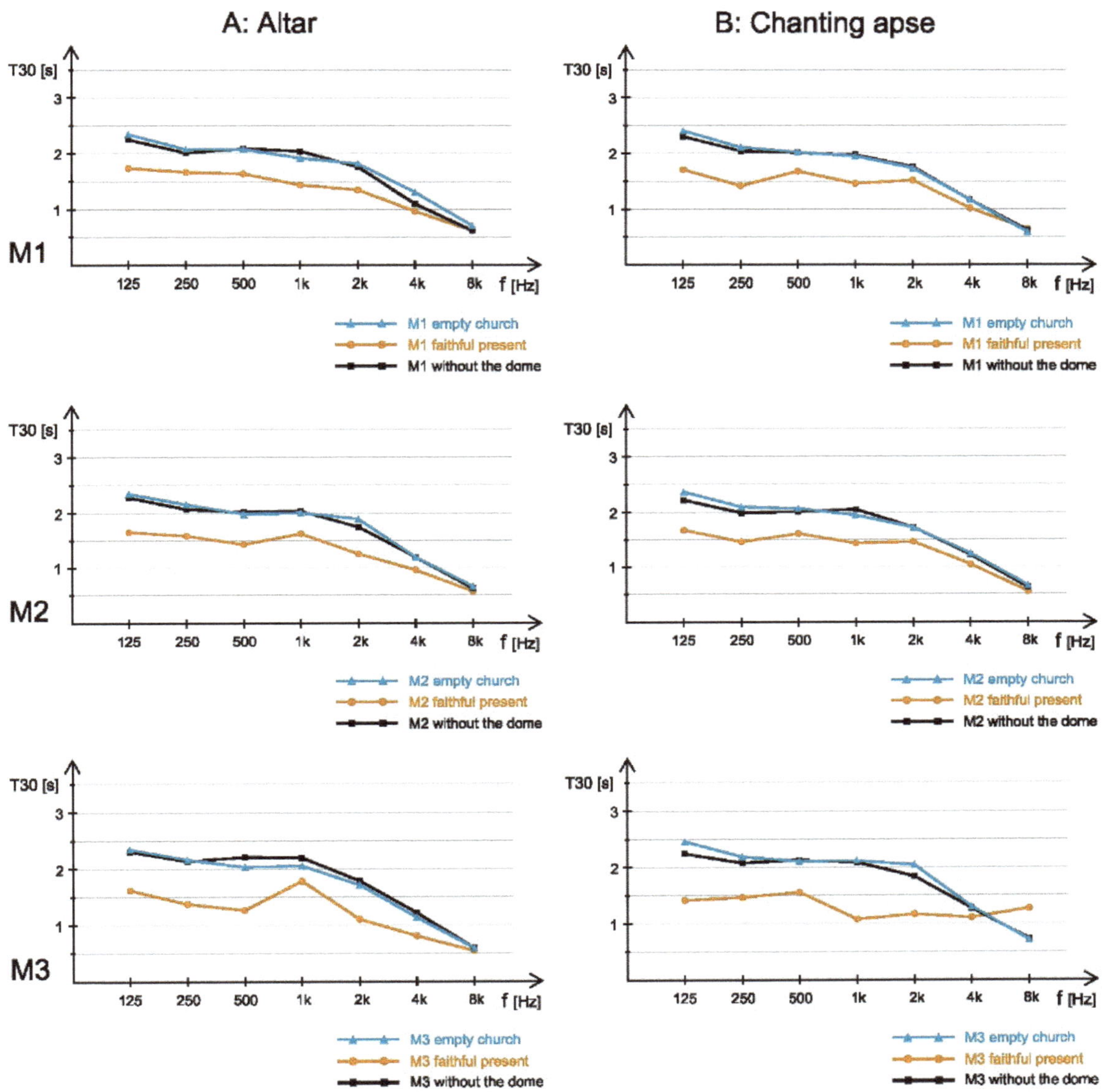

Figure 7. T_{30} octave values for three analyzed cases (case 1—blue, 2—orange, 3—black line) at M1, M2 and M3 measurement points for two sound excitation positions (altar and chanting apse).

In order to examine if the change of height of iconostasis affected Lazarica church acoustics, it is simulated in the acoustic model the lower iconostasis as it was in the time of church building (2.2 m high). Iconostasis was simulated as a single wooden plate made of 6 mm thick plywood. In the middle of the iconostasis is one arch door opening 90/230 cm. When two cases are compared—empty church with iconostasis as it is today, and empty church with iconostasis as it was in medieval times—no significant differences occur, as shown in Table 2 for the STI parameter and on Figure 8 for the T_{30} octave values.

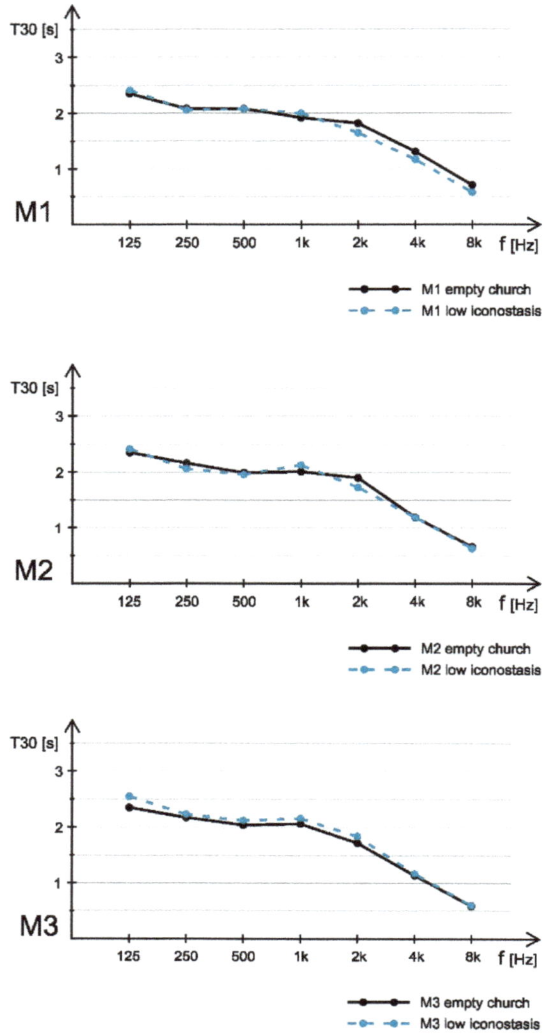

Figure 8. T_{30} octave values for cases of two different iconostasis height at M1, M2 and M3 measurement points.

4. Discussion

The basic parameter that describes church acoustics is reverberation time and its comparison with optimal reverberation time. There are several recommendations in a form of empirical relations for optimal reverberation time for two different purposes, speech and music, and only those derived for speech are of interest for the Orthodox tradition. The reason for this is that there was no instrumental music in Orthodox religious practice, and all sound excitations are exclusively made by the human voice, either as a speech or as a monophonic chanting, which is slow and monotonous, retaining in such way all the basic characteristics of speech. On the other hand, all of those recommendations were derived for churches built in the Western tradition, without any reference to the Orthodox medieval tradition. For example, Perez-Minana relations, which is especially adapted for churches, indicate that optimum reverberation time for churches of this volume (Lazarica volume 717 m^3) should be under 1s @500 Hz [25], and yet Lazarica church has reverberation time that is more than twice as long (Table A1, Appendix A). A reason for such a difference lays in different geometry and interior design materials.

Speech intelligibility is another acoustic property that is essential for churches. In situ measurements indicate that intelligibility under the dome (M1 position) highly depends of the position of the sound source. In the case of excitation from the altar, the STI parameter has a value of 0.491, and in the case of excitation from the chanting apse the same parameter has a value of 0.673.

Such a situation can be explained with specific relations between chanting apse and M1 position—a person standing under the dome is practically in front of the chanting apse, having the highest "direct to reflected" sound ratio, which directly impact speech intelligibility, while there is a physical barrier (iconostasis) between the sound source in the altar and the M1 position. At the other two positions (M2 and M3) such differences cannot be observed, which indicates that speech intelligibility does not depend on the position of the sound source.

Overall speech intelligibility inside the church can be rated as fair inside the naos (M1 and M2 position), and poor inside the narthex area (M3), as show in Table 3. As already stated, the only exception of such a description is the case when listener is standing right in front of the speaker. Although the Orthodox service requires the speech intelligibility, it also has the opposite demand to emphasize the spiritual experience of the chanting art. Therefore, besides the speech intelligibility, also the richness and specific nature of reflected sound is of a great importance.

Table 3. Intelligibility rating based on the STI value.

Intelligibility Rating	STI
Excellent	greater than 0.75
Good	0.60–0.75
Fair	0.45–0.60
Poor	0.30–0.45
Bad	less than 0.30

Measured values of the EDT and T_{30} parameters at M1 and M2 positions are very similar, as can be observed in Figure 6, which indicates at steady decay of energy inside the naos. On the other hand, in the narthex, EDT is consistently longer than the reverberation time for about 20% at the average. The naos and narthex are two different rooms linked with the door opening, which makes them acoustically coupled spaces. In the case of acoustically coupled spaces, the sound energy decay curve in the receiving room (room without the sound source - in this case narthex) can be expected to have a fairly slow early decay rate, followed by a reverberation tail which is determined by the space with longer reverberation time (in this case it is naos) [26] (pp. 445–448). Comparison of the in situ measurement results for the M2 and M3 positions showed that T_{30} is practically the same, but EDT is longer in the narthex area. One should have in mind that EDT correlates with reverberance, as audience perceive it during a music play or a speech. Also, both EDT and T_{30} parameters have different values for each measurement points, which indicates that the sound field inside the church cannot be considered as diffused and homogenous.

As for the sound field simulation conducted in EASE software, beside the case of empty church, three more cases were analysed: (1) church with people inside, (2) empty church without the dome, and (3) empty church with the lower iconostasis. The acoustic analysis of the empty church was important in order to calibrate acoustic response of the simulated model, and T_{30} parameter was used for this process. Although it was impossible to achieve perfect matching of this parameter gained through simulation with measurement results for all three positions, a satisfactory degree of T_{30} match has been achieved. At the same time, matching of the STI parameter was significantly lower, especially for the case of the excitation from the altar. It turns out that the iconostasis has much greater influence on decreasing speech intelligibility than was predicted by the simulation. One possible way to interpret such a case is that the influence of the iconostasis (as a wooden obstacle between sound source and the rest of the church) is not described precisely enough in the process of simulation. This kind of analysis is beyond the scope of this paper, but it could be of interest for some further investigation.

In the case of the presence of people in the church, the results are completely in line with the expectations. Namely, increased absorption decreased the overall reflected energy which results in consistent lowering of EDT and T_{30}. Influence of the people on church acoustics was simulated with increased absorption of the floor area, with absorption coefficients that describes human absorption.

At the same time, STI values increased, once again pointing to the fact that the presence of the people raises the level of speech intelligibility (Figure 5).

In order to investigate the effect of the dome on the overall acoustic response within the empty church, the initial model was modified by "closing" the dome and replacing it with a flat ceiling on the top of the arches that support the dome (marked with blue dashed line in the church section, Figure 2). It can be noticed that the STI parameter in the case of the model with flat ceiling has even lower values than in the initial case with the dome (although minimal, this difference exists). This implies that the dome does not have a bad influence on overall speech intelligibility, although such a thing may be expected at first glance because of the concave shape of the dome. Considering time parameters, it was expected that flat ceiling case has lower EDT and T_{30} values because of decreased overall church volume. However, although such a trend exists, it is not fully consistent. At some frequencies those parameters have greater values in the case of a flat ceiling, but their regularity is difficult to be predicted from the corresponding curve. What is possibly interesting is that differences between those two cases are the least pronounced for the M1 position, which is directly under the dome. Such result obtained by the simulation could indicate that dome has greater acoustic influence on the positions inside the church that are further from the dome. Although not definite, this conclusion can indicate some further directions for dedicated investigation of the dome influence on the overall church acoustics.

5. Concluding Remarks

In the last period of the Serbian medieval state, church building and chanting art reached their apex [16]. The goal of this paper was to examine the acoustics of Lazarica church, as a typical representative of a Morava architectural style, with a layout that combines a compressed inscribed cross and a triconch. Acoustic properties were investigated through the simulation process conducted in the EASE acoustic software. The acoustic model was calibrated upon the on-site measurement of acoustic impulse response, by defining the absorption coefficient of the wall material. The sound source positions were characteristic for the Holy Liturgy—the altar (behind the iconostasis) and the chanting apse (where is chanted and preached from). Three receiver positions were analysed: M1 under the dome, M2 on the longitudinal axe of naos, and M3 in the narthex. The results obtained by the measurements were achieved by an excitation from a directed loudspeaker, in order to simulate a "real life" situation (sound field excitation by the human voice), which should be kept in mind when it comes to repeatability and reproducibility of measurement results. Due to the Orthodox religious practice and the related organization of the church interior space, several cases were of interest for the acoustic analysis, such as the acoustic effect of (1) the central dome, (2) faithful presence, and (3) the medieval iconostasis (because it was half the height than the one standing today). Acoustic requirements in medieval Orthodox church were for the speech intelligibility, as well as the experience of monophonic chanting. Therefore, three acoustic parameters were tracked: reverberation time (T_{30}), early decay time (EDT) and speech transmission index (STI).

Regarding speech intelligibility, the acoustic analysis showed that it is increased with the presence of people, as expected, but lowered when the dome is replaced with a flat ceiling. It is highly dependable on the listener position inside the church, which might indicate on inhomogeneous sound field. The experience of monophonic chanting is correlated with T_{30} and EDT. The simulation showed that the presence of the faithful in the church causes the decrease of T_{30} for about 0.5 s, comparing to the case of the empty church (average rate of 25%). In all the examined cases, it is observed that there is no significant deviation between the T_{30} and EDT parameters, which indicates uniform sound energy decay. Closing of the dome with a flat ceiling did not show any significant impact on T_{30}. The simulation showed that the height of the iconostasis has no significant influence on the acoustics of Lazarica church.

Due to the lack of written records on medieval Serbian Orthodox church design and in what range the aural environment was considered, it is valuable to analyse their acoustics in order to better understand the complex relation between architecture, religious practice and acoustics. The examination of Lazarica church acoustics, presented in this paper, contributes to this goal.

Author Contributions: Conceptualization Z.Đ.; methodology Z.Đ. and D.N.; modelling and examination of the acoustic model U.A. and D.N.; formal analysis D.N.; writing Z.Đ. and D.N.

Funding: This research received no external funding.

Acknowledgments: The paper is a result of research on the project Theory and practice of science in society: multidisciplinary, educational and intergenerational perspectives (number 179048), financed by the Ministry of Education, Science and Technological Development of Republic of Serbia. The authors are grateful to Stefan Dimitrijević, who worked on the pilot acoustic model of Lazarica church that was presented on the conference TAKTONS 2015 in Novi Sad, Serbia.

Conflicts of Interest: The authors declare no conflict of interest.

Appendix A

Table A1. Lazarica church in situ acoustic measurements: Octave values of T30 and EDT for three measurement points (M1, M2 and M3) and two sound source positions (A: Altar and B: Chanting apse).

In situ	M1				M2				M3			
	A: Altar		B: Chanting apse		A: Altar		B: Chanting apse		A: Altar		B: Chanting apse	
Hz\parameter	EDT	T_{30}	EDT	T_{30}	EDT	T_{30}	EDT	T_{30}	EDT	T_{30}	EDT	T_{30}
125	2.50	2.78	2.32	2.40	1.82	2.67	2.10	2.39	2.74	2.47	2.63	2.74
250	2.84	2.01	2.37	2.40	2.49	2.55	2.19	2.03	2.83	2.41	2.62	2.51
500	2.13	2.04	2.33	2.11	2.40	2.22	2.31	2.05	2.80	2.26	2.55	2.22
1000	2.14	1.94	2.21	2.05	2.11	2.13	2.21	1.93	2.76	2.10	2.65	2.04
2000	1.80	1.80	2.01	1.83	1.96	1.90	1.89	1.86	2.30	1.90	2.45	1.85
4000	1.10	1.33	1.47	1.32	1.25	1.35	1.26	1.33	1.46	1.39	1.73	1.43
8000	0.76	0.79	0.01	0.72	0.64	0.80	0.80	0.78	0.89	0.87	1.09	0.89

Table A2. Lazarica church acoustic model examination: Octave values of T_{30} and EDT for receiver point under the dome (M1) and two sound source positions (A: Altar and B:Chanting apse) in three cases – when church is empty, when faithful are present, when the central dome is excluded, and when the iconostasis is 2.2 m high as in medieval times.

M1	Empty church				Faithful present				Without the dome				Low iconostasis			
	A: Altar		B: Chanting apse		A: Altar		B: Chanting apse		A: Altar		B: Chanting apse		A: Altar		B: Chanting apse	
Hz\param.	EDT	T_{30}	EDT	T_{30}	EDT	T_{30}	EDT	T_{30}	EDT	T_{30}	EDT	T_{30}	EDT	T_{30}	EDT	T_{30}
125	2.45	2.35	2.17	2.41	1.99	1.74	1.67	1.71	2.33	2.25	2.10	2.30	2.40	2.41	2.36	2.43
250	2.14	2.08	2.05	2.11	1.53	1.67	1.60	1.42	1.87	2.01	1.79	2.04	2.02	2.06	2.07	2.12
500	1.83	2.08	1.94	2.02	1.64	1.64	1.77	1.68	2.06	2.09	1.95	2.01	2.07	2.08	1.98	2.04
1000	1.94	1.92	1.80	1.95	1.28	1.44	1.47	1.46	1.87	2.04	1.79	1.97	1.62	2.00	2.06	1.97
2000	1.46	1.82	1.40	1.74	1.58	1.35	0.95	1.52	1.52	1.76	1.65	1.76	1.10	1.65	1.71	1.64
4000	1.21	1.31	0.74	1.17	0.64	0.97	0.77	1.02	1.22	1.10	0.93	1.17	0.95	1.17	0.70	1.15
8000	0.63	0.71	0.52	0.59	0.04	0.62	0.53	0.64	0.86	0.62	0.55	0.62	0.65	0.58	0.55	0.56

Table A3. Lazarica church acoustic model examination: Octave values of T_{30} and EDT for receiver point in naos (M2) and two sound source positions (A: Altar and B:Chanting apse) in three cases – when church is empty, when faithful are present, when the central dome is excluded, and when the iconostasis is 2.2 m high as in medieval times.

M2	Empty church				Faithful present				Without the dome				Low iconostasis			
	A: Altar		B: Chanting apse		A: Altar		B: Chanting apse		A: Altar		B: Chanting apse		A: Altar		B: Chanting apse	
Hz\param.	EDT	T_{30}	EDT	T_{30}	EDT	T_{30}	EDT	T_{30}	EDT	T_{30}	EDT	T_{30}	EDT	T_{30}	EDT	T_{30}
125	2.53	2.35	2.36	2.37	1.90	1.66	1.77	1.68	2.32	2.28	2.24	2.22	2.55	2.41	2.43	2.37
250	2.17	2.16	2.06	2.10	1.67	1.59	1.45	1.46	1.97	2.07	1.90	1.99	2.15	2.06	2.02	2.18
500	2.08	1.99	1.99	2.07	1.31	1.43	1.50	1.61	2.01	2.02	1.86	2.01	1.96	1.96	2.12	1.91
1000	2.20	2.01	2.04	1.95	1.53	1.63	1.43	1.43	1.88	2.04	1.83	2.05	1.69	2.13	1.74	2.02
2000	1.45	1.90	1.58	1.73	1.56	1.25	1.44	1.46	1.78	1.75	1.81	1.73	1.23	1.73	1.42	1.96
4000	1.06	1.19	0.84	1.24	1.02	0.96	0.60	1.04	1.12	1.19	1.26	1.21	0.93	1.19	1.27	1.17
8000	0.48	0.66	0.50	0.66	0.37	0.57	0.47	0.55	0.62	0.62	0.56	0.62	0.60	0.63	0.47	0.61

Table A4. Lazarica church acoustic model examination: Octave values of T_{30} and EDT for receiver point in the narthex (M3) and two sound source positions (A: Altar and B:Chanting apse) in three cases – when church is empty, when faithful are present, when the central dome is excluded, and when the iconostasis is 2.2 m high as in medieval times.

M3	Empty church				Faithful present				Without the dome				Low iconostasis			
	A: Altar		B: Chanting apse		A: Altar		B: Chanting apse		A: Altar		B: Chanting apse		A: Altar		B: Chanting apse	
Hz\param.	**EDT**	**T_{30}**	**EDT**	**T_{30}**	**EDT**	**T_{30}**	**EDT**	**T_{30}**	**EDT**	**T_{30}**	**EDT**	**T_{30}**	**EDT**	**T_{30}**	**EDT**	**T_{30}**
125	3.01	2.35	2.67	2.46	2.11	1.63	2.29	1.42	2.88	2.31	2.80	2.24	2.93	2.55	2.68	2.44
250	2.65	2.17	2.69	2.19	1.43	1.38	1.71	1.47	2.54	2.14	2.36	2.08	2.71	2.23	2.69	2.16
500	2.65	2.04	2.63	2.11	1.39	1.27	1.90	1.56	2.26	2.21	2.53	2.13	2.51	2.12	2.39	2.20
1000	2.36	2.06	2.82	2.12	1.39	1.79	1.78	1.08	2.12	2.20	2.60	2.09	2.43	2.16	2.45	2.12
2000	2.00	1.72	2.36	2.05	1.02	1.11	2.19	1.17	2.04	1.79	2.13	1.84	1.63	1.84	2.16	1.95
4000	1.22	1.14	1.72	1.30	1.22	0.81	1.21	1.11	1.22	1.22	1.67	1.27	0.83	1.17	1.45	1.25
8000	0.40	0.59	1.02	0.71	0.20	0.54	0.89	1.27	0.54	0.59	1.00	0.73	0.64	0.60	0.68	0.69

References

1. Crnčević, D. Architectural work, context, meaning: Approaches to the study of architecture of Moravian Serbia. *Hist. Rev.* **2008**, *LVII*, 93–106.
2. Đorđević, Z. Intangible tangibility: Acoustical heritage in architecture. *Struct. Integr. Life* **2016**, *16*, 59–66.
3. Suárez, R.; Alonso, A.; Sendra, J.J. Archaeoacoustics of intangible cultural heritage: The sound of the Maior Ecclesia of Cluny. *J. Cult. Herit.* **2016**, *19*, 567–572. [CrossRef]
4. Girón, S.; Álvarez-Morales, L.; Zamarreño, T. Church acoustics: A state-of-the-art review after several decades of research. *J. Sound Vib.* **2017**, *411*, 378–408. [CrossRef]
5. Alvarez-Morales, L.; Martellotta, F. A geometrical acoustic simulation of the effect of occupancy and source position in historical churches. *Appl. Acoust.* **2015**, *91*, 47–58. [CrossRef]
6. Zamarreño, T.; Girón, S.; Galindo, M. Assessing the intelligibility of speech and singing in Mudejar-Gothic churches. *Appl. Acoust.* **2008**, *69*, 242–254. [CrossRef]
7. Brezina, P. Measurement of intelligibility and clarity of the speech in romanesque churches. *J. Cult. Herit.* **2015**, *16*, 386–390. [CrossRef]
8. Alonso, A.; Suárez, R.; Sendra, J. The Acoustics of the Choir in Spanish Cathedrals. *Acoustics* **2018**, *1*, 35–46. [CrossRef]
9. Navarro, J.; Sendra, J.J.; Muñoz, S. The Western Latin church as a place for music and preaching: An acoustic assessment. *Appl. Acoust.* **2009**, *70*, 781–789. [CrossRef]
10. Martellotta, F.; Cirillo, E.; Della Crociata, S.; Gasparini, E.; Preziuso, D. Acoustical reconstruction of San Petronio Basilica in Bologna during the Baroque period: The effect of festive decorations. *J. Acoust. Soc. Am.* **2008**, *123*, 3607. [CrossRef]
11. Palazzo-Bertholon, B.; Valière, J.-C. *Archéologie du son: Les dispositifs de pots acoustiques dans les édifices anciens*; Bulletin Monumental, Suppl. 5; Picard: Paris, France, 2012; ISBN 978-2-901837-41-1.
12. Đorđević, Z.; Penezić, K.; Dimitrijević, S. Acoustic vessels as an expression of medieval music tradition in Serbian sacred architecture. *Muzikologija* **2017**, *1*, 105–132.
13. Elicio, L.; Martellotta, F. Acoustics as a cultural heritage: The case of Orthodox churches and of the "Russian church" in Bari. *J. Cult. Herit.* **2015**, *16*, 912–917. [CrossRef]
14. Antonopoulos, S.; Gerstel, S.E.J.; Kyriakakis, C.; Raptis, K.T.; Donahue, J. Soundscapes of Byzantium. *Speculum* **2017**, *92*, S321–S335. [CrossRef]
15. Gerstel, S.E.J.; Kyriakakis, C.; Raptis, K.T.; Antonopoulos, S.; Donahue, J. Soundscapes of Byzantium: The Acheiropoietos Basilica and the Cathedral of Hagia Sophia in Thessaloniki. *Hesperia J. Am. Sch. Class. Stud. Athens* **2018**, *87*, 177–213. [CrossRef]
16. Peno, V.; Obradović, M. On the chanting space and hymns that were sung in it. Searching for chanting-architectural connections in the middle ages. *Muzikologija* **2017**, *23*, 145–173. [CrossRef]
17. Mourjopoulos, J.; Papadakos, C.; Kamaris, G.; Chryssochoidis, G.; Kouroupetoglou, G. Optimal acoustic reverberation evaluation of Byzantine chanting in churches. In Proceedings of the International Computer Music Conference, Athens, Greece, 14–20 September 2014; pp. 14–20.
18. Ristić, V.; Lazarica, I.K.G. *Republički Zavod Za Zaštitu*; Spomenika Kulture Beograd: Beograd, Serbia, 1989.

19. Korać, V. *Između Vizantije i Zapada: Odabrane Studije o Arhitekturi*; Prosveta [u.a.]: Beograd, Serbia, 1987; ISBN 978-86-80371-01-6.
20. Stevović, I. Arhitektura Moravske Srbije: Lokalna graditeljska škola ili epilog vodećih tokova poznovizantijskog graditeljskog stvaranja. *Zb. Rad. Viz. Inst.* **2006**, *XLIII*, 231–253.
21. Korać, V.; Šuput, M. *Arhitektura Vizantijskog Sveta*; Zavod za Udžbenike: Beograd, Serbia, 2010; ISBN 978-86-17-16798-9.
22. Ristić, V. Restauracija Lazarice Pere, J. Popovića iz 1904–1908 godine. *Saopštenja* **1983**, *15*, 129–146.
23. Anđelković, B. Liturgija i unutrašnji poredak hrama—kanonski osnovi crkvenog graditeljstva. In *Tradicija i Savremeno Srpsko Crkveno Graditeljstvo*; Stojkov, B., Manević, Z., Eds.; Institut za arhitekturu i urbanizan Srbije: Beograd, Serbia, 1995; pp. 31–61.
24. Stefanović, D. *Old Serbian Music—Examples of 15th Century Chant*; Institute of Musicology SASA: Belgrade, Serbia, 1975.
25. Fernandez, M.; Recuero, M. Data Base Design for Acoustics: The Case of Churches. *Build. Acoust.* **2005**, *12*, 31–40. [CrossRef]
26. Kurtović, H. *Osnovi Tehničke Akustike*; Naučna knjiga: Beograd, Serbia, 1990; ISBN 978-86-23-21072-3.

6

Applicability of Acoustic Concentration Measurements in Suspensions of Artificial and Natural Sediments Using an Acoustic Doppler Velocimeter

Oliver Chmiel *, Ivo Baselt and Andreas Malcherek

Institute of Hydromechanics and Hydraulic Engineering, University of the German Armed Forces Munich, 85579 Neubiberg, Germany; ivo.baselt@unibw.de (I.B.); andreas.malcherek@unibw.de (A.M.)

* Correspondence: oliver.chmiel@unibw.de

Abstract: For the investigation of turbulence and particles, interaction measurement systems are required, which are able to measure velocity and concentration fluctuations simultaneously. Acoustic Doppler Velocimeters (ADV) are widely used for velocity and turbulence measurements in natural and artificial flows. Based on the acoustic sonar theory, a model is presented that correlates the ADV's Signal-to-Noise Ratio (SNR) and the suspended solids concentration of several natural (Ems Estuary, Lake Eixendorf, Lake Altmühl) and artificial sediments (Chinafill, quartz powder, bentonite, metakaolin) for the range 0.001 g/L–50 g/L. Within the presented method, the sound absorption in water and on particles is considered in a continuous approach for sampling frequencies up to 100 Hz. The widely-used log-linear relation between the SNR and the concentration, which is only valid for low concentrations, was extended for the high concentration regime. Measurement results show a similar behavior of the SNR with respect to varying suspended solid concentrations for different sediments. However, the analysis of the fit parameters shows systematic differences depending on the type of sediment. It is concluded that the proposed model is applicable as well for laboratory use as for measurements in rivers and estuaries. Finally, we discuss the reliability of the results and the methodology with regard to measurements in rivers, lakes, and estuaries.

Keywords: acoustic doppler velocimetry (ADV); signal-to-noise ratio (SNR); sonar theory; suspended solids concentration; quartz powder; particle sound absorption; Urick-coefficient

---

1. Introduction

ADV probes provide the opportunity to measure in situ the three-dimensional flow velocity in liquids with a high temporal resolution (up to 200 Hz). This allows evaluating the velocity fluctuations in terms of the turbulent behavior in the flow. The three-dimensional velocity fluctuations $u'_i = u_i - \bar{u}_i$ are described as the deviation of the mean value $\bar{u}_i$. The intensity of turbulence is described by the turbulent kinetic energy $k = \frac{1}{2}\left(\overline{u'_i u'_i}\right)$, which consists of the diagonal contributions of the Reynolds stress tensor, e.g., [1].

Especially in natural surface waters like rivers, channels, lakes, and estuaries, near-bottom suspended solids of different concentrations are common. For instance, in the Ems Estuary in the German Bight, Suspended Solid Concentrations (SSC) were measured up to 300 g/L [2]. Cellino demonstrated that SSC influence the flow behavior within a stream, in particular the vertical structure of turbulence. It was shown that SSC can lead to turbulence damping and decreasing eddy viscosity [3].

The common method to describe the interaction between concentration and turbulence is to quantify the turbulent Schmidt-number $Sc_t = \frac{\nu_t}{K_t}$ with ν_t being the turbulent viscosity and K_t being the turbulent diffusivity. The physical behavior of both quantities is the same, as the diffusivity of particles increases with turbulence. The turbulent Schmidt-number is often used for calibrating experimental and numerical data. Depending on the type of particles, it is given as $Sc_t = 0.3–1.5$ [4].

Applying the eddy viscosity hypothesis, ν_t can be formulated by means of the extra diagonals of the Reynolds stress tensor $\overline{u'w'}$ and the vertical gradient of the mean velocity $\frac{\partial \bar{u}}{\partial z}$ in the main stream direction:

$$\overline{u'w'} = -\nu_t \frac{\partial \bar{u}}{\partial z} \tag{1}$$

Following the same principle, K_t is described by the fluctuations of the concentration c' and the vertical gradient of the mean concentration.

$$\overline{c'w'} = -K_t \frac{\partial \bar{c}}{\partial z} \tag{2}$$

In order to investigate the interaction of turbulence and particle concentration, it is necessary to measure the velocity and the concentration fluctuations simultaneously. Further, to detect as many turbulent eddies as possible, the measuring frequency should be as high as possible.

Salehi et al. summarized three methods for measuring the particle concentrations in currents [5]: (1) the extraction of samples at different positions with a suction device and the external analysis of the samples taken; (2) determination of the particle concentration through scattering and reflection of light (OBS); and (3) the concentration measurement by means of the reflection and scattering of an acoustic signal. The intensity of sound is attenuated by the particles within the measurement volume. Furthermore, the reflected signal is superimposed by noise, which is taken into account by the Signal-to-Noise Ratio (SNR).

As mentioned in Equation (2), for the investigation of the diffusivity coefficient K_t, measurements of the concentration fluctuations are necessary. Applying the suction method, for instance, no information about the concentration and velocity fluctuations at the same time and location are obtained. Optical measurement devices provide a wide spectrum of recordable concentrations; however, they are not able to measure the velocity. The investigations of Fugate et al. show higher accuracy for SSC measurements in estuaries when using acoustic devices rather than optical instruments [6]. Acoustic devices, like ADV probes, are used to measure the three-dimensional flow velocities with a sampling frequency f up to 200 Hz. Additionally, the SNR is recorded, which is affected by the particle concentration in the measurement volume. If it is possible to find a unique function for the SNR with respect to the concentration, then it is possible to measure velocity, turbulence, and concentration synchronously.

The foundation of the acoustic method is the sonar theory, which was first applied for small particles by Urick in 1948 [7]. Urick investigated the absorption behavior of kaolin and fine silt while using a quartz crystal to send acoustic pulses. After ADV measurements became a standard technique for velocity measurements, the acoustic scattering theory again was applied for concentration measurements [8]. Thorne et al. gave a summary of the advantages of acoustic methods for concentration measurements since optical devices reached their working range at this time [9]. Jay et al. described difficulties when calibrating the SNR for natural rivers, since the reflected signal was strongly dependent on the particle's material, shape, and size [10]. Hoitink et al. were able to calibrate the SNR for concentrations up to 1 g/L [11]. However, neither study, from Jay et al. and Hoitink et al., considered the sound absorption on particles; that is why their methods are only valid up to approximately 1 g/L.

The acoustic method for concentration measurement can be applied at laboratory conditions with a homogeneous particle distribution and known sediment properties. Ha et al. and Salehi et al. presented laboratory experiments for calibrating the reflected signal intensity and the particle concentration [5,12]. Nevertheless, they neglected the sound attenuation on particles, although it is a crucial parameter. Decrop et al. presented an experimental setup to measure concentrations up to

10 g/L with an ADV probe while considering sound attenuation on particles. Their measurements were conducted with a measurement frequency of 25 Hz [13]. To investigate the interaction of turbulent concentration and velocity fluctuations, higher sampling frequencies are required to resolve turbulent flow behavior.

In order to investigate the interaction of turbulence and concentration, different experimental studies were realized. Best et al. conducted a laboratory experiment to measure velocities and concentrations with a Laser Doppler Anemometer (LDA). However, the application of their optical method was restricted to low concentrations [14]. Cellino used a self-constructed acoustic particle flux profiler for measuring the concentration and velocity with a frequency of 16 Hz [15]. A part of his research was to investigate the decay of turbulence with increasing suspended sediment concentration in a laboratory flume [3].

Moreover, Hurther et al. showed the development of an Acoustic Concentration and Velocity Profiler (ACVP), which was able to measure the two-dimensional velocity field and the concentration over a vertical profile of 25 cm [16]. This device was applied by Revil-Baudard et al. for laboratory experiments on investigations of sheet-flow processes [17]. They reported about maximum concentration measurement frequencies of 4.9 Hz.

The necessity of simultaneously measuring flow velocity and concentration is presented by means of the Ems Estuary (Germany, North Sea) as an example. In the Ems Estuary, simultaneous measurements of flow velocity and concentration are needed in order to describe the dynamics of fluid mud layers. Fluid mud is a high concentrated mud suspension, which has huge impacts on the maintenance of rivers and estuaries, since navigation can be impeded [18]. Recently, Becker et al. showed concentration and velocity measurements of the fluid mud layers in the Ems Estuary and highlighted the unsteady and complex behavior of the mud layers with tide. Within this measurement campaign, velocity and concentration were measured by different devices. Furthermore, no turbulence measurements were presented [19]. Having a measurement device that would be able to measure velocity, concentration, and turbulence simultaneously, the unsteady dynamics of such fluid mud layers could be investigated in more detail.

With developing measuring techniques, more accurate turbulence results are expected when using measurement devices with sampling frequencies as high as possible. In this case, small eddies are able to be resolved, which could not be recognized from the measurement device in former studies. Therefore, a Vectrino Profiler (Nortek, Norway) was used with a sampling frequency of 100 Hz. The Nortek Vectrino Profiler was launched in 2010 and consists of a single central transceiver and four passive receivers at an angle of 30° towards the transceiver. A detailed description of the probe was given by Craig et al. and by Thomas et al. [20,21].

Within this paper, a method is presented to model the concentration of different suspended materials by means of the SNR with a modern ADV probe (Vectrino Profiler). The acoustic model is based on the well-known sonar theory approach, and it is applied for artificial and natural river sediments. The authors show how sound attenuation is considered in water and on sediment particles within the sonar theory-based model. In experimental studies for cohesive sediment dynamics, quartz powder, metakaolin, and Chinafill are often used [13,22,23]. Therefore, the following artificial sediments were investigated: quartz powder, bentonite, metakaolin, and Chinafill. Afterwards, the results were compared with those of natural sediments from the Ems Estuary on the German Coast, Lake Altmühl, and Lake Eixendorf, both located in Bavaria, Germany. The experimental setup and the proceeding measurements are presented. Furthermore, the authors describe the analysis of the results of different sediments in order to specify the model's applicability in nature. Furthermore, it is shown that concentrations can be measured up to 50 g/L depending on the type of sediment.

This paper describes a method for simultaneously measuring velocity, turbulence, and concentration, which can be applied in laboratory investigations, as well as in rivers, estuaries, or lakes. The characteristics of the method are presented in detail, and indications are given for practical use in nature. The results of this study provide a contribution for further investigations of turbulence and diffusivity equally in the laboratory and for use in nature.

2. Acoustics of the ADV Probe

In the following, the theoretical aspects are described for sound and particles' interaction in water. First, the particle diameters are verified, which are applicable to the acoustic backscatter theory. Afterwards, the authors present the implementation of the general sonar theory using the backscatter intensity. The aim of this part is to derive an equation to relate the suspension concentration to an acoustic quantity, which can be measured by the ADV probe.

2.1. Rayleigh Scattering and the Intensity of the Backscatter Signal

The working principle of ADV devices is based on Rayleigh scattering, which occurs when the diameter of the particle is smaller than the wavelength of the emitted signal. In that case, the energy transfer towards the particle can be neglected. For Rayleigh scattering, the condition:

$$k_w r < 1 \tag{3}$$

has to be fulfilled, with $k_w = \frac{2\pi}{\lambda}$ as the wave number, λ as the wavelength, and r as the particle radius [24]. Decrop et al. argued that a sufficiently high backscatter intensity only can be achieved for $k_w r > 0.05$ [13].

In this study, we used different sediments with mean particle diameters 7 μm $\leq d_{50} \leq$ 20 μm. The ADV's emitter frequency was F = 10 MHz, which lead to $0.147 \leq k_w r \leq 0.419$ (Table 1). Therefore, the condition was fulfilled to apply the principles of Rayleigh scattering here.

Table 1. The different particle diameters used to investigate the validity of Rayleigh scattering.

Sediment/Location	d_{50} (μm)	$k_w r$ (−)
Quartz powder	18	0.377
Bentonite	7	0.147
Metakaolin	9	0.189
Chinafill	9	0.189
Ems 2012	10	0.209
Ems 2015	18	0.377
Lake Eixendorf	20	0.419
Lake Altmühl	12	0.251

The intensity S_v of the backscatter signal increases with the reflecting particle amount n_b inside the measurement volume and their scattering cross-section σ_{bs}. Assuming unidirectional reflection, Medwin and Clay presented the scattering cross-section as:

$$\sigma_{bs} = 4\pi r^6 k_w^4 \mathrm{Y} \tag{4}$$

The factor Y is a material parameter, which is determined as:

$$\mathrm{Y} = \left(\frac{\Delta E - 1}{3\Delta E}\right)^2 + \frac{1}{3}\left(\frac{\Delta\rho - 1}{2\Delta\rho + 1}\right)^2 \tag{5}$$

where $\Delta\rho = \rho_s / \rho_w$ is the density ratio of the particle and water and $\Delta E = E_s / E_w$ is the elasticity ratio. The backscatter signal can be described in terms of n_b and σ_{bs} as the volume backscattering coefficient:

$$s_v = n_b \sigma_{bs} \tag{6}$$

which is a measure of the backscatter strength from particles. The backscatter strength increases with the number of reflecting particles and with the size of the scattering cross-section. That means that an increase of s_v is equivalent to a greater backscatter intensity, which is received by the ADV probe. The backscatter strength can be expressed in decibel units as [11,24]:

$$S_v = 10\, log_{10}\,(n_b \sigma_{bs}) \tag{7}$$

Assuming spherical particles with a mass $M = 4/3r^3\pi\rho_s$, the concentration:

$$C = M\, n_b \tag{8}$$

can be calculated from the number of particles inside the scattering volume. Combining Equations (4), (6) and (8) leads to:

$$S_v = 10\, log_{10}\left(\frac{3r^3 k_w^4 Y}{\rho_s} C\right) \tag{9}$$

which describes the backscatter intensity as a function of the concentration.

The backscatter intensity S_v must not be mistaken with the ADV receiver intensity. S_v is rather the backscatter intensity in the vicinity of the particle. On the way back to the ADV receiver, the signal intensity is exposed to further losses.

2.2. *Application of Sonar Theory*

The ADV transmitter emits an acoustic signal, which will be reflected on particles in the water, but it is only partly received by the probe's receiver. This process can be described by the principles of active sonar, which are illustrated in Figure 1. All quantities within the sonar theory are described in dB.

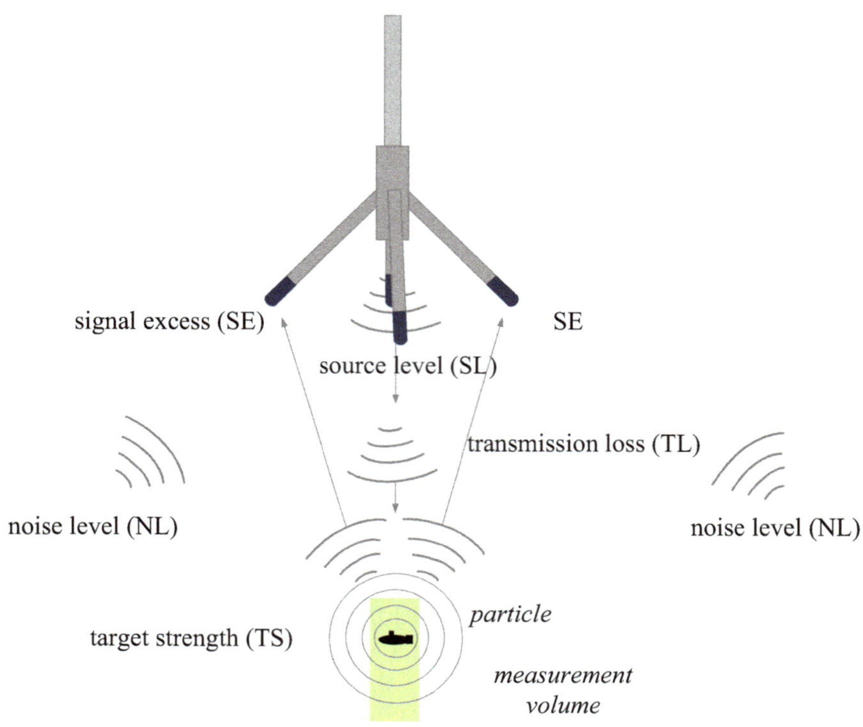

Figure 1. Sound intensities, which are of importance to state the sonar balance equation.

The emitted signal intensity (Signal Level (SL)) is exposed to Transmission Losses (TL) in both traveling directions. Additionally, the reflected signal is influenced by the specifics of the particles (e.g., shape, size, elasticity), which are accounted for in the Target Strength (TS). The ambient sound, which is not related to the ADV probe, is called noise and is described in NL as the noise level. Within sonar theory, SE (Signal Excess) is the received backscatter intensity inclusive of noise. In order to detect the reflected signal, SE has to overcome at least the noise intensity (Noise Level (NL)). The sonar balance equation as presented in similar form in Hoitink et al. and in Salehi et al. reads [5,11]:

$$SE - NL = S_v - 40\, log_{10}\,(R) - 2\alpha R + C_{ADV} \tag{10}$$

$(SE - NL)$ is the backscatter intensity under exclusion of noise (it is also called the Signal-to-Noise Ratio (SNR)). The target strength within the measurement volume is the backscatter intensity S_v. C_{ADV} describes the influence of additional device-specific settings. The transmission losses $2\text{TL} = 40\, log_{10}\,(R) + 2\alpha R$ are accounted for in both directions, while R is the distance from the transmitter to the measurement volume [25]. The effect of sound absorption is considered by:

$$\alpha = \alpha_w + \alpha_p \tag{11}$$

as the sum of the occurring sound absorption losses in water α_w and on particles α_p.

The ADV device, which is used in the experiment, has an emitter frequency of $F = 10$ MHz. For this emitter frequency, the coefficient for absorption losses in water is given as $\alpha_w = 26.9$ dB/m [26]. Unfortunately, the particle absorption coefficient α_p is not a constant. It depends on the acoustic properties of the transmitted signal and the properties of the particles. Urick presented an approach for the particle absorption coefficient in granular suspensions, which is valid for homogeneous particle distributions [7]. This approach was successfully applied by Hoitink et al. and Salehi et al. when conducting ADCP (Acoustic Doppler Current Profiler) and ADV measurements [5,11]. Therefore, Urick's formula for α_p was used within this paper, while the constant term is called the Urick-coefficient Ur [m^2/kg]. Ur is calculated by sediment and material constants, namely the particle radius r, the solid density ρ_s, the specific density ratio $\Delta\rho = \rho_s/\rho_w$, the molecular viscosity ν_m, the emitter frequency F, and its wave number k_w:

$$\alpha_p = \underbrace{\left(\frac{k_w^4\, r^3}{96\rho_s} + \frac{k_w s\,(\Delta\rho - 1)^2}{2\rho_s\left(s^2 + (\Delta\rho + \delta)^2\right)}\right)\frac{20}{ln\,(10)}}_{Ur} C \tag{12}$$

The following auxiliary variables are introduced:

$$s = \frac{9}{2\beta r}\left(1 + \frac{2}{\beta r}\right)$$

$$\delta = \frac{1}{2}\left(1 + \frac{9}{\beta r}\right)$$

$$\beta = \sqrt{\frac{\pi F}{\nu_m}}$$

It is highlighted that next to sedimentological and device-specific parameters, Equation (12) is a function of the concentration C.

2.3. *How Particles Influence the SNR*

The intensity of the backscatter signal, which is received by the ADV, depends on the sediment concentration. Two effects have to be considered simultaneously. First, an increasing particle concentration will be accompanied by an increase of the overall scattering cross-section, if a homogeneous particle distribution is ensured over the volume. This leads to rising SNR values.

Second, while α_p is a function of the concentration, increasing sound absorption causes a decrease of the SNR.

Combining Equations (9) and (10), splitting $2\alpha R = 2\alpha_p R + 2\alpha_w R$, introducing the transformation parameter Π_1, and solving for the concentration leads to:

$$log_{10}(A\,C) = 4log_{10}(R) + \frac{1}{10}\,\Pi_1\,SNR - \frac{C_{ADV}}{10} + \frac{1}{5}\alpha_w R + \frac{2R}{10}\,\alpha_p \tag{13}$$

with:

$$A = \frac{3r^3 k_w^4 Y}{\rho_s}$$

and:

$$\Pi_1\,SNR = SE - NL$$

In order to quantify the dependency of the SNR and the concentration, we present an experiment to measure the SNR in suspensions of pre-defined concentrations. Afterwards, the measurements will be compared to the results gained by the sonar theory model.

3. Experimental Investigation

After deriving the relation between the concentration and the SNR, within this section, the authors present the experimental setup for calibrating the SNR to the concentration of suspension flows. After discussing the suitability of the used ADV device, the exemplary results are presented. On behalf of simplicity, the description of the experimental setup is restricted only to one sediment type, namely quartz powder. Results of the resuming sediments are shown in Section 4.

3.1. Experimental Setup

Figure 2 illustrates the experimental setup consisting of a rectangular box, which was B_1 = 0.96 m wide and L_1 = 0.80 m long. The box was mounted on a steel bottom. The water depth in the box was H_1 = 0.30 m. Two opposite installed stirrers guaranteed homogeneously-mixed suspensions. A Nortek Vectrino Profiler ADV probe (sampling frequency f = 100 Hz) was mounted on a dynamometer wagon and adjusted in a way that the probe was box-centered and able to be translated vertically.

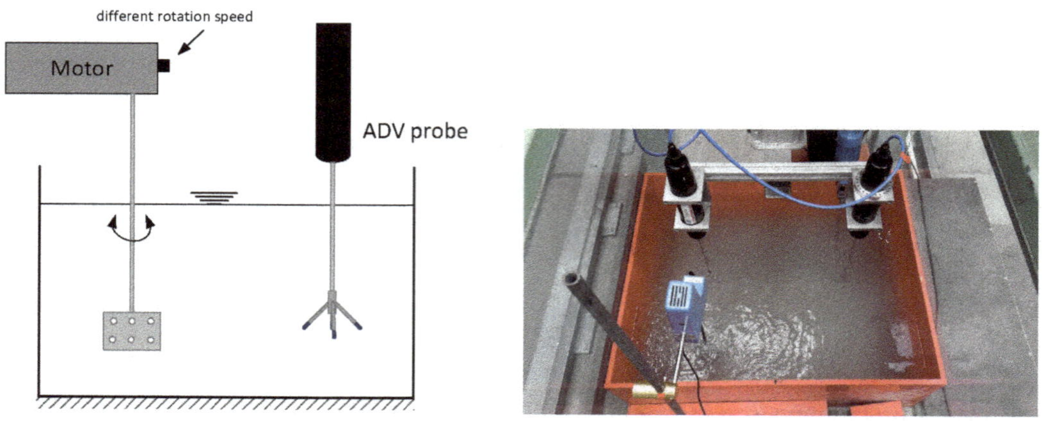

Figure 2. The experimental setup is shown for calibrating the ADV's SNR and the concentration. The sketch is showing the stirrer and the ADV probe in side view, and the photograph is showing the experiment in top view.

Before adding the sediment, an initial reference SNR was measured to account for the prevailing contamination in the water used. The sediment was weighed with a high precision electronic scale in order to prepare the desired concentrations. The concentration was stepwise increased from

$C = 0.001$ g/L–50 g/L. To prevent the aggregation of particles, we initiated a strong turbulence by the stirrers and added the powder through a sieve. To control vertical homogeneous mixing conditions, measurements were conducted at four depths, equally separated.

3.2. *Suitability of the ADV Probe*

To determine the most unambiguous relation between the SNR and the concentration, firstly, a clear SNR signal was needed, and secondly, the deviations within the four receiver signals had to be as small as possible. Figure 3 shows the box plot evaluation of the SNR signal for three repeated measurements in a low concentrated quartz powder suspension.

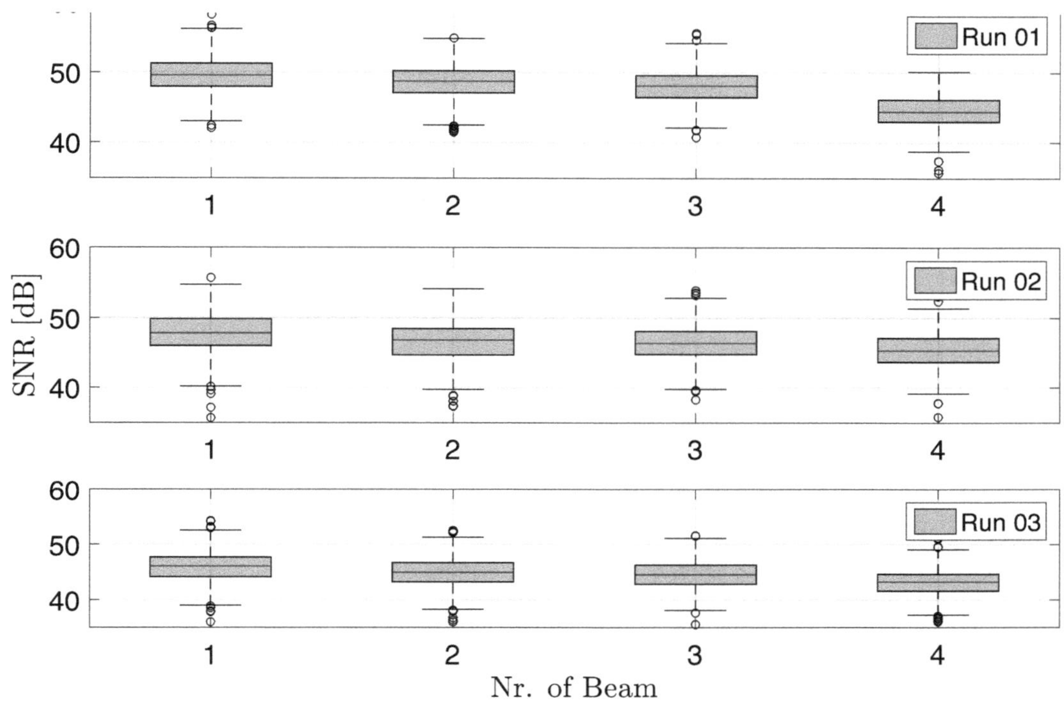

Figure 3. A box plot evaluation is shown of the SNR data for three repeated measurements in a low concentrated quartz powder suspension. Only negligible deviations of the median of each beam are recognizable.

The scattering of the Interquartile Range (IQR) was similar for all four beams and for three repetitions, within the range of 2–3 dB. It is noticeable that Beam 4 was measuring lower values than the remaining receivers. On the contrary, Beam 1 was providing slightly greater values.

In the case of Run 03, this behavior is shown in Figure 4, where the SNR signals are presented as histograms. While fitting Gaussian curves to the histograms, the mean μ and the standard deviation σ were calculated. It shows that Beams 2 and 3 provided similar mean values $\mu_2 = 44.96$ dB and $\mu_3 = 44.61$ dB, which is in between $\mu_1 = 46.03$ dB and $\mu_4 = 43.08$ dB. Nevertheless, when averaging the data of all four receivers, the equal deviations of Beam 1 and Beam 4 were eliminated. The standard deviations of the SNR distributions were in the range of $\sigma = 2.5$–2.7 dB, which corresponded to the values obtained by the box plot evaluation.

Due to the narrow range of the IQR and the few detected outliers in the box plots, as well as similar mean values and low standard deviations, it was assumed that this ADV probe was satisfactorily suitable for calibration measurements.

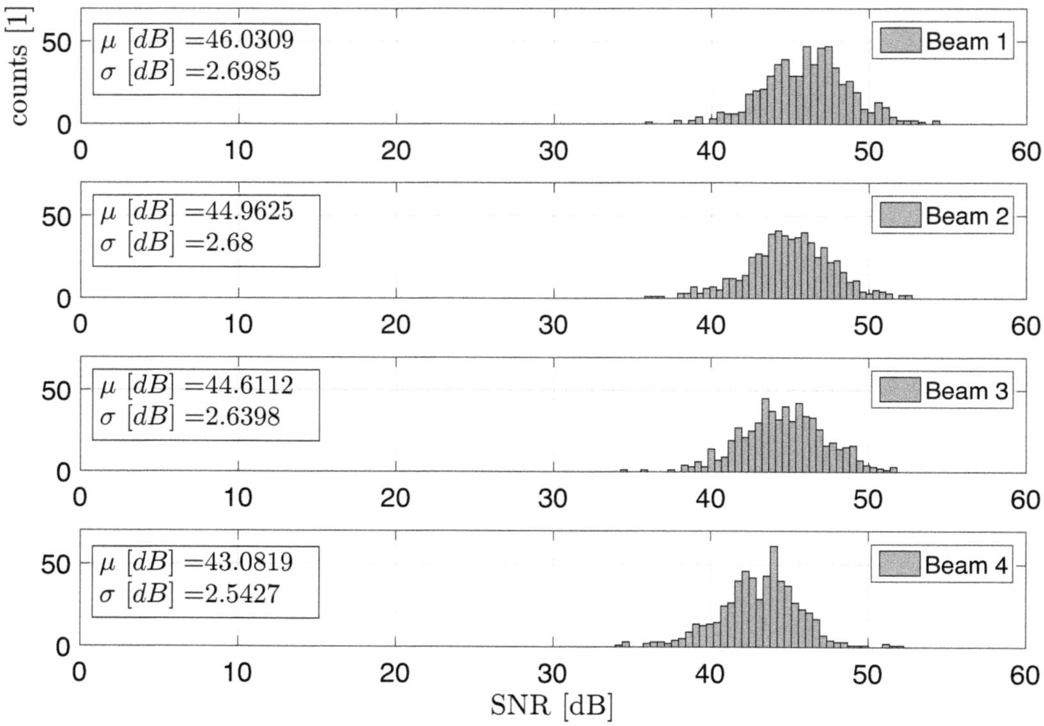

Figure 4. The evaluation of the frequency of measured SNR data is described in histograms (Run 03). Histograms are given for each receiver (beam) with corresponding mean and standard deviation values.

3.3. Acoustic Feedback: Based on Quartz Powder Measurements

After averaging the SNR data of the four receivers, the SNR was plotted against the quartz powder concentration in semi-logarithmic scale. While the SNR of the four receivers was already averaged, Figure 5 illustrates the dependency of SNR on the concentration for two different measurement heights above the bottom (z_1 = 19 cm, z_2 = 15 cm). Although the measurements at different heights fluctuated in a range of about 5 dB, no systematic deviations related to the z-coordinate were noticeable. Therefore, we concluded that the prepared suspensions were well mixed, and a homogeneous concentration was apparent in the experimental box.

In semi-logarithmic scale, a characteristic linear shape can be observed up to $C \approx 1$ g/L. For higher concentrations, no linear dependency can be regarded any more. On the basis of the quartz powder measurements, we propose the following subdivision of the chart in Figure 5 into three acoustically-characteristic regions. (I) The whispering zone: The whispering region is characterized by few echoes returning from the measurement volume due to low particle concentration. Within this region, SNR data reacted sensitively to small deviations of the particle amount. The log-linear behavior between the SNR and the concentration was limited in this region since the ADV probe always detected a minimum SNR value. (II) The audibility zone: Within this region, a clear echo intensity was received. The absorption losses on sediment particles were considered to be low. It is common to neglect α_p for concentrations up to $C \approx 1$ g/L [5]. (III) The muffling zone: Due to a high concentration, the acoustic backscatter signal was intense. In this region, the log-linear shape provided no suitable data approximation. Still, the overall scattering coefficient was increasing with concentration, which led to an increase of SNR. At the same time absorption, losses on particles were becoming more dominant. This resulted in a vertex after which the SNR decreased again. When absorption losses overcome the increase of the scattering coefficient, SNR data will decrease as a matter of fact.

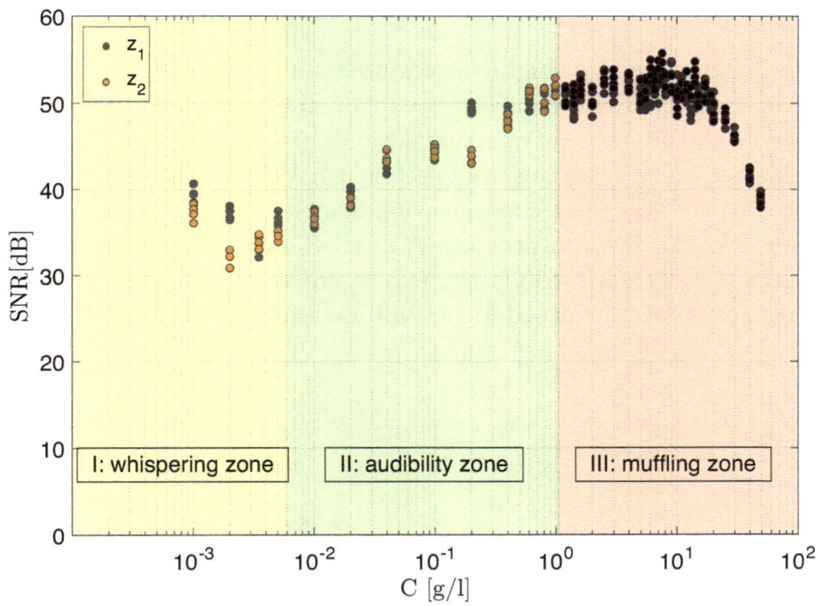

Figure 5. SNR measurements are shown for two different measurement heights as a function of the quartz powder concentration. Measurement results of each receiver are summarized. For better interpretation, the curve is divided into three characteristic areas (I, II, III).

The continuous flattening of the data points (with increasing concentration) in Figure 5 was due to the increasing influence of α_p. Not knowing at which concentration α_p would be negligible or not, it is proposed to consider both the absorption coefficients α_w and α_p irrelevant with respect to the concentration. Since α_p is a function of C itself (Equation (12)), it will be automatically reduced in the case of low C. Therefore, a steady function SNR = f(C) has to be found.

4. Modeling the Signal-to-Noise Ratio

In this section, we describe the methodology of fitting the sonar equation parameters (Equation (13)) into the measurements. Doing so, the influence of α_p is considered. Furthermore, we show how the sonar model is applied to compute the concentration by means of the SNR. In the end, we evaluate and compare the results of different kinds of sediment.

4.1. Application of the Sonar Equation

When rearranging Equation (13), the SNR is described as:

$$SNR = \frac{10}{\Pi_1}\left(log_{10}\left(A\,C\right) - 4\,log_{10}\left(R\right) + \frac{C_{ADV}}{10} - \frac{1}{5}R\,\alpha_w - \frac{2\,R}{10}\,\alpha_p\right) \tag{14}$$

with the three unknown parameters: Π_1, $\alpha_p = \Pi_2 \cdot Ur_{analyt.} \cdot C$ and C_{ADV}. Π_2 is interpreted as an empirical correction factor of $Ur_{analyt.}$.

Figure 6 describes the experimental data in comparison to the parametrized sonar function, which is based on Equation (14). The parameterization of the sonar equation was done with a non-linear least-squares method in the MATLAB® Curve-Fitting-Tool. It can be seen that the fit function and the measurements are in good agreement. The unknown parameters were found as Π_1 = 1.712, Π_2 = 4.34, and C_{ADV} = 42.08 dB. We mention that the analytical solution of the Urick-coefficient gave $Ur_{analyt.}$ = 1.727 m^2/kg, which deviated in comparison to $Ur_{sonar} = \Pi_2 \cdot Ur_{analyt.}$ = 7.43 m^2/kg. Furthermore, we mention that Π_1 is a transformation coefficient of the SNR (see Equation (13)), which is expected to be close to 1.

When considering the region of high concentrations, the sonar model describes a flattening followed by a decrease of SNR. This behavior is explained by the absorption coefficient α_p in the sonar equation. Even if the influence of α_p remains small in low concentrations, the more particles are in the flow, the higher is the scattering cross-section σ_{bs}, which increases the influence of α_p. At a certain concentration, the influence of α_p becomes dominant and leads to a signal damping, resulting eventually in a decrease of the SNR function. Especially, SNR measurements at higher concentrations depict this effect. A sudden decrease of SNR data was regarded for all tested sediments. The "turning point" of SNR differed with the type of sediment, thus it was expected to be material dependent.

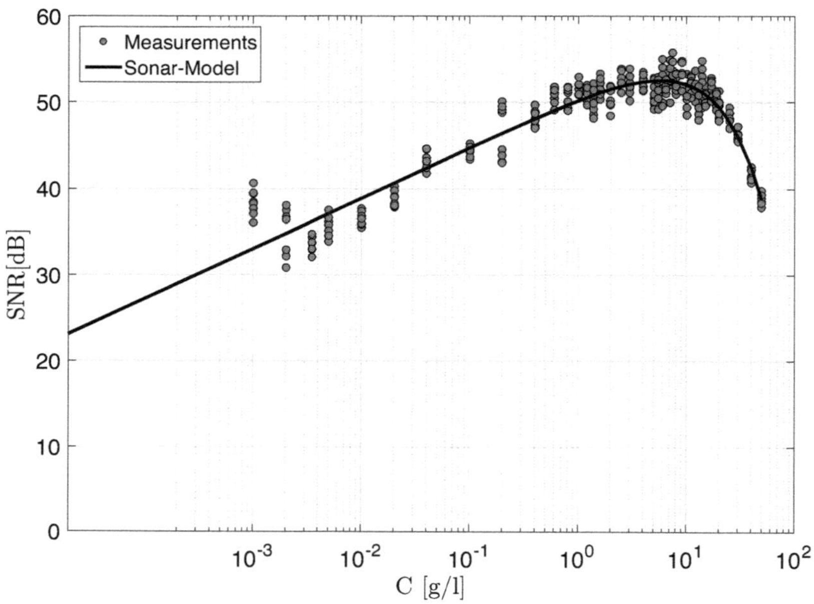

Figure 6. The measured SNR data are shown as a function of quartz powder concentration. The fit function is given for the evaluation of the sonar equation (Equation (14)). This model predicts a decreasing SNR in the high concentration regime and a log-linear development in the low concentration regime.

4.2. *Evaluation of the Sonar Model for Different Sediments*

Next, with respect to quartz powder, we conducted measurements of three other artificial sediments, namely bentonite, metakaolin, and Chinafill. All artificial sediments were used as dry powders so that no drying or sediment preparation was necessary. In order to testify to our methodology for natural sediments, we repeated the measurements for coastal sediments from the Ems Estuary (North Sea, Germany) and for freshwater sediments from Lake Altmühl and Lake Eixendorf (Bavaria, Germany). Before using the natural sediments, they needed to be dried, crushed, and sieved (<25 mm) in order to gain the necessary powder.

Dry densities of the sediments were measured with a gas pycnometer (Pycnomatic ATC, Porotec), and mean particle diameters were calculated from grain size distributions measured with a particle sizer (partica LA-950, Horiba). A list of the resulting material constants $Ur_{analyt.}$, Y, and A is given in Table 2.

Table 2. The different particle parameters used to calculate the *Ur*-coefficient and the material parameter *A*.

Sediment	r (μm)	ρ_s (kg/m^3)	C_U (-)	$Ur_{analyt.}$ (m^2/kg)	Y (-)	A (m^{-1})
Quartz powder	9	2650	3.806	1.727	0.129	0.329
Bentonite	3.5	2600	3.90	4.066	0.129	0.019
Metakaolin	4.5	2215	2.01	2.788	0.123	0.047
Chinafill	4.5	2673	2.15	3.264	0.129	0.041
Ems 2012	5	2525	3.20	2.846	0.128	0.058
Ems 2015	9	2450	4.16	1.649	0.127	0.348
Lake Eixendorf	10	2500	4.61	1.545	0.127	0.470
Lake Altmühl	6	2463	6.49	2.357	0.127	0.103

After repeating the calibration experiment for different sediments, the sonar model was fitted to the measurements. The results for the artificial sediments (quartz powder, bentonite, metakaolin, and Chinafill) are plotted in Figure 7. It can be seen that the sonar model was capable of reproducing the experimental datasets.

In the case of artificial sediments, it is recognizable that there was some deviation between the model and the experimental data in the log-linear regime (II: audibility zone). However, with increasing concentration, the deviations become smaller. When the concentration was sufficiently high, a decrease of the SNR was observed. The conditions to pass from increasing to decreasing SNR response (III: muffling zone) differed with the type of sediment.

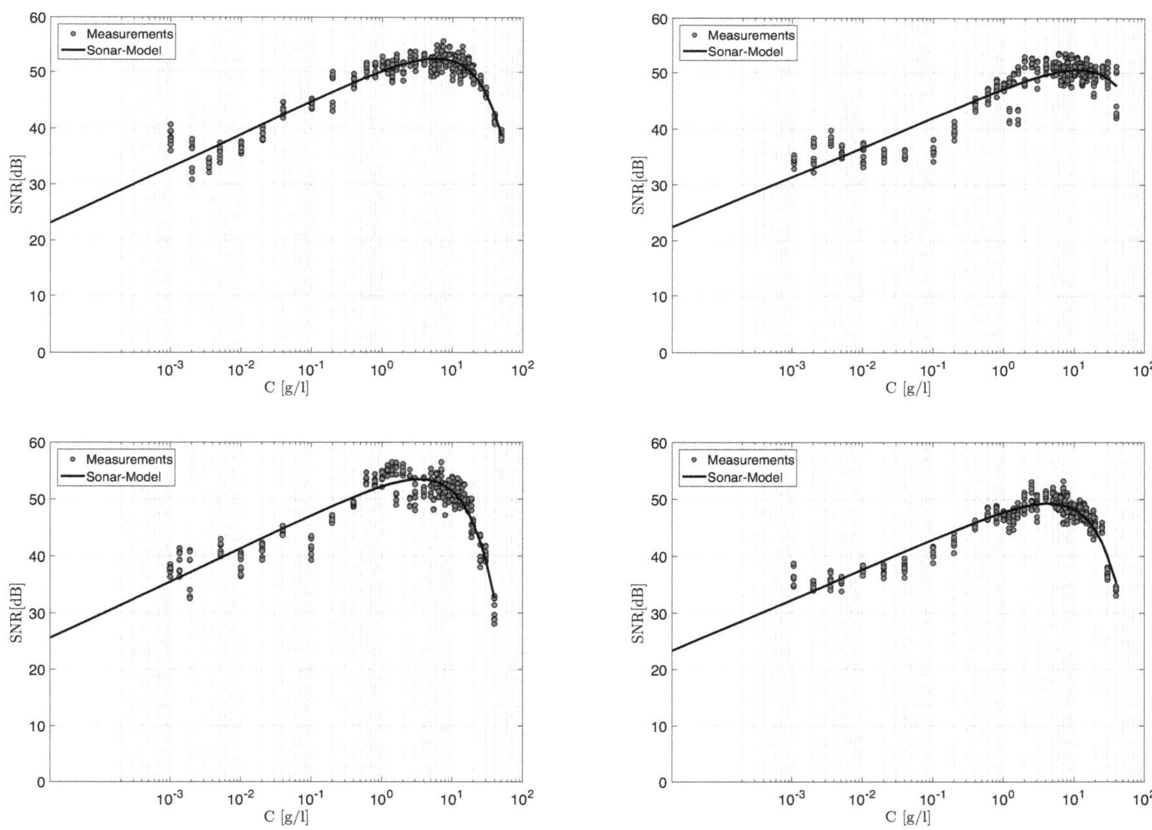

Figure 7. The measured SNR data are shown as a function of concentration for different artificial sediments: (**upper left**) quartz powder; (**upper right**) bentonite; (**lower left**) metakaolin; (**lower right**) Chinafill. The fit function is given for the evaluation of the sonar equation (Equation (14)). This model predicts a decreasing SNR in the high concentration regime and a log-linear development in the low concentration regime.

In comparison to the artificial sediments, the natural sediments from Lake Eixendorf, Lake Altmühl, and the Ems Estuary are presented in Figure 8. The resulting fit parameters are summarized in Table 3 for all measurements. Additionally, R^2 is given as the square of the correlation between the fitted and the measured values.

Even for natural sediments where the composition of sediment is far more complex compared to artificial sediments, our sonar model was able to reproduce the experimental data. Furthermore, it seems that natural sediments rather follow a log-linear approach in the low concentration regime (II: audibility zone) than artificial sediments. For natural sediments, the slope of the modeling results in the log-linear regime does not differ as much as for the artificial sediments. In the opinion of the authors, this effect is not surprising since natural sediments likely consist of similar sediment partitions.

For natural sediments, the vertices of the modeled SNR were located very close to each other, whereas the variation was greater for artificial sediments. Out of this, we conclude that there exists a sediment- and device-specific critical concentration, which separates the increasing and decreasing SNR response.

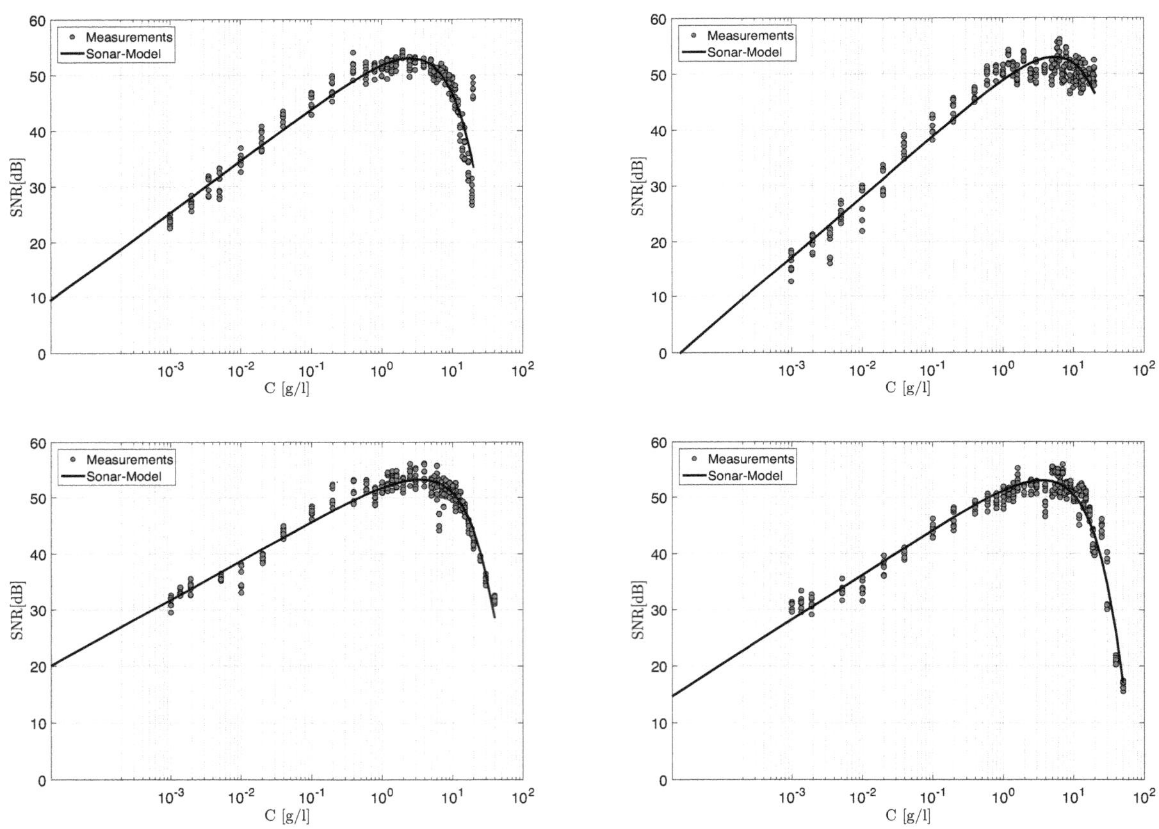

Figure 8. The measured SNR data are shown as a function of concentration for different artificial sediments: upper left: Lake Eixendorf: upper right: Lake Altmühl; lower left: Ems 2012; lower right: Ems 2015. The fit function is given for the evaluation of the sonar equation (Equation (14)). This model predicts a log-linear evolution in the low concentration regime and a decreasing SNR in the high concentration regime.

Table 3. Fit parameter of the sonar model for different sediments. R^2 as a statistical parameter is the square of the correlation between the fitted and the measured values.

Material/Location	Π_1 (-)	Π_2 (-)	C_{ADV} (dB)	R^2 (%)
Quartz powder	1.712	4.34	42.08	96.34
Bentonite	1.890	0.94	57.06	88.91
Metakaolin	1.719	4.57	54.98	91.02
Chinafill	1.887	3.08	55.54	92.99
Ems 2012	1.441	4.47	38.84	96.32
Ems 2015	1.253	7.40	20.62	95.69
Lake Eixendorf	1.070	11.06	10.97	88.34
Lake Altmühl	0.921	3.37	6.27	96.59

5. Discussion

We analyzed the fit parameters Π_2 and C_{ADV} and separated the data into artificial sediments (quartz powder, bentonite, metakaolin, and Chinafill) and natural sediments (Ems Estuary, Lake Eixendorf, and Lake Altmühl). Knowing that the mentioned artificial sediments are natural materials, but artificially processed, we use this terminus to distinguish between the sediments extracted from nature. When reading the fit parameters in Table 3, differences between Π_2 and C_{ADV} can be observed. Furthermore, it can be seen that Π_2 and C_{ADV} can be separated into artificial and natural sediments. This is shown in Figure 9 for Π_2 and in Figure 10 for C_{ADV}. Π_2 is described through $\alpha_{p,fit}/\alpha_{p,analyt.}$, which is the coefficient of the fitted and the analytically-calculated particle absorption coefficient. A value close to one would imply perfect agreement between the modeled sound absorption on particles and the value based on Urick's formula in Equation (12).

Artificially-processed sediments are characterized by a very homogeneous grain size distribution, which is not or to a negligible extent contaminated by other sediments or particle diameters. The grading curve therefore is steeper than for sediments extracted from nature. The uniformity coefficient $C_U = d_{60}/d_{10}$ is a dimensionless number to describe the uniformity of a sediment sample. The diameters d_{60} and d_{10} correspond to 60% and 10% sieve passage, respectively. The sediment samples' uniformity coefficients C_U are shown in Table 2. The grading curves of the artificial sediments result in lower C_U-values than for natural sediments. The formula of Urick (Equation (12)), which is implemented in the model, is based on the assumption of homogeneous particle diameters [7]. In Salehi and Strom, it is mentioned that the particle size distribution also affects α_p [5]. In the case of ideal sediment, where all particles have the same size, C_U would be equal to one. This was not the case, no matter which sediment we used. However, in Figure 9, it is shown that $\alpha_{p,fit}/\alpha_{p,analyt.}$ is lower when using artificial sediments with lower C_U values. However, as the data for "Ems 2012" show, C_U is not the only parameter that leads to a decrease of $\alpha_{p,fit}/\alpha_{p,analyt.}$. "Ems 2012" with $C_U = 3.2$ results in $\alpha_{p,fit}/\alpha_{p,analyt.} = 4.47$, which is greater than the values of quartz powder and bentonite.

Figure 10 presents the fit parameter C_{ADV} with respect to C_U. Our results show that C_{ADV} is not only a device-specific parameter, which can be kept as constant for every material. We conclude that it is rather a combined device- and sediment-specific parameter. The data for C_{ADV} again can be separated into artificial and natural sediments and were in the range of C_{ADV} = 42–58 dB for quartz powder, bentonite, Chinafill, and metakaolin. On the contrary, C_{ADV} data for natural sediments were in the range of C_{ADV} = 6–39 dB.

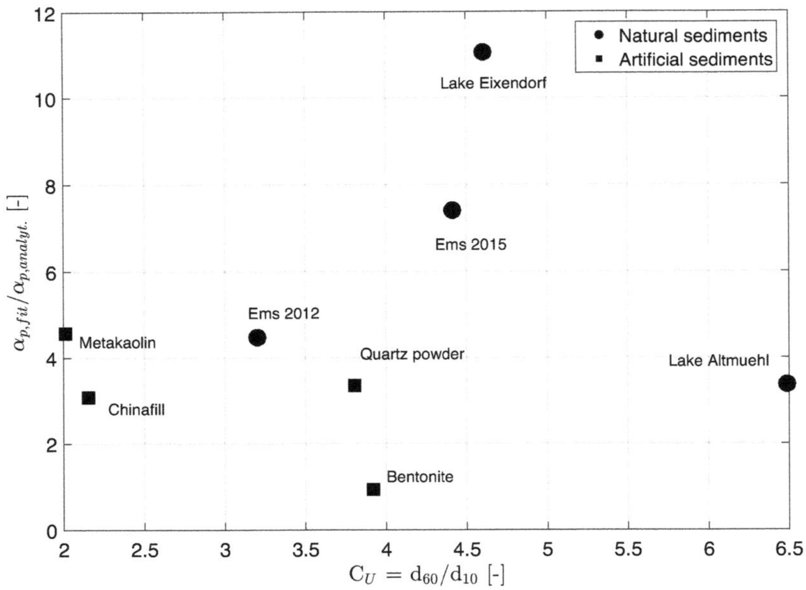

Figure 9. Artificial sediments have lower uniformity coefficients C_U than natural sediments. This results in lower values of $\alpha_{p,fit}/\alpha_{p,analyt.}$, which means that the agreement increases towards the formula of Urick.

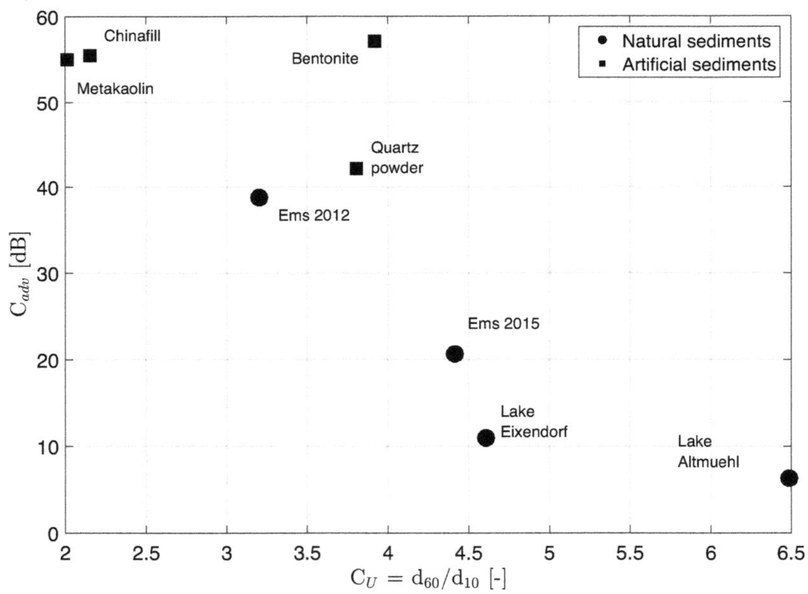

Figure 10. Artificial sediments have lower uniformity coefficients C_U than natural sediments. As a result, the sonar model predicts greater C_{ADV} values.

To summarize, our results show that the sonar model can be applied for different sediments without losing its qualitative behavior. We also showed that the model parameters clearly indicate the variation of sediment type. It was shown that the sediment absorption coefficients α_p and C_{ADV} not only rely on the sediment diameter, but also on the grain size distribution.

Practical Use and Reliability

In order to assess the practical use, applicability, and reliability of measurements in rivers and estuaries, for instance, we highlight some main characteristics of the method.

Due to the model's dependence on the particle grain size distribution, the user in the field would need to specify the characteristic sediment that occurs at the measurement location. When Urick's sound absorption coefficient is used, the mean particle radius and the solid density are needed.

Next, with respect to the sediment-specific parameters, knowledge about the measurement device (here: ADV) is necessary. In general, the sonar model can be also applied for ADCP measurements (e.g., [11,27]), though within this study, we only discuss ADV measurements. The ADV probe (Nortek, Vectrino Profiler) that was used has a dynamic range of 60 dB. In comparison, the Vectrino+ (Nortek) has a dynamic range of 25 dB. Both probes have an emitter frequency of 10 MHz. When repeating our measurements with the Vectrino+ probe, we could not find any trend between the SNR and the concentration. Therefore, it can be concluded that a certain dynamic range is of importance to resolve the SNR response with respect to the concentration. Furthermore, it is mentioned that the presented parameters are only valid for the Vectrino Profiler with a dynamic range of 60 dB.

Following the SNR function to concentrations lower than 0.1 mg/L, one will perceive that in this case, Equation (14) predicts negative SNR values. In particular for $C = 3 \times 10^{-6}$ g/L, the function crosses the axis of ordinate at SNR = −10 dB. Following the principles of the sonar theory, the SNR has to be positive for particle detection [25]. Negative values of the SNR would imply that noise is more dominant compared to the particle backscatter signal. Thus, the region of negative SNR values should be neglected for the interpretation of the model function based on sonar theory. Additionally, it can be observed that the experimental SNR data seem to reach a minimum threshold value for concentrations close to zero. This is explained by the SNR definition, which is implemented in the software of the measuring device. Within the ADV probe, the SNR is calculated by the difference of the backscatter signal intensity, which is here called Amp_b (dB), and the noise intensity, Amp_n (dB):

$$SNR = Amp_b - Amp_n \tag{15}$$

Each of the two intensities is calculated by the received counts as:

$$Amp_i = 20\ log_{10}\left(\frac{counts_i + 1}{counts_{max}}\right) \tag{16}$$

with $counts_{\ max}$ = 8000 for the used ADV probe [28]. This formulation leads to a minimum intensity Amp_{min} = −78.062 dB for no detected counts and to a maximum intensity Amp_{max} = 0 dB in the case of reaching the maximum counts. When inserting Equation (16) into Equation (15), the SNR is written as:

$$SNR = 20\ log_{10}\left(\frac{counts_b + 1}{counts_n + 1}\right) = 20\ log_{10}\left(\frac{counts_p + counts_n + 1}{counts_n + 1}\right) \tag{17}$$

where $counts_b$ includes the counts related to particle backscattering $counts_p$, as well as the noise $counts_n$. In the case of no backscatter intensity (absolutely no scattering particles in the measurement volume), the minimum SNR would be SNR_{min}= 0 dB and fully independent of noise intensity. Indeed, this is a theoretical threshold, because there are always some scattering particles in the measurement volume. Nevertheless, there is a minimum threshold value for the SNR, which cannot be reproduced by the log-linear sonar model.

Assuming well-mixed conditions in the experimental tank, depth and bottom material are not expected to influence the SNR. As the name signal-to-noise ratio says, it describes the sound intensity of the ratio of the received signal and the background noise. In practice, just before measuring, the ADV probe records the noise intensity. Later during the active measurements, this intensity is subtracted from the absolute received signal (amplitude). Of course, depth and bottom material can influence the intensity of noise. However, the SNR is not influenced by this. This makes the SNR a reliable and universal measure for different environments. However, it has to be noted that the device's technique to determine the SNR is only valid for steady noise intensity. When conducting long-term measurements with frequent changes of noise (ships, wind, etc.) during the data acquisition, the SNR will be falsified.

In our experiments, we tried to avoid flocculation of particles while the suspension was well mixed. In nature, especially in estuaries, various floc populations are observed, e.g., [29]. Sound attenuation on flocs is expected to be different than on single primary particles. Flocs are a connection of solids and organic substances, which vary in size, density, and elasticity. In order to calibrate the probe for flocculated suspensions, advanced experimental setups are needed where flocs will not be destroyed during the experiment.

Summing up, the sonar model is able to reproduce the experimental data for a number of different sediments. It shows the log-linear increase of SNR with concentration, as well as the vertex behavior and the decrease of SNR for higher concentrations. This particular behavior of the SNR (increase and decrease) expects attention and special treatment when applying such a model for concentration measurements in nature. One SNR value can be assigned to two concentrations, which is a disadvantage of the model and which has to be taken into account. It reduces the measurement range of this method either to the low concentration regime (positive slope of SNR) or to the high concentration regime (negative slope of SNR). While only using the ADV probe for concentration measurements, the user will not be able to assess the appropriate concentration. Especially in nature, the usage of a second concentration measurement device (e.g., OBS, LISST, tuning fork, etc.) would be necessary to specify the concentration regime [6,11,30]. Then, it is possible to utilize an ADV to measure the velocity, the turbulence, and the concentration simultaneously.

6. Conclusions

In the scope of the experimental investigations, a clear relation was found between the SNR and the concentration of several sediment suspensions. The method was found to be valid for maximum concentrations $C_{max} = 50$ g/L, as is shown for sediments from the Ems Estuary. In higher concentrations, correlation values between two received signals dropped below 70 %, that is why those measurement results were discarded within the evaluation [31]. While combining scattering and sonar theory, an advanced model was presented for computing the concentration out of the SNR for natural and artificial sediments. The derived model parameters are only valid for the ADV probe and the sediments that we used in this experiment. However, it was shown that the methodology is valid for different sediments, both natural or artificial ones, and it can be applied for further investigations of turbulent suspension flows. Since this measurement technique allows temporal high resolved (f = 100 Hz) non-invasive concentration measurements, it is meant to be a basis for further experimental investigations of turbulence and diffusivity, either in nature or at the laboratory scale.

Author Contributions: Conceptualization, O.C. and I.B.; methodology, O.C. and I.B.; software, O.C.; validation, O.C., I.B. and A.M.; formal analysis, O.C.; investigation, O.C.; resources, I.B., A.M.; data curation, O.C. and I.B.; writing, original draft preparation, O.C.; writing, review and editing, O.C., I.B. and A.M.; visualization, O.C.; supervision, I.B.; project administration, A.M.; funding acquisition, A.M.

Funding: The work presented in this paper is part of a joint research project MudEstuary. The MudEstuary project is funded by the German Coastal Engineering Research Council (KFKI) within the German Federal Ministry of Education and Research (BMBF) through the project management of Projektträger Jülich (PTJ) under Grant Numbers 03KIS112 and 03KIS113. The funding is gratefully acknowledged.

Conflicts of Interest: The authors declare no conflict of interest. The founding sponsors had no role in the design of the study; in the collection, analyses, or interpretation of data; in the writing of the manuscript; nor in the decision to publish the results.

References

1. Nezu, I.; Nakagawa, H. Turbulence in Open-Channel Flows. *J. Fluid Mech.* **1994**, *269*, 373–374. [CrossRef]
2. Schrottke, K. *Dynamik fluider Schlicke im Weser und Ems-Ästuar—Untersuchung und Analysen zum Prozessverständnis*; BAW/BfG Kolloqium: Karlsruhe, Germany, 2006.
3. Cellino, M. Experimental Study of Suspension Flow in Open Channels. Ph.D. Thesis, Ecole Polytechnique Federale de Lausanne, Lausanne, Switzerland, 1998.
4. Pope, S.B. *Turbulent Flows*; Cambridge University Press: Cambridge, UK, 2001.
5. Salehi, M.; Strom, K. Using Velocimeter Signal to Noise Ratio as a Surrogate Measure of Suspended Mud Concentration. *Cont. Shelf Res.* **2011**, *31*, 1020–1032. [CrossRef]
6. Fugate, D.C.; Friedrichs, C.T. Determining concentration and fall velocity of estuarine particle populations using ADV, OBS and LISST. *Cont. Shelf Res.* **2002**, *22*, 1867–1886. [CrossRef]
7. Urick, R. The absorption of sound in suspensions of irregular particles. *J. Acoust. Soc. Am.* **1948**, *20*, 283–289. [CrossRef]
8. Lohrmann, A.; Cabrera, R.; Kraus, N.C. Acoustic-Doppler Velocimeter (ADV) for Laboratory Use. In Proceedings of the Fundamental and Advancements in Hydraulic Measurements and Experimentation, Buffalo, NY, USA, 1–5 August 1994; pp. 351–365.
9. Thorne, P.D.; Hanes, D.M. A review of acoustic measurement of small-scale sediment processes. *Cont. Shelf Res.* **2002**, *22*, 603–632. [CrossRef]
10. Jay, D.A.; Orton, P.; Kay, D.J.; Fain, A.; Baptista, A.M. Acoustic determination of sediment concentrations, settling velocities, horizontal transports and vertical fluxes in estuaries. In Proceedings of the IEEE Sixth Working Conference on Current Measurement, San Diego, CA, USA, 13 March 1999; pp. 258–263.
11. Hoitink, A.; Hoekstra, P. Observations of Suspended Sediment from ADCP and OBS Measurements in a Mud-dominated Environment. *Coast. Eng.* **2005**, *52*, 103–118. [CrossRef]
12. Ha, H.; Hsu, W.Y.; Maa, J.Y.; Shao, Y.; Holland, C. Using ADV backscatter strength for measuring suspended cohesive sediment concentration. *Cont. Shelf Res.* **2009**, *29*, 1310–1316. [CrossRef]
13. Decrop, B.; De Mulder, T.; Toorman, E.; Sas, M. New methods for ADV measurements of turbulent sediment fluxes–application to a fine sediment plume. *J. Hydraul. Res.* **2015**, *53*, 317–331. [CrossRef]
14. Best, J.; Bennet, S.; Bridge, J.; Leeder, M. Turbulence Modulation and Particle Velocities over Flat Sand Beds at Low Transport Rates. *J. Hydraul. Eng.* **1997**, *123*, 1118–1129. [CrossRef]
15. Cellino, M. Ultrasonic Measurements of Instantaneous Velocity and Suspended Concentration in Open-Channel Flow. In Proceedings of the Third International Symposium on Ultrasonic Doppler Methods for Fluid Mechanics and Fluid Engineering Co-organized by EPFL and PSI, Lausanne, Switzerland, 9–11 September 2002.
16. Hurther, D.; Thorne, P.D.; Bricault, M.; Lemmin, U.; Barnoud, J.M. A multi-frequency Acoustic Concentration and Velocity Profiler (ACVP) for boundary layer measurements of fine-scale flow and sediment transport processes. *Coast. Eng.* **2011**, *58*, 594–605. [CrossRef]
17. Revil-Baudard, T.; Chauchat, J.; Hurther, D.; Barraud, P.A. Investigation of sheet-flow processes based on novel acoustic high-resolution velocity and concentration measurements. *J. Fluid Mech.* **2015**, *767*, 1–30. [CrossRef]
18. McAnally, W.H.; Friedrichs, C.; Hamilton, D.; Hayter, E.; Shrestha, P.; Rodriguez, H.; Sheremet, A.; Teeter, A. Management of Fluid Mud in Estuaries, Bays, and Lakes. I: Present State of Understanding on Character and Behavior. *J. Hydraul. Eng.* **2007**, *133*, 9–22. [CrossRef]
19. Becker, M.; Maushake, C.; Winter, C. Observations of Mud-Induced Periodic Stratification in a Hyperturbid Estuary. *Geophys. Res. Lett.* **2018**, *45*, 5461–5469. [CrossRef]
20. Craig, R.G.; Loadman, C.; Clement, B.; Rusello, P.J.; Siegel, E. Characterization and testing of a new bistatic profiling acoustic Doppler velocimeter: The vectrino-II. In Proceedings of the 2011 IEEE/OES 10th Current, Waves and Turbulence Measurements (CWTM), Monterey, CA, USA, 20–23 March 2011; pp. 246–252.
21. Thomas, R.; Schindfessel, L.; McLelland, S.; Creëlle, S.; De Mulder, T. Bias in mean velocities and noise in variances and covariances measured using a multistatic acoustic profiler: The Nortek Vectrino Profiler. *Meas. Sci. Technol.* **2017**, *28*, 075302. [CrossRef]
22. Bruens, A. *Communications on Hydraulic and Geotechnical Engineering—Entraining Mud Suspensions*; Report No. 03-1; Delft University of Technology: Delft, The Netherlands, 2003.

23. Mietta, F.; Chassagne, C.; Winterwerp, J. Shear-induced flocculation of a suspension of kaolinite as function of pH and salt concentration. *J. Colloid Interface Sci.* **2009**, *336*, 134–141. [CrossRef] [PubMed]
24. Medwin, H.; Clay, C.S. *Fundamentals of Acoustical Oceanography*; Academic Press: Cambridge, MA, USA, 1997.
25. Lerch, R.; Sessler, G.; Wolf, D. *Technische Akustik: Grundlagen und Anwendungen*; Springer: Berlin/Heidelberg, Germany, 2009.
26. Lohrmann, A. *Monitoring Sediment Concentration with Acoustic Backscattering Instruments*; Nortek Technical Note; Nortek: Bologna, Italy, 2001; p. 9.
27. Guerrero, M.; Szupiany, R.N.; Latosinski, F. Multi-frequency acoustics for suspended sediment studies: An application in the Parana River. *J. Hydraul. Res.* **2013**, *51*, 696–707. [CrossRef]
28. AS Nortek. *VECTRINO Velocimeter User Guide (Rev. C)*; AS Nortek: Carlsbad, CA, USA, 2004.
29. Dyer, K.; Manning, A. Observation of the size, settling velocity and effective density of flocs, and their fractal dimensions. *J. Sea Res.* **1999**, *41*, 87–95. [CrossRef]
30. Groposo, V.; Mosquera, R.L.; Pedocchi, F.; Vinzón, S.B.; Gallo, M. Mud Density Prospection Using a Tuning Fork. *J. Waterw. Port Coast. Ocean Eng.* **2015**, *141*, 04014047. [CrossRef]
31. Ali, A.; Lemckert, C.J. A traversing system to measure bottom boundary layer hydraulic properties. *Estuar. Coast. Shelf Sci.* **2009**, *83*, 425–433. [CrossRef]

7

The Effects of Adding Pyramidal and Convex Diffusers on Room Acoustic Parameters in a Small Non-Diffuse Room

Hassan Azad [1,*,†], Julie Meyer [2,†], Gary Siebein [1] and Tapio Lokki [2]

[1] School of Architecture, University of Florida, Gainesville, FL 32611, USA
[2] Department of Computer Science, Aalto University, FI-02150 Espoo, Finland
* Correspondence: h.azad@ufl.edu
† These authors contributed equally to this work.

Abstract: This paper presents an investigation of the effects of relatively large-scale pyramidal and convex-shaped diffusers on the acoustical properties of a small non-diffuse rectangular room. Room impulse responses (RIRs) were measured in various room configurations to extract the early decay time (EDT), reverberation time (T_{20}), early-to-late arriving sound ratio (C_{50}), and clarity (C_{80}). The difference between the parameters measured in the empty room were chosen to be the reference, and those measured in other room configurations was calculated. Statistical analysis of the measurement results supplements the investigation to determine whether the coverage and type of diffusers contribute significantly to the variation of the acoustical parameters. The results show that adding diffusers in the room generally decreases EDT as well as T_{20}, and increases C_{50} as well as C_{80} for both diffuser types. The statistical analysis shows that the coverage of diffusers significantly contributes to the variation of the acoustical parameters in most conditions (octave band, diffuser type). The effect of the diffuser shape is only significant for some of the conditions (at 4 kHz, the number of diffusers). The data presented demonstrate that in a small non-diffuse room the reverberation can be controlled efficiently by redirecting the sound energy towards the most absorbing surfaces.

Keywords: pyramidal diffuser; convex diffuser; non-diffuse room; acoustical parameters

---

1. Introduction

Sound waves reflected back from diffusing surfaces are temporally and spatially dispersed. Therefore, the inclusion of such surfaces in an enclosed space plays a key role in its acoustic quality. For example, these types of surfaces have proven to be effective at preventing echoes in auditoriums and concert halls [1], and at reducing coloration in small rooms used for sound reproduction [2]. Another application example is the reduction of focusing effects of concave surfaces [3].

While research on the quantification and characterization of the surface scattering is well documented (e.g., two international standards focusing on methods to measure the directional uniformity of the surface scattering have been published [4,5]), fewer studies focus on the effects of these surfaces on the acoustical properties of enclosed spaces. Several of these studies employed scale models and in-situ measurements to determine the effects of diffusing surfaces on the sound field. Jeon et al. [6] investigated the effects of wall diffusing elements on the diffuseness of sound-fields in both a 1:10 scale model hall and in a real recital hall. Among the results, it was found that diffusive surfaces enhanced effective absorption and therefore decreased mainly early decay time

(EDT) for both halls. However, other objective acoustical parameters did not show any consistent tendencies with respect to the presence of diffusers. In another work, Jeon and Kim [7] measured EDT, reverberation time (T_{30}), and clarity (C_{80}), in two scale models (1:50 and 1:25) of a concert hall. These parameters were measured for eight configurations of the hall created using hemispherical diffusers with different geometrical height and a reference configuration without diffusers. It was concluded that the acoustical parameters were inconsistently changing across the configurations as a result of adding diffusers. Choi [8] studied the effects of periodic type diffusers on classroom acoustics by adding different amounts of absorptive and diffusing materials to one or more of five surfaces of a 1:10 scale model classroom. The comparison of the most successful treatment with a typical acoustical treatment of a classroom led to early-to-late arriving sound ratio (C_{50}) values increased by more than one just noticeable difference (JND) while EDT and T_{30} values decreased by more than one JND. Shtrepi et al. [9] measured the acoustic scattering effects produced by a lateral diffusive wall in a small variable-acoustics hall. Four objective acoustic parameters were compared between a diffusive and a reflective condition of the wall. It was found that, as the acoustic scattering increased, the values of EDT and T_{30} increased while these of C_{80} and definition (D50) decreased.

Computer simulations have also been performed to determine the effects of diffusing surfaces on the sound field. Shtrepi et al. [10] investigated both objectively and perceptually the effects of different scattering coefficients applied to the walls and ceiling of a simulated concert hall, using three prediction models based on geometrical acoustics. The authors reported that the values of the analyzed acoustic parameters (T_{30}, EDT, C_{80}, and sound strength G) depend mainly on the source-to-receiver distance and on the scattering coefficient variation, rather than on the distance from the considered lateral wall. An increase in EDT and a decrease in C_{80} have also been observed for increasing scattering coefficient values for all three software, while no similar trend was observed for the other parameters. Table 1 shows a summary of the findings of the previous studies in regard to the change of the acoustical parameters as a result of the addition of diffusing surfaces in the room.

Despite recent advances in understanding why and where diffusers should be applied, there is a need to further investigate the right placement, amount, or shape of these surfaces as it still seems to be guided by "guess-work" in the design process of rooms [16]. One difference that causes this research to stand out from the previous similar studies is the large number (35) of different measured scattered sound fields or configurations that allowed for a more in-depth analysis of the results and more solid conclusions. Additionally, rather than using a metric characterizing the scattering produced by the diffusers, the effects produced by the inclusion of the diffusers in the room on its acoustical parameters are investigated by changing the placement, number, and type (shape) of the diffusers. Furthermore, the differences in acoustical parameters between comparable room configurations and a reference, that is the empty room (i.e., without diffusers), are statistically analyzed. Another feature of the present work is the non-diffuse nature of the empty room characterized by a distribution of hard, thus reflective, materials on the walls and absorptive materials on the ceiling and floor. The aim of this work is to investigate the effect of the inclusion of diffusing surfaces with respect to their placement, amount, and shape in small non-diffuse spaces. As such, the scope of the present paper is restricted to this type of room as the volume of the room as well as the distribution of hard and absorptive materials will affect how the diffusers impact the sound field.

The remaining sections of this paper are organized as follows. Section 2 describes the room impulse response (RIR) measurement procedure along with the measured room configurations and the two types of diffuser used. Section 3 presents the acoustical parameter measurement results. In Section 4, the results are analyzed both in terms of differences with respect to the empty room and statistically to compare the effect of the number of diffusing elements and their shape on the measured acoustical parameters. Section 4 discusses these results before concluding on the study in Section 5.

Table 1. Summary of the findings of the background studies. The +, -, and ○ signs indicate that with the addition of the diffusers the values were increased, decreased, and inconsistently changed, respectively. The results consistent with those of the present paper are indicated with the marker *. This table is adapted from [10].

Reference (Year)	Room	Diffuser Shape	Method	EDT	RT	C_{50}	C_{80}
Suzumura et al. [11] (2000)	Concert Hall	Column	1:10 Scale Model		○		
Chiles [12] (2004)	Concert Hall	Hemisphere/Batten	1:25 Scale Model	+	- *		
Jeon and Kim [7] (2008)	Concert Hall	Hemisphere	1:25 Scale Model	○	○		○
	Concert Hall	Hemisphere	1:50 Scale Model	- *	- *		
Ryu and Jeon [13] (2008)	Concert Hall	Hemisphere	1:10 Scale Model	- *	- *		○
Kim et al. [14] (2011)	Concert Hall	Hemisphere	1:50 Scale Model 1	- *	- *		+ *
	Concert Hall	Hemisphere	1:50 Scale Model 2	- *	- *		+ *
Kim et al. [15] (2011)	Concert Hall	Wave-shape	1:25 Scale Model	+			
	Concert Hall	Convex	In-situ 1	- *	○		-
	Concert Hall	Stepped	In-situ 2	+	- *		
Choi [8] (2013)	Classroom	Periodic	1:10 Scale Model	- *	- *	+ *	
Jeon et al. [6] (2015)	Concert Hall	Polygon	1:10 Scale Model 1	- *	- *		+ *
	Chamber Hall	Saw-tooth	1:10 Scale Model 2	- *	○		-
Shtrepi et al. [10] (2015)	Concert Hall	Simulated [a]	Ray-based 1	+	- *		-
	Concert Hall	Simulated [a]	Ray-based 2	+	- *		○
	Concert Hall	Simulated [a]	Ray-based 3	+	+		-
Shtrepi et al. [9] (2016)	Multi-purpose	Triangle	In-situ	+	+		-

[a] Different scattering coefficients were applied to the ceiling and walls to model scattered reflections.

2. Impulse Response Measurements

2.1. Measurement Procedure

Impulse response measurements were carried out according to ISO 3382-2:2008 [17] in an unfurnished room of volume $V = 25.2\ \text{m}^2$ having no other use than testing purposes. All measurements were conducted with no people inside the room in a few sessions spread over several days. In order to achieve an appropriate coverage in the room and to account for influences likely to cause differences in reverberation time (RT) throughout the room, one omni-directional class 1 microphone (NTi Audio M2230) placed at a height of 1.2 m was moved between five different and evenly distributed positions across the room. The output of the signal recorded by the microphone was taken by a signal recorder for later analysis. The distance from any microphone position to any reflecting surface of the room and to the source was set to be 1.0 m and 0.84 m, respectively, which complies with the minimum requirement set in the ISO standard [17]. To make sure that all measurement positions were independent, the minimum distance between two microphones was set to be 0.84 m. Because the room used for conducting the measurements has no other special use than testing purposes, a single source position was used (height was set at 1.5 m). To improve the signal-to-noise ratio, each source-receiver measurement was repeated four times for each room configuration, thus resulting in a total of $4 \times 5 = 20$ impulse responses to compute for each room configuration. The source and receiver positions, as well as the room dimensions are illustrated in Figure 1.

The source signal fed to the dodecahedron loudspeaker (Larson Davis, Type BAS001) was a logarithmic sweep of a duration of 4.5 s from 100 Hz to 8 kHz that was generated using a personal laptop. A 100 ms fade-in and fade-out was applied to smooth the amplitude of the excitation signal fed to the loudspeaker. A silence of 3 s was also added in the excitation signal between each sweep repetition to allow for the room response to be fully recorded. Finally, a click was introduced 1 s before the start of the sweep so that the source and recorded signals could be synchronized for the calculation of the impulse response taking into account any delays introduced by the measurement chain.

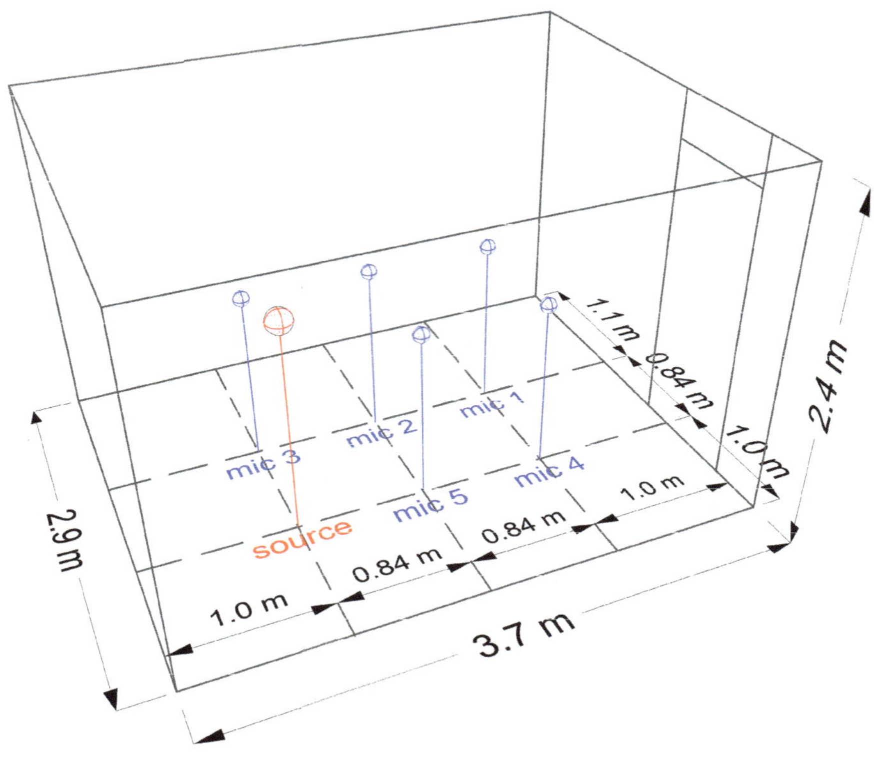

Figure 1. The 3D model of the empty room indicating the room dimensions and the impulse response measurement setup.

Impulse responses were calculated after all measurement data were collected from the signals recorded at the microphone positions. To obtain the impulse responses, the microphone recordings were first synchronized with the full source excitation signal (including the click at its beginning). This synchronization was done by aligning visually the two channels corresponding to the emitted and recorded signals, thus removing any delays between the full source excitation signal and the recorded signals. After synchronization, recorded signals were convolved with the inverse sweep to result in the impulse response [18].

2.2. Measured Configurations

As part of a bigger study, a total of 69 room configurations, including the empty room serving as the reference, were measured following the procedure described in Section 2.1. These configurations were sketched out such that the effects of scattered reflections on the room acoustic parameters were investigated in regard to i. the placement of the diffusers, i.e., the distribution pattern of the diffusers over the wall surfaces; ii. the coverage of the diffusers, i.e., the number of diffusers added in the room; iii. the type of diffuser between a pyramidal and a convex-shaped diffuser acquired from a manufacturer (AVL Systems, Inc., Ocala, FL, USA); iv. the size of the diffusers, either "full-scale" (1.2 m × 1.2 m) or "half-scale" (0.6 m × 0.6 m) diffusers; v. the combination of different types of diffusers. In the present paper, the investigation solely focuses on the contribution of the placement, coverage, and type of the diffusers in the change of the room acoustic parameters, thus reducing the number of configurations to analyze from 69 to 35 (including the empty room).

In the empty room (reference configuration), the structure of the wall assembly is comprised of two layers of 13 mm plasterboards on frame with a cavity of 10 cm in between filled with mineral wool. A 80 cm × 200 cm solid wooden door located on one of the walls discontinues the homogeneous structure of the walls. The floor is covered with loop pile tufted carpet (1.4 kg/m^2) with no underlay, and the ceiling is made of acoustical tiles (USG Radar Ceramic 5/8"). As such, while the wall finishes are made of reflective surfaces, the floor and ceiling are absorptive, resulting in a non-diffuse room. The absorption coefficients of the aforementioned materials are reported in Table 2.

Table 2. Absorption coefficients of the room surfaces and diffusers materials, and the maximum change in the total absorption area of the room with respect to the empty room (in %).

Room Surface	Area (m^2)	One Octave Band (Hz)				
		250	500	1000	2000	4000
3 *Convex*/3 *Pyramid*	4.9/4.6	0.16	0.09	0.05	0.04	0.04
Ceiling [a]	10.7	0.28	0.38	0.60	0.76	0.77
Floor [3]	10.7	0.08	0.17	0.33	0.59	0.75
Walls [3]	29.4	0.12	0.08	0.06	0.06	0.05
Door [3]	1.7	0.10	0.06	0.08	0.10	0.10
$A_{3\text{diff}}$		7.7	8.4	11.8	16.3/16.2	17.8
A_{ref}		7.5	8.3	11.8	16.3	17.9
Maximum percentage change		2.6	0.6	0.5	0.7	0.3

[a] Data adopted from the manufacturer (USG) https://www.usg.com/content/dam/USG_Marketing_Communications/united_states/product_promotional_materials/finished_assets/usg-ceilings-systems-catalog-en-SC2000.pdf.

The maximum number of diffusers added in the room was set to three for both the pyramidal and convex-shaped diffusers, and only one size, the "full-scale", was used. The 3D models and dimensions of the pyramidal and convex-shaped diffusers, from now referred to as pyramid and convex diffusers, respectively, are shown in Figure 2. The pyramid and convex diffusers have a volume of 0.17 m^3 and 0.20 m^3, respectively, and rounded up to one decimal their volume is the same (0.2 m^3). Their surface area, as seen from inside the room, is also approximately the same (1.5 m^2 for pyramid and 1.6 m^2 for convex). Additionally, both diffuser types are made of the same finishes and similar materials as confirmed by visual inspection and discussion with the manufacturer. Given these observations and since the absorption coefficients of the diffusers were not provided by the manufacturer for the size in use, these were found from other manufacturers' databases producing the same diffuser types. The average absorption coefficients across four different manufacturers' databases [19–22] for the pyramidal diffuser (maximum standard deviation (SD) across octave bands of 0.02) was used to estimate the absorption area of both diffuser types, and is reported in Table 2.

The location, especially in height, of the diffusers on the walls was constrained by the placement of racks to mount the diffusing geometries. The area centroid point of the diffusers was located at approximately half of the height of the room, i.e., ~1.3 m from the floor. The mounting clips attached to the back of the diffusers also constrained the convex diffuser to be oriented horizontally in all configurations. As for the pyramid diffusers, the clips constrained the positioning on the wall to some specific orientations visible in Figure 3 and on the 3D models of the configurations (see Tables A1–A6).

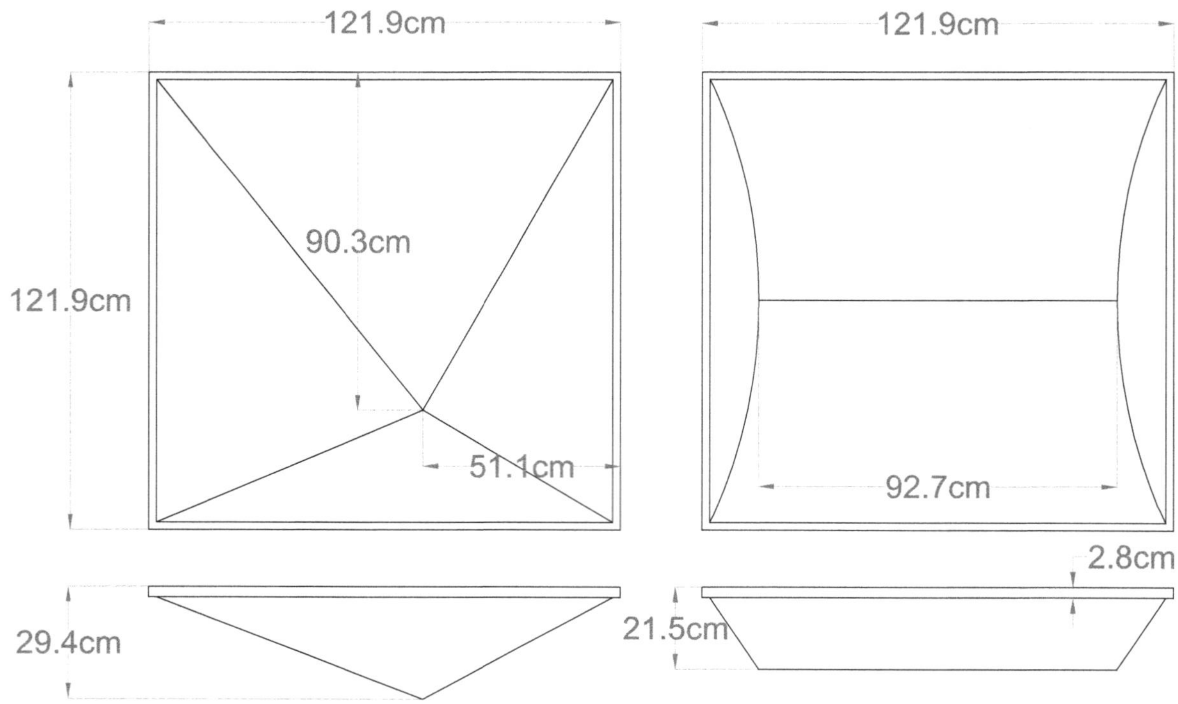

Figure 2. Dimensions and front (**top**) and top (**bottom**) views of the 3D models of the pyramid (**left**) and convex (**right**) diffusers.

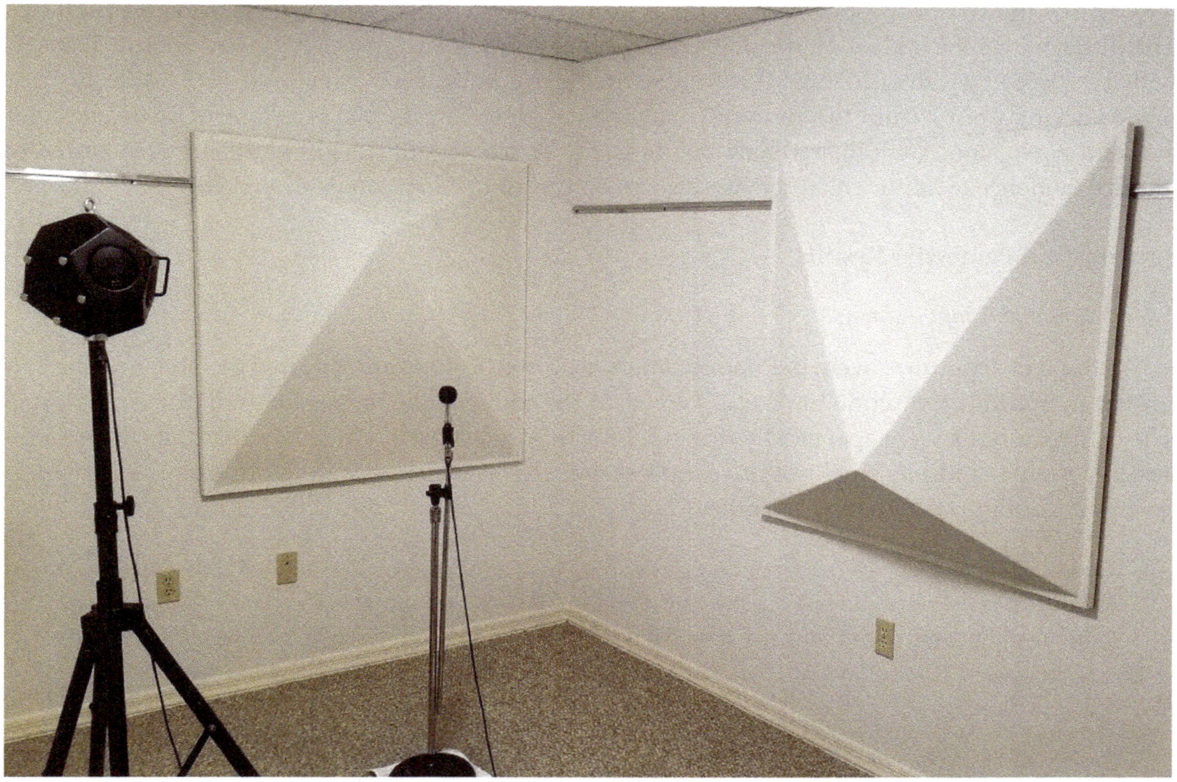

Figure 3. Photo of the measurement setup for one of the configurations of study (2Pi) with receiver at location 3. The source and receiver height is 1.5 and 1.2 m, respectively.

2.3. Assumptions

Three assumptions were made prior to analyzing the data. First, the maximum change in the volume of the room as a result of mounting three diffusers is considered negligible since it is of 2.4% with respect to the empty room.

Second, the maximum change in the total absorption area of the room due to the mounted diffusers is also considered negligible. This assumption is based on the results calculated using Equation (1) and presented in Table 2. As can be seen in Table 2, the maximum change in the total absorption area of the room varies from 0.3% to 2.6% across all octave bands, which can be considered negligible.

$$\text{Percentage change} = 1 - \frac{A_{\text{3diff}}}{A_{\text{ref}}}, \quad (1)$$

where A_{3diff} and A_{ref} denote the absorption area of the room containing three diffusers and the empty room, respectively. A_{3diff} and A_{ref} have been calculated using the definition of the absorption area given by $A = \sum_i \alpha_i S_i$, where i is the material in the room and α_i and S_i are its absorption coefficient and surface area, respectively. It should be noted that this formula applies only in a diffuse sound field and we are assuming a roughly diffuse sound field in this context.

Third, the net absorption due to the gap between the diffusers and the walls is neglected. The diffusers were installed on the walls using the mounting clips fixed at the back of the diffusers, with a measured width of 10 mm. This resulted in the slightly slanted installation of the diffusers with a gap varying from 10 mm to 0 mm in size over a length of 1.05 m, that is a slope of <1%. The small size of the gap along with the reduction of its size to zero across the height of the diffusers provided enough evidence to assume that the net absorption due to this gap can be neglected.

3. Acoustical Parameter Results

3.1. Groups and Configurations

A total of six groups were defined for analysis. These groups are reported in Table 3 along with the configurations they contain. The 3D views of the configurations are shown in the Appendix A. The configurations were specifically designed for this investigation in a way that the effects of the coverage, type, and placement of the diffusers were to be easily considered by comparing certain groups of configurations with each other.

Table 3. Groups and their corresponding configurations. The naming of the configurations corresponds to Integer: number of diffusers in the configuration, P: pyramid diffuser, C: convex diffuser, letter: location and pattern of the installation on the walls of the room (a, m: clustered; k, l, n, o: distributed).

#	Group Name	Configurations
1	1P	1Pa, 1Pb, 1Pc, 1Pd, 1Pe
2	2P	2Pa, 2Pf, 2Pg, 2Ph, 2Pi, 2Pj
3	3P	3Pa, 3Pk, 3Pl, 3Pm, 3Pn, 3Po
4	1C	1Ca, 1Cb, 1Cc, 1Cd, 1Ce
5	2C	2Ca, 2Cf, 2Cg, 2Ch, 2Ci, 2Cj
6	3C	3Ca, 3Ck, 3Cl, 3Cm, 3Cn, 3Co

3.2. Acoustical Parameters

From the measured impulse responses, four acoustical parameters, namely EDT, T_{20}, C_{50} and C_{80}, were calculated according to the standard [17]. More specifically, these acoustical parameters were calculated from the decay curves corresponding to the backward integrated squared IRs. Measurements were averaged across microphone positions per octave band by finding the individual acoustical parameters for all the decay curves and taking the mean value. The calculated acoustical parameters per octave band are reported in Table 4 for the empty room and Tables A1–A6 for the configurations with added diffusers, that can be found below and in the Appendix A, respectively.

Table 4. Mean and SD over the five source-receiver combinations of the early decay time (EDT), T_{20}, C_{50}, and C_{80} for the empty room.

Freq. (Hz)	EDT (s)	T_{20} (s)	C_{50} (dB)	C_{80} (dB)
250	0.50 (0.08)	0.55 (0.06)	4.8 (1.8)	9.8 (1.5)
500	0.33 (0.05)	0.31 (0.03)	9.5 (0.9)	15.8 (1.1)
1000	0.36 (0.07)	0.59 (0.07)	8.7 (0.8)	12.7 (1.2)
2000	1.00 (0.05)	1.17 (0.04)	2.7 (0.7)	4.9 (0.7)
4000	1.01 (0.06)	1.19 (0.02)	5.0 (1.2)	6.5 (1.1)

4. Results Analysis

The data was analyzed using two approaches. In the first approach, referred to hereafter as amount change, the amounts of variation of objective acoustical parameters due to the addition of the diffusers to the empty room are analyzed for the different groups reported in Table 3. In the second approach, a series of statistical tests (Kruskal–Wallis tests) were carried out to examine the significance of difference in the acoustical parameters between the groups of configurations with the intention to analyze the effects of the type and coverage of the diffusers. As a way of data reduction with the caution of not losing important data, three frequency ranges were defined: a low frequency range comprising center frequencies ranging from 250 to 500 Hz, a mid frequency range including 1 k and 2 kHz octave bands, and a high frequency range including the center frequency 4 kHz. The choice of splitting the data into these three frequency ranges was intentionally made prior to analyzing it such that no bias was introduced in the data analysis while the frequency-dependency of the results could be maintained. For each defined frequency range, the data was averaged across corresponding octave bands, thus resulting in a single value for each of the three defined (low, mid, and high) frequency ranges.

4.1. Amount Change

In this approach, the mean value of each acoustical parameter (average across microphone positions) measured in each room configuration was subtracted from this of the empty room. The resulting difference representing the amount of change of each parameter for the configurations of each group with respect to the reference is expressed in seconds for EDT and T_{20}, and in decibels for C_{50} and C_{80}. These amounts of change are shown in Figures 4 and 5 per octave band and discussed in the following paragraphs.

From the plots in Figures 4 and 5, it can be seen that the change in the acoustical parameters with respect to the reference generally resulted in the reduction of both EDT and T_{20} and in the increase of both C_{50} and C_{80}. Over 34 configurations, only five resulted in an increase of EDT, two resulted in an increase of T_{20}, and six in a reduction of C_{50}, at the center frequency of 500 Hz only. The maximum amount of change across these configurations was only of 0.03 s for EDT, 0.01 s for T_{20}, and 0.5 dB for C_{50}.

The amounts of change in the acoustical parameters with respect to the reference were more significant in the mid and high frequency ranges than in the low frequency range. More specifically, the reduction of EDT varied from 0 to 0.17 s in the low frequency range while it varied from 0.08 to 0.78 s and from 0.48 to 0.78 s in the mid and high frequency ranges, respectively. A similar trend was observed for T_{20} for which the reduction fluctuated between 0 to 0.21 s in the low frequency range, from 0.19 to 0.80 s in the mid frequency range and from 0.45 to 0.80 s in the high frequency range. As for C_{50}, it increased from 0.1 to 4.1 dB in the low frequency range, while it increased from 2.0 to 9.8 dB and from 2.1 to 6.8 dB in the mid and high frequency ranges, respectively. Similarly to C_{50}, C_{80} increased from 0.4 to 4.4 dB, from 3.3 to 12.4 dB, and from 3.4 to 9.9 dB in the low, mid and high frequency ranges, respectively.

Studying the effect of the diffusers placement on the acoustical parameters is not straightforward. To consider the effect of the placement only, an introductory analysis comparing the two configurations having the maximum and minimum values of change in the acoustical parameters for each octave band in each group was made. The latter showed that the maximum change due to the placement was 0.09 s for EDT (groups 1PF at 4 kHz, 2PF at 2 kHz, 2CF and 3CF at 500 Hz), 0.17 s for T_{20} (group 3CF at 2 kHz), 2.1 dB for C_{50} (groups 1PF at 250 Hz and 3CF at 2 kHz), 2.9 dB for C_{80} (group 3PF at 2 kHz). This introductory analysis also suggested that the configurations with one diffuser with the greatest distance between the source and the diffuser (those named with the letters e and c) have the lowest amounts of change in the acoustical parameters. Another brief analysis focusing on configurations with three diffusers was made to investigate if differences exist on the amount change of the acoustical parameters between configurations with a clustered diffuser installation design (those named with the letters a and m) and these having a distributed diffuser installation design (configurations named with the letters k, l, n, and o). However, the results from this brief analysis did not show any clear differences between the effect of the two defined diffuser installation designs (clustered vs. distributed) on the amount change of the acoustical parameters.

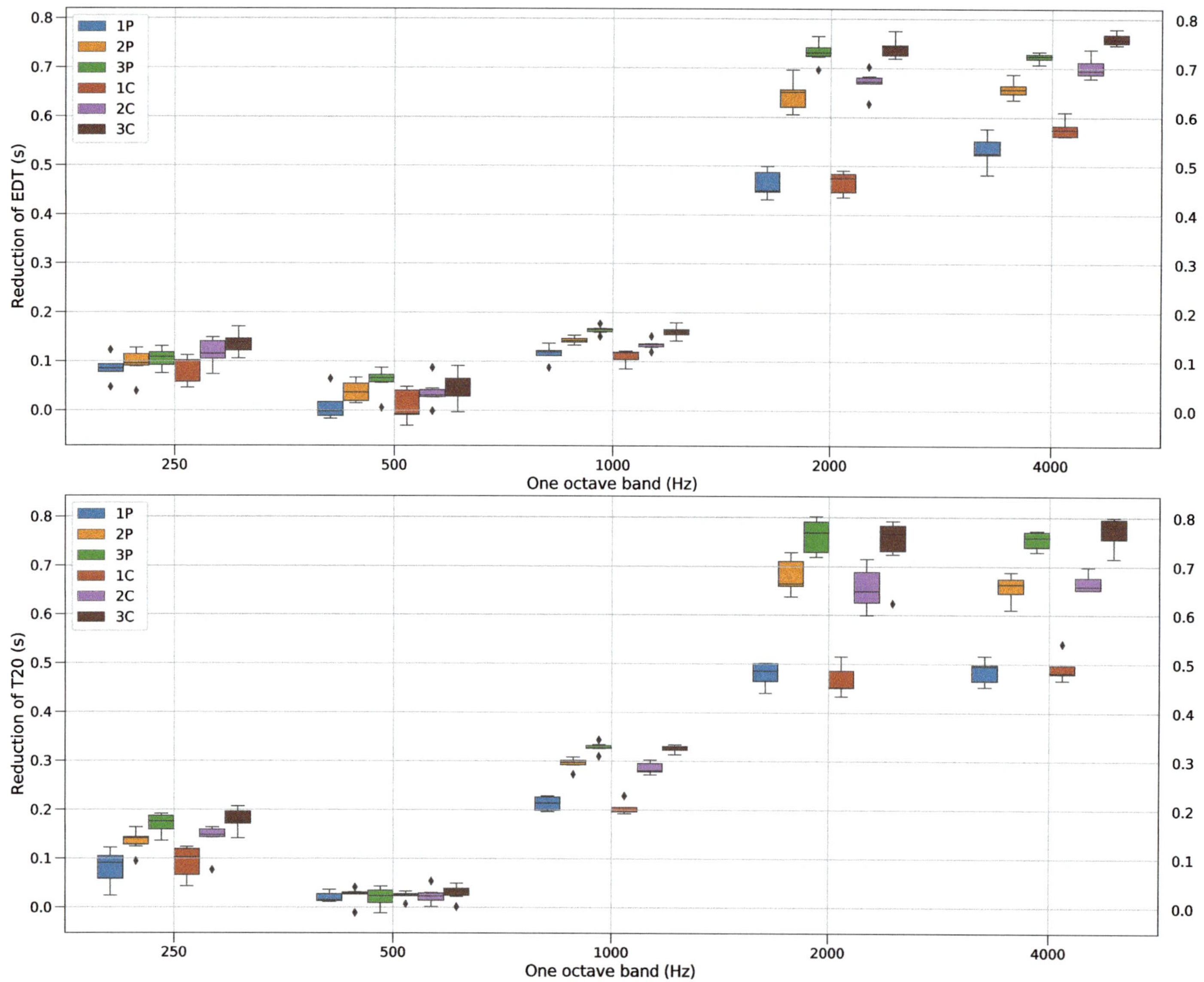

Figure 4. Reduction of early decay time (EDT) (**top**) and T_{20} (**bottom**) with respect to the reference per octave band for all the groups.

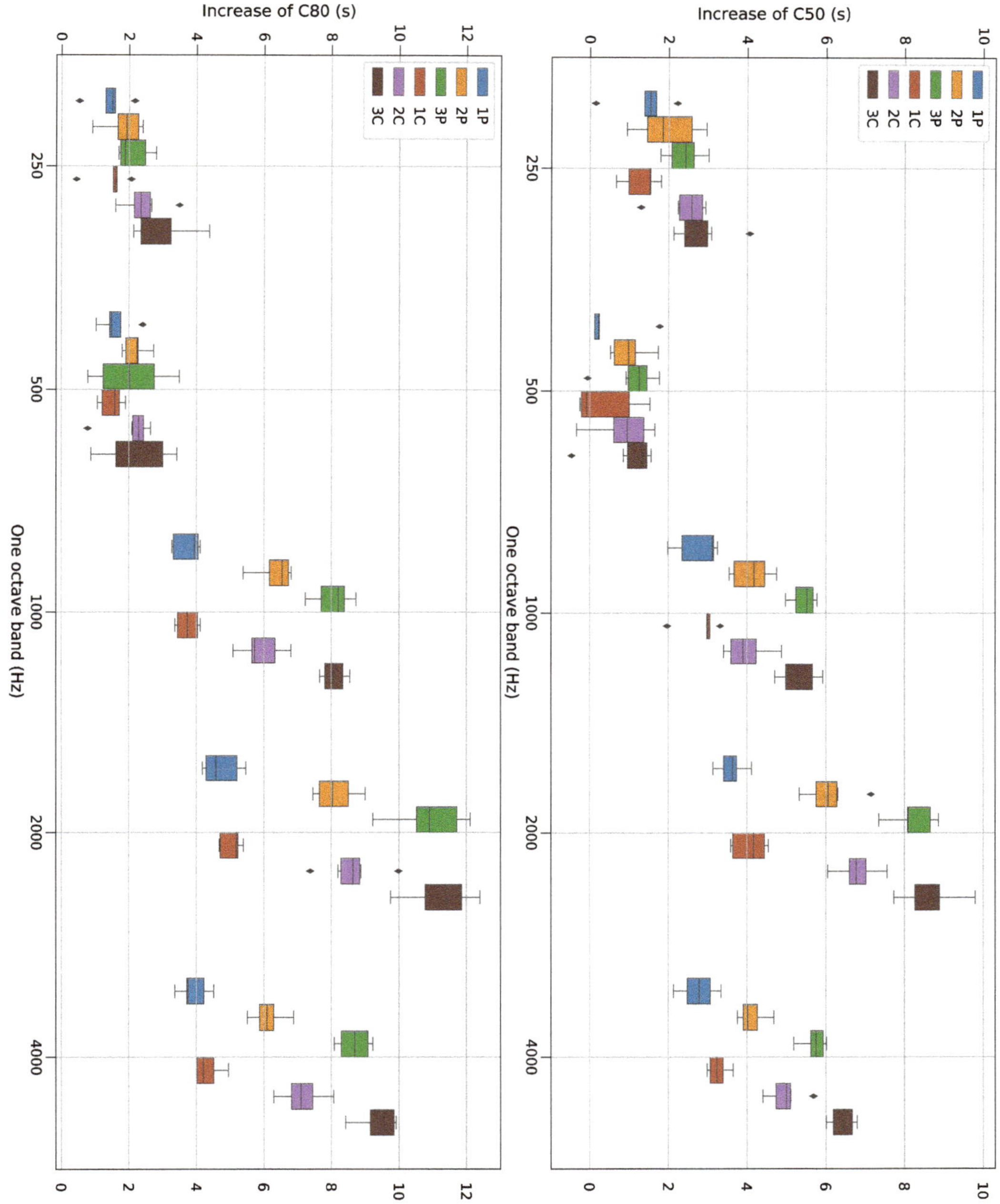

Figure 5. Increase of C_{50} **(top)** and C_{80} **(bottom)** with respect to the reference per octave band for all the groups.

Because a maximum of three diffusers only were added in the room, the effect of placement on the change of the acoustical parameters was neglected when considering the effect of the coverage or type. Considering the coverage of the diffusers, the amounts of change averaged across both pyramid and convex diffuser types were increased as the number of diffuser varied from one to three for all the four acoustical parameters and for all the three frequency ranges. In particular, as the number of diffuser increased from one to three, EDT decreased by 0.04, 0.16, and 0.19 s in the low, mid, and high frequency range, respectively. T_{20} decreased even more significantly than EDT with reduction values of 0.05, 0.2, and 0.27 s in the low, mid, and high frequency range, respectively. C_{50} and C_{80} were increased as the number of diffuser varied from one to three by 0.9 and 0.8 dB in the low, 3.6 and 5.3 dB in the mid, and 3.1 and 4.9 dB in the high frequency range, respectively.

Considering the type of the diffusers, the amounts of change were slightly higher for the convex type compared to the pyramidal type in all the measured acoustical parameters and at most of the center frequencies. More specifically, for EDT and T_{20}, the differences in the amount of reduction between the two diffuser types varied from 0 s in the 500 Hz octave band to 0.05 s in the 2 kHz octave band. The amount of increase between the two diffuser types for C_{50} and C_{80} ranged from 0 dB in the 500 Hz octave band to 1.1 dB in both 2 and 4 kHz octave bands.

4.2. Statistical Analysis

A series of 60 statistical hypothesis tests were performed to examine the significance of the effect of the coverage and type of the diffusers on the measured acoustical parameters. More precisely, this statistical analysis is done to determine if significant differences exist between groups 1P, 2P, and 3P (and between 1C, 2C, and 3C similarly) to investigate the effect of coverage, and between groups 1P vs. 1C, 2P vs. 2C, and 3P vs. 3C to investigate the effect of the diffuser type. In other words, for analyzing the effect of the coverage and excluding the effect of the type, the data points for the configurations with pyramid diffusers were grouped separately from the ones for the configurations including convex diffusers. As a result, two tests per frequency range were performed for each acoustical parameter, resulting in an aggregate of 24 statistical tests. Alternatively, in order to analyze the effect of the type of the diffusers, the data points were grouped such that the configurations with one, two, and three diffusers were separately studied to exclude the effect of coverage, thus resulting in three groups to analyze. For each of these groups, one test per frequency range was performed for each of the four acoustical parameters, resulting in a total of 36 statistical tests.

For each of the aforementioned 60 tests, the data points were examined for the conformity with the assumptions of the analysis of variance (ANOVA), namely the normality of the residuals and homogeneity of the variance, to be able to confirm if ANOVA is a suitable candidate for the purpose of the presented statistical analysis. The Shapiro–Wilk and Levene tests [23,24] were performed to verify if the former and the latter assumptions were met, respectively. The results demonstrated that among the 60 tests, 35 (58.33%) were not complying with the normality assumption. As for the homogeneity assumption, this percentage was 21.66%. Consequently, the Kruskal–Wallis test was selected as an appropriate substitute to the ANOVA test. The results of the Kruskal–Wallis tests are reported in Tables 5 and 6 for the analysis of the coverage and the type of the diffusers, respectively.

The results for the analysis of the coverage of the diffusers showed that for EDT, significant differences were observed between the configurations of all the groups in all the frequency ranges except for the group containing pyramid diffusers in the low frequency range. A same analysis for T_{20}, C_{50}, and C_{80} confirmed the existence of significant differences between all the groups only in the mid and high frequency ranges. In other words, these results indicated that increasing the number of either pyramid or convex diffusers from one to two or three had significant effect on the values of all the four acoustical parameters of the study in the mid and high frequency ranges.

The results for the analysis of the type of the diffusers indicated that the values of EDT for the configurations with one, two, and three pyramid diffusers were significantly different from the values for the corresponding convex-contained configurations only in the high frequency range. The same analysis for the values of T_{20} showed a significant difference only for the configurations with three diffusers and only in the high frequency range. For C_{50} and C_{80}, significant differences were observed only between the configurations with two and three pyramid diffusers and their corresponding convex-contained configurations and only in the high frequency range.

Table 5. Results of the Kruskal–Wallis tests for the analysis of the coverage characteristics (significant *p*-values at the 5% significance level are indicated in **bold**).

Acoustical Parameter	Freq. Range	Groups	Chi-Squared	df	*p*-Value
EDT	Low	1P-3P	3.70	2	0.158
		1C-3C	8.18	2	**0.017**
	Mid	1P-3P	31.21	2	**<0.001**
		1C-3C	39.44	2	**<0.001**
	High	1P-3P	62.20	2	**<0.001**
		1C-3C	57.39	2	**<0.001**
T_{20}	Low	1P-3P	3.57	2	0.167
		1C-3C	5.73	2	0.057
	Mid	1P-3P	44.27	2	**<0.001**
		1C-3C	44.57	2	**<0.001**
	High	1P-3P	74.47	2	**<0.001**
		1C-3C	72.51	2	**<0.001**
C_{50}	Low	1P-3P	2.10	2	0.350
		1C-3C	4.48	2	0.107
	Mid	1P-3P	43.01	2	**<0.001**
		1C-3C	47.43	2	**<0.001**
	High	1P-3P	49.56	2	**<0.001**
		1C-3C	50.00	2	**<0.001**
C_{80}	Low	1P-3P	0.82	2	0.665
		1C-3C	3.03	2	0.220
	Mid	1P-3P	46.96	2	**<0.001**
		1C-3C	52.20	2	**<0.001**
	High	1P-3P	68.45	2	**<0.001**
		1C-3C	66.63	2	**<0.001**

Table 6. Results of the Kruskal–Wallis tests for the analysis of the type characteristics (significant p-values at the 5% significance level are indicated in **bold**).

Acoustical Parameter	Freq. Range	Groups	Chi-Squared	df	p-Value
EDT	Low	1P, 1C	0.03	1	0.874
		2P, 2C	0.38	1	0.537
		3P, 3C	0.57	1	0.456
	Mid	1P, 1C	0.03	1	0.863
		2P, 2C	0.16	1	0.692
		3P, 3C	0.12	1	0.725
	High	1P, 1C	5.56	1	**0.018**
		2P, 2C	12.98	1	**<0.001**
		3P, 3C	16.78	1	**<0.001**
T_{20}	Low	1P, 1C	0.10	1	0.754
		2P, 2C	0.02	1	0.881
		3P, 3C	0.49	1	0.483
	Mid	1P, 1C	0.39	1	0.533
		2P, 2C	1.11	1	0.291
		3P, 3C	0.18	1	0.674
	High	1P, 1C	0.63	1	0.426
		2P, 2C	1.09	1	0.297
		3P, 3C	6.52	1	**0.011**
C_{50}	Low	1P, 1C	0.08	1	0.775
		2P, 2C	0.14	1	0.711
		3P, 3C	0.14	1	0.711
	Mid	1P, 1C	0.26	1	0.608
		2P, 2C	0.41	1	0.522
		3P, 3C	0.10	1	0.753
	High	1P, 1C	2.29	1	0.130
		2P, 2C	8.27	1	**0.004**
		3P, 3C	5.39	1	**0.020**
C_{80}	Low	1P, 1C	0.05	1	0.820
		2P, 2C	0.23	1	0.635
		3P, 3C	0.49	1	0.482
	Mid	1P, 1C	0.08	1	0.777
		2P, 2C	0.03	1	0.871
		3P, 3C	0.16	1	0.690
	High	1P, 1C	1.93	1	0.165
		2P, 2C	12.23	1	**<0.001**
		3P, 3C	6.06	1	**0.014**

4.3. Discussion

Although the statistical analysis revealed that the change of the acoustical parameters with respect to the type of diffuser (pyramid or convex) is significant in the high frequency range for all acoustical parameters, the number of diffusers also seem to play a role since the significance is mainly achieved for groups containing more than one diffuser. Moreover, the differences observed in the change of the acoustical parameters due to the change of the diffuser type cannot be fully conclusive since their respective absorption area differs by 0.1 m^2 for a single diffuser added to the room and by 0.3 m^2 for three diffusers. However, both types seem to have the same effect on the change of the acoustical parameters (i.e., a reduction of EDT and T_{20}, and an increase of C_{50} and C_{80}).

Introducing even a single diffuser to any walls of such a non-diffuse room is enough to significantly change the reverberation time in the room as seen in the measurement results and confirmed by the statistical analysis. This is because when such a room is empty, the sound keeps bouncing between the walls for a considerable amount of time while the sound field is mainly in the lateral plane. However, when a diffuser is introduced on a wall, the sound energy is directed down to the floor and up to the ceiling which are highly absorptive surfaces. The role of diffusers in this particular room is therefore much greater than in a room in which all boundaries have the same material. Consistently, as the number of added diffuser increases, the faster the sound energy is directed towards absorptive boundaries (floor/ceiling), which is in fact observed in the measurement results.

The changes in the acoustical parameters observed in the measurement results are in line with the trends observed in previous studies investigating scale models when switching from a reflective to a diffuse condition [6,14]. However, these results also differ from other publications [9,10]. Finally, it could be generally concluded that the effects produced by scattering surfaces on the acoustic parameters depend on several factors such as the amount of diffusion already present in the space, the position of the scattering surfaces, the position and amount of the absorptive surfaces, as well as the presence of mirrored reflective surfaces. Furthermore, the size and shape of the space are important parameters influencing the sound field diffusivity [14,25].

5. Conclusions and Future Work

This paper presented an investigation on the effects created by the inclusion of relatively large-scale pyramidal and convex-shaped diffusers in an empty non-diffuse room of small size on four acoustical parameters (EDT, T_{20}, C_{50}, and C_{80}) measured in the room. A total of 35 room configurations were measured such that the inclusion of the diffusers was controlled and analyzed in terms of their placement, number and shape.

The measurement results showed that adding the diffusers in the room generally decreases EDT and T_{20}, and increases C_{50} and C_{80} for both diffuser types. The results also showed that the control of reverberation is possible in small non-diffuse rooms by directing the sound energy towards absorbing surfaces. The statistical analysis showed that adding the diffusing elements in the room significantly changed the sound field at most octave bands regardless of the diffuser shape. The statistical analysis also showed that the influence of the diffuser type on the change of the room acoustic parameters was significant at high frequencies for groups with different number of diffusing elements. This conclusion is supported by all the presented measurement results and analysis as the convex-shaped diffusers caused larger impact on the sound at high frequencies. It is worth mentioning that from the presented results, which are restricted to the measured space and conditions, only tendencies are clear and quantitative rules for room acoustical parameters in non-diffuse sound fields cannot be extrapolated to other cases.

The contributions made by this paper can be divided into two items. Firstly, a large body of measurement data is presented in the Appendix A that can be used by researchers for further (statistical, e.g., regression) analysis in the area of sound diffusers and room acoustics. Secondly, although the study of the effects of diffusers' installation on the room acoustical parameters has already been done in several studies, this paper takes this approach into a more in-depth level by increasing the number of tests and configurations and choosing a room that had no prior assigned use that allowed for an abstract and systematic approach in this area of study.

There have been some areas of future challenges that are of interest to the authors of this paper. Using bigger rooms with various volumetric shapes, playing with the diffusivity of the room by changing the amount of absorption area, and incorporating the type of use of the rooms into the study as one more variable, are among those future studies. With the growing interest in the area of wave-based time domain modeling, e.g. finite-difference time-domain (FDTD) in room acoustics in recent years as reflected in [26,27], an additional area of future research using the FDTD method as a complementary technique to in-situ measurement can be considered for similar investigations to the present paper, thus offering more flexibility in testing numerous configurations. This area of future research was previously investigated by the present paper's authors in similar studies [28,29].

Author Contributions: Conceptualization, H.A.;Formal analysis, H.A. and J.M.; Investigation, H.A. and J.M.; Measurement, H.A.; Writing—Original draft preparation, H.A. and J.M.; Writing—Review and editing, T.L.; Supervision, G.S.

Funding: The first author Hassan Azad has received funding from the University of Florida Jeremy J Hill Memorial Fund over the course of the experiments carried out in this research. The co-author Julie Meyer is part of a project that has received funding from the European Union's Horizon 2020 research and innovation programme under the Marie Skłodowska-Curie grant agreement No 721536.

Acknowledgments: The authors are grateful to Siebein Associates Inc. for providing the equipment and the physical space for conducting the measurements. The manufacturer AVL Systems, Inc. is also thanked for providing the diffusers used in this work.

Conflicts of Interest: The authors declare no conflict of interest.

Abbreviations

The following abbreviations are used in this manuscript:

RIR	Room impulse response
EDT	Early decay time
RT, T_{20}	Reverberation time
C_{50}	Early-to-late arriving sound ratio
C_{80}	Clarity
SD	Standard deviation
ANOVA	Analysis Of variance

Appendix A. Acoustical Parameters

Table A1. 3D view, mean and SD over the five source-receiver combinations of the EDT, T_{20}, C_{50}, and C_{80} for group 1P.

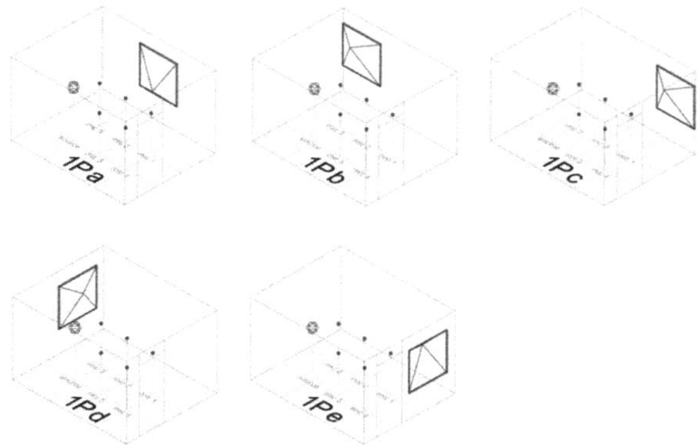

Freq.	Cnfg	EDT (s)	T_{20} (s)	C_{50} (dB)	C_{80} (dB)
250 Hz	1Pa	0.42 (0.07)	0.43 (0.05)	6.2 (1.5)	11.4 (2.2)
	1Pb	0.41 (0.02)	0.46 (0.03)	6.5 (1.7)	11.2 (1.2)
	1Pc	0.41 (0.08)	0.44 (0.05)	6.3 (2.4)	11.4 (1.4)
	1Pd	0.45 (0.03)	0.49 (0.04)	4.9 (1.3)	10.4 (1.1)
	1Pe	0.38 (0.08)	0.52 (0.04)	9.6 (2.1)	14.0 (1.3)
500 Hz	1Pa	0.34 (0.10)	0.30 (0.04)	9.8 (2.9)	16.8 (1.7)
	1Pb	0.31 (0.06)	0.29 (0.01)	9.7 (1.0)	17.2 (1.0)
	1Pc	0.35 (0.05)	0.28 (0.02)	9.7 (0.9)	17.5 (1.2)
	1Pd	0.33 (0.03)	0.27 (0.02)	9.8 (0.7)	17.3 (2.0)
	1Pe	0.27 (0.04)	0.29 (0.02)	13.2 (0.6)	18.8 (1.2)
1000 Hz	1Pa	0.24 (0.01)	0.39 (0.02)	11.0 (0.9)	16.0 (0.5)
	1Pb	0.27 (0.04)	0.40 (0.04)	10.7 (0.8)	16.0 (1.2)
	1Pc	0.23 (0.05)	0.38 (0.02)	11.8 (0.7)	16.7 (1.2)
	1Pd	0.22 (0.04)	0.37 (0.01)	12.0 (0.7)	16.6 (0.6)
	1Pe	0.24 (0.05)	0.36 (0.02)	11.8 (1.3)	15.9 (1.1)
2000 Hz	1Pa	0.55 (0.07)	0.69 (0.02)	5.8 (1.4)	9.1 (1.3)
	1Pb	0.51 (0.06)	0.71 (0.03)	6.8 (0.7)	10.4 (0.9)
	1Pc	0.57 (0.06)	0.73 (0.03)	6.1 (1.0)	9.2 (1.1)
	1Pd	0.55 (0.06)	0.67 (0.04)	6.3 (1.6)	9.5 (1.2)
	1Pe	0.50 (0.04)	0.67 (0.03)	6.4 (1.2)	9.5 (0.8)
4000 Hz	1Pa	0.53 (0.05)	0.73 (0.02)	7.1 (1.0)	9.9 (0.8)
	1Pb	0.49 (0.06)	0.74 (0.01)	7.4 (1.1)	10.2 (1.0)
	1Pc	0.44 (0.05)	0.70 (0.02)	8.3 (0.6)	11.0 (0.9)
	1Pd	0.49 (0.07)	0.70 (0.02)	7.7 (0.9)	10.3 (1.1)
	1Pe	0.46 (0.06)	0.68 (0.04)	5.8 (0.9)	8.0 (1.0)

Table A2. 3D view, mean and SD over the five source-receiver combinations of the EDT, T_{20}, C_{50}, and C_{80} for group 2P.

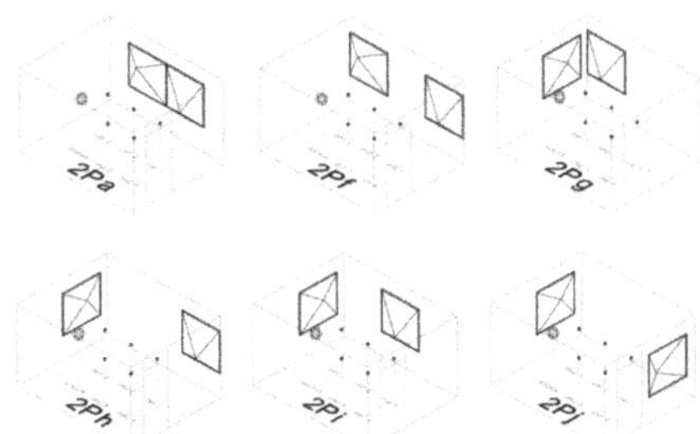

Freq.	Cnfg	EDT (s)	T_{20} (s)	C_{50} (dB)	C_{80} (dB)
	2Pa	0.38 (0.07)	0.41 (0.03)	7.6 (2.5)	12.3 (1.7)
	2Pf	0.37 (0.06)	0.41 (0.04)	7.8 (2.6)	12.2 (1.5)
250 Hz	2Pg	0.41 (0.04)	0.42 (0.06)	6.6 (1.5)	12.0 (1.3)
	2Ph	0.40 (0.02)	0.40 (0.05)	6.1 (1.5)	11.6 (1.2)
	2Pi	0.46 (0.04)	0.38 (0.04)	5.7 (1.0)	10.8 (1.7)
	2Pj	0.41 (0.05)	0.45 (0.07)	6.7 (1.5)	11.5 (1.6)
	2Pa	0.29 (0.09)	0.28 (0.02)	10.7 (2.9)	18.0 (1.1)
	2Pf	0.30 (0.04)	0.28 (0.02)	10.1 (0.7)	17.6 (2.6)
500 Hz	2Pg	0.27 (0.05)	0.28 (0.03)	11.3 (0.5)	18.5 (1.6)
	2Ph	0.32 (0.03)	0.27 (0.03)	10.1 (0.9)	18.1 (2.8)
	2Pi	0.32 (0.08)	0.28 (0.03)	10.4 (2.2)	18.0 (0.9)
	2Pj	0.26 (0.01)	0.32 (0.02)	10.7 (0.6)	17.6 (1.1)
	2Pa	0.22 (0.04)	0.29 (0.03)	13.1 (1.5)	19.3 (1.7)
	2Pf	0.22 (0.02)	0.29 (0.02)	13.5 (0.7)	19.5 (1.5)
1000 Hz	2Pg	0.21 (0.03)	0.32 (0.01)	12.3 (1.1)	18.1 (1.6)
	2Ph	0.21 (0.02)	0.28 (0.02)	13.2 (0.6)	19.4 (0.7)
	2Pi	0.20 (0.02)	0.30 (0.02)	12.6 (0.8)	18.8 (0.5)
	2Pj	0.22 (0.02)	0.30 (0.02)	12.3 (1.0)	19.1 (0.6)
	2Pa	0.34 (0.04)	0.51 (0.03)	8.6 (1.2)	12.6 (1.0)
	2Pf	0.30 (0.04)	0.51 (0.05)	9.8 (0.9)	13.9 (1.1)
2000 Hz	2Pg	0.39 (0.06)	0.51 (0.02)	8.0 (1.1)	12.6 (0.9)
	2Ph	0.35 (0.03)	0.44 (0.02)	9.0 (0.5)	13.3 (0.6)
	2Pi	0.34 (0.05)	0.45 (0.02)	8.9 (1.3)	13.5 (0.8)
	2Pj	0.39 (0.08)	0.54 (0.02)	8.4 (1.5)	12.4 (1.4)
	2Pa	0.36 (0.02)	0.54 (0.02)	8.7 (0.9)	12.5 (1.0)
	2Pf	0.32 (0.01)	0.52 (0.02)	9.6 (0.8)	13.4 (0.8)
4000 Hz	2Pg	0.38 (0.01)	0.58 (0.02)	8.9 (0.6)	12.3 (0.7)
	2Ph	0.35 (0.02)	0.51 (0.01)	9.3 (0.6)	12.8 (0.4)
	2Pi	0.35 (0.03)	0.52 (0.02)	9.1 (1.0)	12.7 (0.9)
	2Pj	0.37 (0.07)	0.55 (0.02)	8.9 (1.5)	12.0 (1.1)

Table A3. 3D view, mean and SD over the five source-receiver combinations of the EDT, T_{20}, C_{50}, and C_{80} for group 3P.

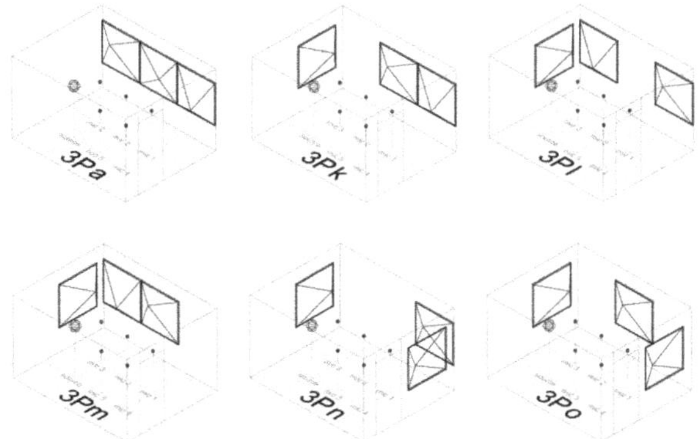

Freq.	Cnfg	EDT (s)	T_{20} (s)	C_{50} (dB)	C_{80} (dB)
	3Pa	0.39 (0.05)	0.39 (0.04)	7.2 (1.7)	11.6 (1.2)
	3Pk	0.42 (0.03)	0.36 (0.06)	6.6 (1.5)	11.7 (1.5)
250 Hz	3Pl	0.37 (0.05)	0.37 (0.06)	7.8 (2.0)	12.7 (1.8)
	3Pm	0.38 (0.06)	0.36 (0.04)	7.3 (1.7)	12.5 (1.5)
	3Pn	0.39 (0.03)	0.41 (0.03)	7.5 (1.6)	11.7 (1.5)
	3Po	0.41 (0.04)	0.37 (0.03)	6.8 (1.4)	11.5 (1.9)
	3Pa	0.33 (0.02)	0.27 (0.02)	9.5 (1.0)	16.8 (1.4)
	3Pk	0.28 (0.09)	0.27 (0.03)	10.5 (1.9)	18.7 (1.0)
500 Hz	3Pl	0.27 (0.05)	0.30 (0.04)	11.3 (1.7)	17.6 (2.1)
	3Pm	0.26 (0.05)	0.26 (0.04)	10.9 (1.7)	19.3 (2.1)
	3Pn	0.26 (0.06)	0.29 (0.04)	10.7 (1.3)	18.0 (1.5)
	3Po	0.24 (0.03)	0.32 (0.01)	11.0 (1.4)	16.6 (1.8)
	3Pa	0.19 (0.02)	0.27 (0.02)	14.1 (1.4)	20.8 (0.6)
	3Pk	0.19 (0.03)	0.25 (0.02)	13.9 (0.7)	21.4 (1.1)
1000 Hz	3Pl	0.19 (0.02)	0.27 (0.03)	14.4 (0.8)	21.0 (1.6)
	3Pm	0.20 (0.04)	0.28 (0.02)	13.7 (0.6)	19.9 (0.6)
	3Pn	0.20 (0.04)	0.27 (0.02)	14.5 (0.8)	20.2 (1.0)
	3Po	0.18 (0.04)	0.26 (0.03)	14.3 (0.7)	21.1 (1.2)
	3Pa	0.27 (0.02)	0.45 (0.05)	10.8 (1.3)	15.5 (1.6)
	3Pk	0.23 (0.03)	0.37 (0.03)	11.5 (1.1)	17.1 (1.5)
2000 Hz	3Pl	0.27 (0.03)	0.38 (0.04)	10.8 (1.1)	16.2 (1.4)
	3Pm	0.27 (0.03)	0.43 (0.04)	10.8 (1.0)	15.5 (1.2)
	3Pn	0.30 (0.04)	0.45 (0.06)	10.0 (1.2)	14.2 (1.1)
	3Po	0.25 (0.04)	0.38 (0.06)	11.5 (1.3)	16.8 (1.6)
	3Pa	0.29 (0.03)	0.44 (0.02)	10.6 (1.1)	15.2 (1.0)
	3Pk	0.28 (0.02)	0.42 (0.02)	11.0 (0.7)	15.7 (0.7)
4000 Hz	3Pl	0.28 (0.02)	0.43 (0.01)	10.8 (0.9)	15.2 (0.7)
	3Pm	0.31 (0.03)	0.46 (0.02)	10.1 (0.9)	14.6 (0.7)
	3Pn	0.29 (0.04)	0.46 (0)	10.6 (0.8)	14.7 (0.6)
	3Po	0.28 (0.04)	0.42 (0.02)	10.9 (1.5)	15.7 (1.2)

Table A4. 3D view, mean and SD over the five source-receiver combinations of the EDT, T_{20}, C_{50}, and C_{80} for group 1C.

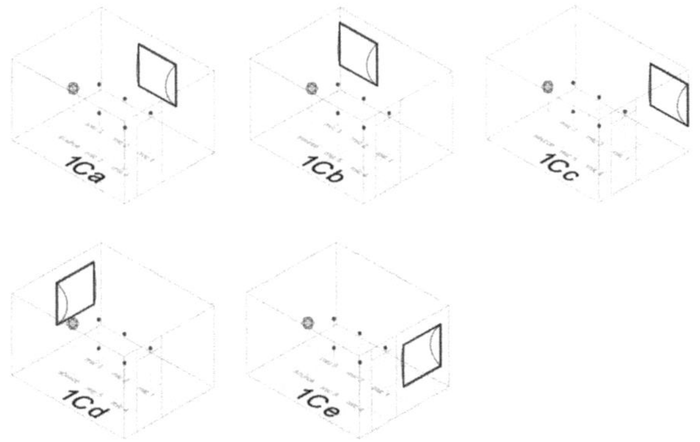

Freq.	Cnfg	EDT (s)	T_{20} (s)	C_{50} (dB)	C_{80} (dB)
250 Hz	1Ca	0.44 (0.07)	0.42 (0.06)	5.8 (1.3)	11.4 (2.1)
	1Cb	0.40 (0.03)	0.45 (0.04)	6.3 (1.5)	11.5 (1.3)
	1Cc	0.40 (0.08)	0.43 (0.04)	6.3 (2.2)	11.4 (1.1)
	1Cd	0.45 (0.04)	0.48 (0.02)	5.5 (1.5)	10.3 (1.2)
	1Ce	0.39 (0.07)	0.51 (0.03)	6.6 (2.0)	11.9 (1.3)
500 Hz	1Ca	0.34 (0.08)	0.28 (0.03)	9.5 (3.1)	17.0 (2.4)
	1Cb	0.34 (0.05)	0.27 (0.02)	9.4 (1.3)	17.4 (1.3)
	1Cc	0.36 (0.04)	0.28 (0.01)	9.3 (0.5)	16.8 (0.6)
	1Cd	0.29 (0.03)	0.28 (0.03)	10.5 (0.4)	17.7 (2.4)
	1Ce	0.28 (0.03)	0.30 (0.03)	11.1 (0.3)	17.5 (1.9)
1000 Hz	1Ca	0.24 (0.02)	0.40 (0.01)	11.7 (1.0)	16.4 (0.6)
	1Cb	0.27 (0.04)	0.40 (0.03)	10.7 (1.0)	16.0 (1.2)
	1Cc	0.24 (0.02)	0.40 (0.04)	11.7 (0.3)	16.7 (1.2)
	1Cd	0.25 (0.03)	0.39 (0.04)	11.8 (1.1)	16.1 (1.2)
	1Ce	0.23 (0.05)	0.36 (0.01)	12.0 (0.5)	16.8 (0.5)
2000 Hz	1Ca	0.52 (0.08)	0.72 (0.03)	7.1 (0.6)	10.3 (0.8)
	1Cb	0.56 (0.07)	0.72 (0.05)	6.3 (1.3)	9.6 (1.1)
	1Cc	0.55 (0.06)	0.74 (0.02)	6.3 (1.0)	9.7 (1.0)
	1Cd	0.51 (0.05)	0.69 (0.02)	6.8 (1.0)	10.2 (1.0)
	1Ce	0.51 (0.11)	0.66 (0.03)	7.2 (1.5)	10.1 (1.5)
4000 Hz	1Ca	0.45 (0.06)	0.71 (0.02)	7.9 (1.2)	10.7 (1.0)
	1Cb	0.44 (0.04)	0.73 (0.02)	8.0 (0.9)	10.5 (0.5)
	1Cc	0.40 (0.04)	0.71 (0.02)	8.6 (0.7)	11.5 (0.8)
	1Cd	0.45 (0.09)	0.70 (0.01)	8.2 (1.2)	10.5 (0.9)
	1Ce	0.43 (0.06)	0.65 (0.02)	8.4 (0.9)	11.0 (1.0)

Table A5. 3D view, mean and SD over the five source-receiver combinations of the EDT, T_{20}, C_{50}, and C_{80} for group 2C.

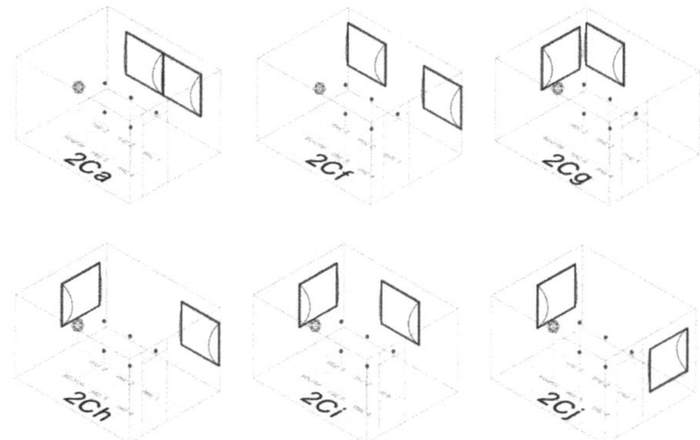

Freq.	Cnfg	EDT (s)	T_{20} (s)	C_{50} (dB)	C_{80} (dB)
	2Ca	0.38 (0.09)	0.41 (0.03)	7.6 (2.2)	12.3 (1.8)
	2Cf	0.35 (0.04)	0.39 (0.05)	7.7 (2.0)	12.5 (1.6)
250 Hz	2Cg	0.35 (0.03)	0.40 (0.04)	7.7 (1.3)	13.3 (1.5)
	2Ch	0.40 (0.04)	0.40 (0.04)	7.0 (1.5)	12.0 (1.4)
	2Ci	0.44 (0.05)	0.38 (0.05)	6.1 (1.4)	11.5 (2.0)
	2Cj	0.39 (0.05)	0.47 (0.05)	7.2 (1.8)	12.1 (1.7)
	2Ca	0.30 (0.08)	0.25 (0.01)	10.1 (2.8)	18.1 (2.0)
	2Cf	0.33 (0.04)	0.28 (0.03)	9.2 (0.6)	16.6 (2.1)
500 Hz	2Cg	0.29 (0.04)	0.30 (0.02)	11.2 (1.0)	18.2 (0.9)
	2Ch	0.30 (0.04)	0.28 (0.05)	10.4 (0.6)	17.9 (1.6)
	2Ci	0.31 (0.08)	0.29 (0.02)	10.6 (2.2)	18.0 (1.0)
	2Cj	0.24 (0.05)	0.31 (0.03)	11.0 (0.8)	18.4 (1.7)
	2Ca	0.22 (0.05)	0.31 (0.02)	12.1 (0.8)	18.4 (0.7)
	2Cf	0.24 (0.02)	0.31 (0.02)	12.2 (1.0)	18.3 (1.1)
1000 Hz	2Cg	0.23 (0.02)	0.32 (0.02)	12.6 (0.7)	17.8 (0.8)
	2Ch	0.22 (0.03)	0.29 (0.03)	13.0 (1.1)	19.2 (1.2)
	2Ci	0.21 (0.03)	0.29 (0.01)	13.6 (0.9)	19.5 (0.7)
	2Cj	0.22 (0.05)	0.31 (0.02)	12.6 (1.8)	18.5 (1.0)
	2Ca	0.33 (0.07)	0.57 (0.03)	9.2 (1.4)	13.1 (1.7)
	2Cf	0.31 (0.03)	0.54 (0.02)	9.8 (0.8)	13.7 (0.9)
2000 Hz	2Cg	0.33 (0.04)	0.50 (0.02)	9.4 (0.6)	13.5 (1.1)
	2Ch	0.32 (0.06)	0.48 (0.06)	9.5 (1.3)	13.8 (1.5)
	2Ci	0.33 (0.05)	0.46 (0.03)	10.2 (1.5)	14.9 (1.0)
	2Cj	0.37 (0.07)	0.55 (0.05)	8.7 (1.3)	12.3 (1.5)
	2Ca	0.32 (0.06)	0.54 (0.02)	9.9 (1.3)	13.7 (1.1)
	2Cf	0.27 (0.03)	0.53 (0.02)	10.6 (1.2)	14.6 (0.9)
4000 Hz	2Cg	0.31 (0.01)	0.54 (0.01)	10.1 (0.4)	13.5 (0.5)
	2Ch	0.33 (0.05)	0.51 (0.03)	9.4 (1.1)	13.3 (1.0)
	2Ci	0.33 (0.05)	0.50 (0.01)	10.1 (1.1)	14.0 (1.1)
	2Cj	0.33 (0.07)	0.54 (0.03)	9.6 (1.3)	12.8 (1.1)

Table A6. 3D view, mean and SD over the five source-receiver combinations of the EDT, T_{20}, C_{50}, and C_{80} for group 3C.

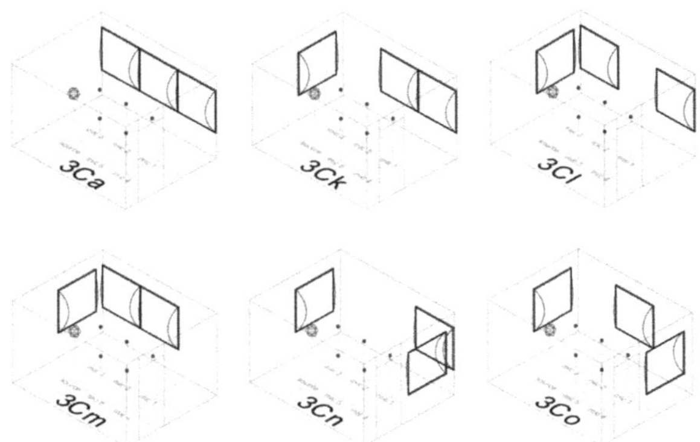

Freq.	Cnfg	EDT (s)	T_{20} (s)	C_{50} (dB)	C_{80} (dB)
250 Hz	3Ca	0.36 (0.03)	0.37 (0.03)	7.5 (1.4)	12.1 (1.1)
	3Ck	0.39 (0.02)	0.37 (0.03)	7.1 (1.1)	12.0 (1.6)
	3Cl	0.33 (0.03)	0.34 (0.04)	8.9 (2.0)	14.2 (2.3)
	3Cm	0.36 (0.05)	0.35 (0.04)	7.4 (1.9)	13.3 (2.0)
	3Cn	0.36 (0.03)	0.41 (0.03)	7.9 (1.0)	12.5 (1.9)
	3Co	0.38 (0.02)	0.38 (0.04)	6.9 (1.3)	12.4 (2.1)
500 Hz	3Ca	0.33 (0.05)	0.28 (0.02)	9.1 (1.4)	16.7 (1.3)
	3Ck	0.31 (0.06)	0.27 (0.03)	10.4 (1.7)	18.9 (1.6)
	3Cl	0.28 (0.04)	0.29 (0.04)	10.9 (0.9)	17.5 (1.6)
	3Cm	0.28 (0.06)	0.26 (0.03)	10.8 (1.7)	18.4 (1.6)
	3Cn	0.27 (0.04)	0.27 (0.03)	11.0 (0.7)	19.2 (2.2)
	3Co	0.24 (0.04)	0.31 (0.03)	11.1 (1.1)	17.4 (1.3)
1000 Hz	3Ca	0.21 (0.02)	0.28 (0.02)	13.6 (0.7)	20.3 (0.6)
	3Ck	0.19 (0.03)	0.26 (0.01)	14.0 (0.6)	20.7 (0.9)
	3Cl	0.20 (0.02)	0.26 (0.02)	14.5 (1.4)	21.1 (1.6)
	3Cm	0.20 (0.02)	0.26 (0.01)	13.4 (0.8)	20.4 (1.0)
	3Cn	0.21 (0.01)	0.27 (0.02)	14.6 (0.6)	20.8 (0.8)
	3Co	0.18 (0.04)	0.26 (0.04)	14.0 (1.2)	21.2 (0.9)
2000 Hz	3Ca	0.25 (0.02)	0.55 (0.06)	11.5 (1.0)	15.7 (1.5)
	3Ck	0.22 (0.03)	0.38 (0.04)	12.5 (0.8)	17.4 (1.0)
	3Cl	0.25 (0.03)	0.39 (0.04)	11.6 (1.2)	16.7 (1.5)
	3Cm	0.28 (0.04)	0.42 (0.05)	10.8 (1.4)	15.9 (1.6)
	3Cn	0.30 (0.03)	0.45 (0.03)	10.4 (1.4)	14.7 (1.2)
	3Co	0.25 (0.04)	0.39 (0.06)	11.5 (1.3)	16.8 (1.5)
4000 Hz	3Ca	0.27 (0.02)	0.48 (0.03)	11.3 (1.1)	15.6 (1.2)
	3Ck	0.23 (0.01)	0.40 (0.01)	11.7 (0.7)	16.3 (0.8)
	3Cl	0.24 (0.03)	0.39 (0.02)	11.8 (0.8)	16.4 (0.6)
	3Cm	0.26 (0.03)	0.43 (0.01)	11.1 (0.9)	15.8 (0.9)
	3Cn	0.29 (0.02)	0.44 (0.02)	11.0 (0.7)	14.9 (0.4)
	3Co	0.25 (0.03)	0.40 (0.02)	11.5 (1.1)	16.4 (0.9)

References

1. D'Antonio, P.; Cox, T. Two Decades of Sound Diffusor Design and Development, Part 1: Applications and Design. *J. Audio Eng. Soc.* **1998**, *46*, 955–976.
2. D'Antonio, P.; Konnert, J.H. The RFZ/RPG approach to control room monitoring. In *Audio Engineering Society Convention 76*; Audio Engineering Society: New York, NY, USA, 1984.
3. Cox, T.; d'Antonio, P. *Acoustic Absorbers and Diffusers: Theory, Design and Application*; Crc Press: Boca Raton, FL, USA, 2016.
4. D'Antonio, P.; Cox, T. AES information document for room acoustics and sound reinforcement systems-characterization and measurement of surface scattering uniformity. *J. Audio Eng. Soc.* **2001**, *49*, 149–165.
5. ISO. *17497-2:2012 Acoustics—Sound-Scattering Properties of Surfaces—Part 2: Measurement of the Directional Diffusion Coefficient in a Free Field*; International Organization for Standardization: Geneva, Switzerland, 2012.
6. Jeon, J.Y.; Jang, H.S.; Kim, Y.H.; Vorländer, M. Influence of wall scattering on the early fine structures of measured room impulse responses. *J. Acoust. Soc. Am.* **2015**, *137*, 1108–1116. [CrossRef] [PubMed]
7. Jeon, J.Y. and Kim, Y.H. Diffuser Design for the New Ifez Arts Center Concert Hall Using Scale Models. In Proceedings of the 7th International Conference on Auditorium Acoustics, Oslo, Norway, 3–5 October 2008; Institute of Acoustics: Milton Keynes, UK, 2008; Volume 30, pp. 357–363.
8. Choi, Y.J. Effects of periodic type diffusers on classroom acoustics. *Appl. Acoust.* **2013**, *74*, 694–707. [CrossRef]
9. Shtrepi, L.; Astolfi, A.; D'Antonio, G.; Guski, M. Objective and perceptual evaluation of distance-dependent scattered sound effects in a small variable-acoustics hall. *J. Acoust. Soc. Am.* **2016**, *140*, 3651–3662. [CrossRef] [PubMed]
10. Shtrepi, L.; Astolfi, A.; Pelzer, S.; Vitale, R.; Rychtáriková, M. Objective and perceptual assessment of the scattered sound field in a simulated concert hall. *J. Acoust. Soc. Am.* **2015**, *138*, 1485–1497. [CrossRef] [PubMed]
11. Suzumura, Y.; Sakurai, M.; Ando, Y.; Yamamoto, I.; Iizuka, T.; Oowaki, M. An evaluation of the effects of scattered reflections in a sound field. *J. Sound Vib.* **2000**, *232*, 303–308. [CrossRef]
12. Chiles, S. Sound Behaviour in Proportionate Spaces and Auditoria. Ph.D. Thesis, University of Bath, Bath, UK, 2004.
13. Ryu, J.K.; Jeon, J.Y. Subjective and objective evaluations of a scattered sound field in a scale model opera house. *J. Acoust. Soc. Am.* **2008**, *124*, 1538–1549. [CrossRef] [PubMed]
14. Kim, Y.H.; Kim, J.H.; Jeon, J.Y. Scale model investigations of diffuser application strategies for acoustical design of performance venues. *Acta Acust. United Acust.* **2011**, *97*, 791–799. [CrossRef]
15. Kim, Y.H.; Kim, J.H.; Rougier, C.; Scmich, I.; Jeon, J.Y. In-situ measurements of diffuse reflections from lateral walls in concert halls. In Proceedings of the 8th International Conference on Auditorium Acoustics, Dublin, Ireland, 20–22 May 2011; Institute of Acoustics: Milton Keynes, UK, 2011.
16. Jeon, J.Y.; Lee, S.C.; Vorländer, M. Development of scattering surfaces for concert halls. *Appl. Acoust.* **2004**, *65*, 341–355. [CrossRef]
17. ISO, EN. *3382-2, 2008, "Acoustics—Measurement of Room Acoustic Parameters—Part 2: Reverberation Time in Ordinary Rooms"*; International Organization for Standardization: Brussels, Belgium, 2008.
18. Farina, A. Simultaneous measurement of impulse response and distortion with a swept-sine technique. In *Audio Engineering Society Convention 108*; Audio Engineering Society: New York, NY, USA, 2000.
19. Wenger Corporation. Traditional Acoustic Panel Series: Diffuser Panels. Available online: https://www.wengercorp.com/Lit/Wenger%20Traditional%20Acoustic%20Panels-Diffusers-TS.pdf (accessed on 19 March 2019).
20. Kinetics Noise Control. Kinetics Geometric Diffusers. Available online: https://kineticsnoise.com/interiors/pdf/geometricDiffusers.pdf (accessed on 19 March 2019).
21. AcousticsFirst Corporation. TechSpecs. p. 25. Available online: http://www.acousticsfirst.com/docs/AF-TECSPECS.pdf (accessed on 19 March 2019).
22. Acoustical Solutions. Technical Data Sheet—Pyramid Sound Diffuser. p. 25. Available online: https://acousticalsolutions.com/wp-content/uploads/2015/01/as-pyramid-sound-diffuser-data-sheet.pdf (accessed on 19 March 2019).

23. Shapiro, S.S.; Wilk, M.B. An analysis of variance test for normality (complete samples). *Biometrika* **1965**, *52*, 591–611. [CrossRef]
24. Levene, H. Contributions to probability and statistics. In *Essays in Honor of Harold Hotelling*; Stanford University Press: Redwood City, CA, USA, 1960; pp. 278–292.
25. Shtrepi, L.; Astolfi, A.; Rychtáriková, M. Influence of a volume scale factor on the scattering coefficient effects on the prediction of room acoustic parameters. In Proceedings of the International Symposium on Room Acoustics, Toronto, ON, Canada, 9–11 June 2013.
26. Hamilton, B.; Bilbao, S. FDTD Methods for 3-D Room Acoustics Simulation with High-order Accuracy in Space and Time. *Ieee/Acm Trans. Audio Speech Lang. Process.* **2017**, *25*, 2112–2124. [CrossRef]
27. Saarelma, J. Finite-Difference TIME-domain Solver for Room Acoustics Using Graphics Processing Units. Master's Thesis, Aalto University, Helsinki, Finland, 2013.
28. Meyer, J.; Lokki, T. Optimization of a diffuser geometry using parametric modeling tools and finite-difference time-domain simulations. *Audit. Acoust.* **2018**, 638–645.
29. Azad, H.; Ketabi, R.; Siebein, G. A Study of Diffusivity in Concert Halls Using Large Scale Acoustic Wave-Based Modeling and Simulation. In Proceedings of the INTER-NOISE and NOISE-CON Congress and Conference, Chicago, IL, USA, 26–29 August 2018; Institute of Noise Control Engineering: Reston, VA, USA, 2018; Volume 258, pp. 5431–5442.

8

Plate-Type Acoustic Metamaterials: Experimental Evaluation of a Modular Large-Scale Design for Low-Frequency Noise Control

Linus Yinn Leng Ang [1,2,*], Yong Khiang Koh [2] and Heow Pueh Lee [1]

[1] Department of Mechanical Engineering, National University of Singapore, 9 Engineering Drive 1, Singapore 117575, Singapore; mpeleehp@nus.edu.sg
[2] Kinetics Design and Manufacturing, ST Engineering Land Systems Ltd., 249 Jalan Boon Lay, Singapore 619523, Singapore; kohyk@stengg.com
* Correspondence: ylang@u.nus.edu or anglinus@stengg.com

Abstract: For industrial applications, the scalability of a finalised design is an important factor to consider. The scaling process of typical membrane-type acoustic metamaterials may pose manufacturing challenges such as stress uniformity of the membrane and spatial consistency of the platelet. These challenges could be addressed by plate-type acoustic metamaterials with an internal tonraum resonator. By adopting the concept of modularity in a large-scale design (or meta-panel), the acoustical performance of different specimen configurations could be scaled and modularly combined. This study justifies the viability of two meta-panel configurations for low-frequency (80–500 Hz) noise control. The meta-panels were shown to be superior to two commercially available noise barriers at 80–500 Hz. This superiority was substantiated when the sound transmission class (STC) and the outdoor-indoor transmission class (OITC) were compared. The meta-panels were also shown to provide an average noise reduction of 22.7–27.4 dB at 80–400 Hz when evaluated in different noise environments—traffic noise, aircraft flyby noise, and construction noise. Consequently, the meta-panel may be further developed and optimised to obtain a design that is lightweight and yet has good acoustical performance at below 500 Hz, which is the frequency content of most problematic noises.

Keywords: large-scale; plate-type acoustic metamaterial; resonator; modular; noise control; sound transmission loss; noise reduction; environmental noise

---

1. Introduction

Membrane-type acoustic metamaterials have been studied extensively over the past decade [1–12] because they could exhibit good acoustical performance at below 1 kHz while being compact and lightweight. To date, the literature provides many different designs—ranging from simple to complex constructions—that may be considered as a benchmark for further development. This class of acoustic metamaterial is usually realised by adhering the edges of a pretensioned membrane to a rigid frame. To pretension the membrane, a mechanism could be designed to stretch the membrane edges uniformly over a prescribed distance [1]. Alternatively, epoxy could be applied along the membrane edges in which stretching is achieved by thermal curing [4]. Evidently, to produce the specimen in large quantities, the assembly process would likely require a huge amount of time and effort. This issue may be addressed by considering either a honeycomb panel [13] or a perforated panel [14] as the frame. The acoustical performance is characterised solely by the resonant behaviour of the membrane in which the frequency bandwidths of good acoustical performance are achieved at around the anti-resonance frequencies, while the frequency bandwidths of poor acoustical performance are obtained at around the resonance frequencies. By varying the dimensions, the material properties, and the pretension of the membrane, the frequency bandwidths of good acoustical performance could be adjusted to coincide with the frequency range of the unwanted noise. Consequently, noise control is achieved. The membrane could also be attached with a highly dense platelet to introduce additional tuning parameters. The list of tuning parameters may be extended. For example, an additional support structure could be implemented to suppress specific eigenmodes of the membrane [8], or a perforated membrane could be considered to introduce the effect of a Helmholtz resonator [2].

Knowing that membrane-type acoustic metamaterials could potentially be considered in the industry for noise control, it is imperative that the finalised small-scale designs are extendable. Of course, it would be ideal if the large-scale design could exhibit the same acoustical performance achieved during the small-scale study. Considering how typical designs are realised, manufacturing challenges may be faced during the scaling process. These challenges, for example, include stress relaxation and stress uniformity of the membrane, and spatial consistency of the platelet on the membrane. Moreover, if numerous platelets were considered, the scaling process would be laborious. Huang et al. [15] reported that the challenges could be faced even for small-scale designs. Hence, avenues should be explored to address the challenges and improve the scalability of membrane-type acoustic metamaterials.

Recently, numerous works [16–20] have shown that a plastic film can be considered in place of the elastic membrane to eliminate the challenges pertaining to stress relaxation and stress uniformity. Since there is negligible pretension in the film, the membrane-type acoustic metamaterial is transformed into another class of acoustic metamaterial known as the plate-type acoustic metamaterial. Despite this transformation, the acoustical performance is still characterised solely by the resonant behaviour of the film. Again, if a platelet was considered to modify the resonant behaviour, the challenge relating to spatial consistency would remain unresolved. To address this challenge, Ang et al. [21,22] proposed that the design could adopt a double-leaf configuration with an internal tonraum resonator (i.e., two cavities connected via an orifice) so that the objective of having more tuning parameters could still be met without attaching platelets.

Over the past five years, researchers [1,23,24] have shown growing interest in large-scale membrane-type acoustic metamaterials despite the aforementioned challenges. However, large-scale plate-type acoustic metamaterials have yet to be studied extensively, evident from the literature in which limited works [20,25,26] can be found. Varanasi et al. [20,25] proposed a large-scale design (1.22×1.22 m^2) that had close resemblance to a plate-type acoustic metamaterial. The design was fabricated from a continuous acrylic panel in which the unit cells were realised by varying the thickness at different regions of the panel. Although numerous configurations were considered, none had dissimilar unit cells on the same panel to study whether they could collectively contribute to the acoustical performance of the panel. Ang et al. [26] filled in the research gap with a large-scale design (994×994 mm^2) that adopted the concept of modularity for customisable acoustical performance. The proof-of-concept design—also referred to as the meta-panel—was realised by slotting specimens with two different configurations (without and with resonator) into the periodic cutouts on a host structure. By varying the distribution percentage of the specimens on the host structure, the acoustical performance of the meta-panel was shown to be collectively contributed by the acoustical performance of each specimen configuration. Additionally, the acoustical performance of the meta-panel was found to be insensitive to how the specimens were oriented or arranged on the host structure. The findings suggest that the meta-panel could potentially be considered in the industry for noise control.

This study aims to justify whether the meta-panel could indeed be viable for noise control in the industry by investigating two selected meta-panel configurations reported previously [26] in greater detail. This investigation compares the acoustical performance with that of two commercially available noise barriers and evaluates the acoustical performance in three different noise environments—traffic noise, aircraft flyby noise, and construction noise. Section 2 presents the design details of the two meta-panel configurations and the experimental methods. Section 3 presents the results. Section 4 discusses how the findings could provide avenues for future research. Section 5 concludes the study based on the observations.

2. Materials and Methods

This section first presents the two selected meta-panel configurations (Section 2.1) before moving on to discuss the experimental methods (Section 2.2). To minimise repetition, only the differences between this study and the previous work [26] are discussed. Nonetheless, crucial information is still provided.

2.1. Meta-Panel Configurations

The two meta-panel configurations were selected because they exhibited the widest frequency bandwidth of good acoustical performance at low frequencies (<500 Hz) compared to that of the other configurations [26]. Both configurations were solely occupied by specimens with resonator (Figure 1a), which were previously investigated in a small-scale study [21,22]. The difference between the two configurations is the material of the host structure in which one was made of aluminium and the other was made of plywood.

The host structure consisted of an inner panel and two outer skins (Figure 1b). The former was designed with 10 × 10 cutouts (22 mm apart) that were used to house the specimens, while the latter served as a flange to constrain the specimens from falling out of the cutouts. To achieve the flange, the 10 × 10 cutouts on the outer skins were 32 mm apart. The inner panel had a thickness (20 mm) that matched that of the specimens, while the outer skins were specified with a thickness (3 mm) so that they would not be too flimsy especially for the case of plywood. Fasteners were used to assemble the inner panel and the outer skins at 32 points. This number of points is inclusive of the eight points used by the handles, which were replaced by countersunk screws during the measurements.

Each specimen was assembled from two Mylar films (127 μm thick) and three printed components. The printed components—two outer frames and one inner frame (Figure 1a)—were produced from polylactide filament via fused deposition modelling (MOMENT Moment S). To assemble the specimen, a layer of adhesive tape (Nitto Denko No. 500) was first applied on each side of the inner frame. The width of the tape (70 mm) was specially customised to be larger than the lengths of the inner frame (60 mm) to avoid discontinuity, which could result in air leakages from the cavities. Each side of the inner frame was then adhered with a film that was simply laid flat on a tabletop. This approach minimised the presence of pretension—an essential attribute of plate-type acoustic metamaterials. The excess tapes and films were then removed by cutting. Finally, one outer frame (applied with a layer of adhesive tape on one side) was adhered to each film to emulate a pinned boundary condition along the film edges. The orifice was specified at the centre of the partition for ease of numerical modelling and realisation. The orifice diameter (5 mm) was specified such that the influence of the viscous boundary layer would be minimal as shown in the previous work [22].

2.2. Experimental Methods

This section first presents how the sound transmission class (STC) and the outdoor-indoor transmission class (OITC) of the meta-panels were determined for comparison with that of two commercially available noise barriers (Sections 2.2.1 and 2.2.2, respectively). Next, this section discusses how the meta-panels were evaluated to determine the noise reduction in three different noise environments (Sections 2.2.3).

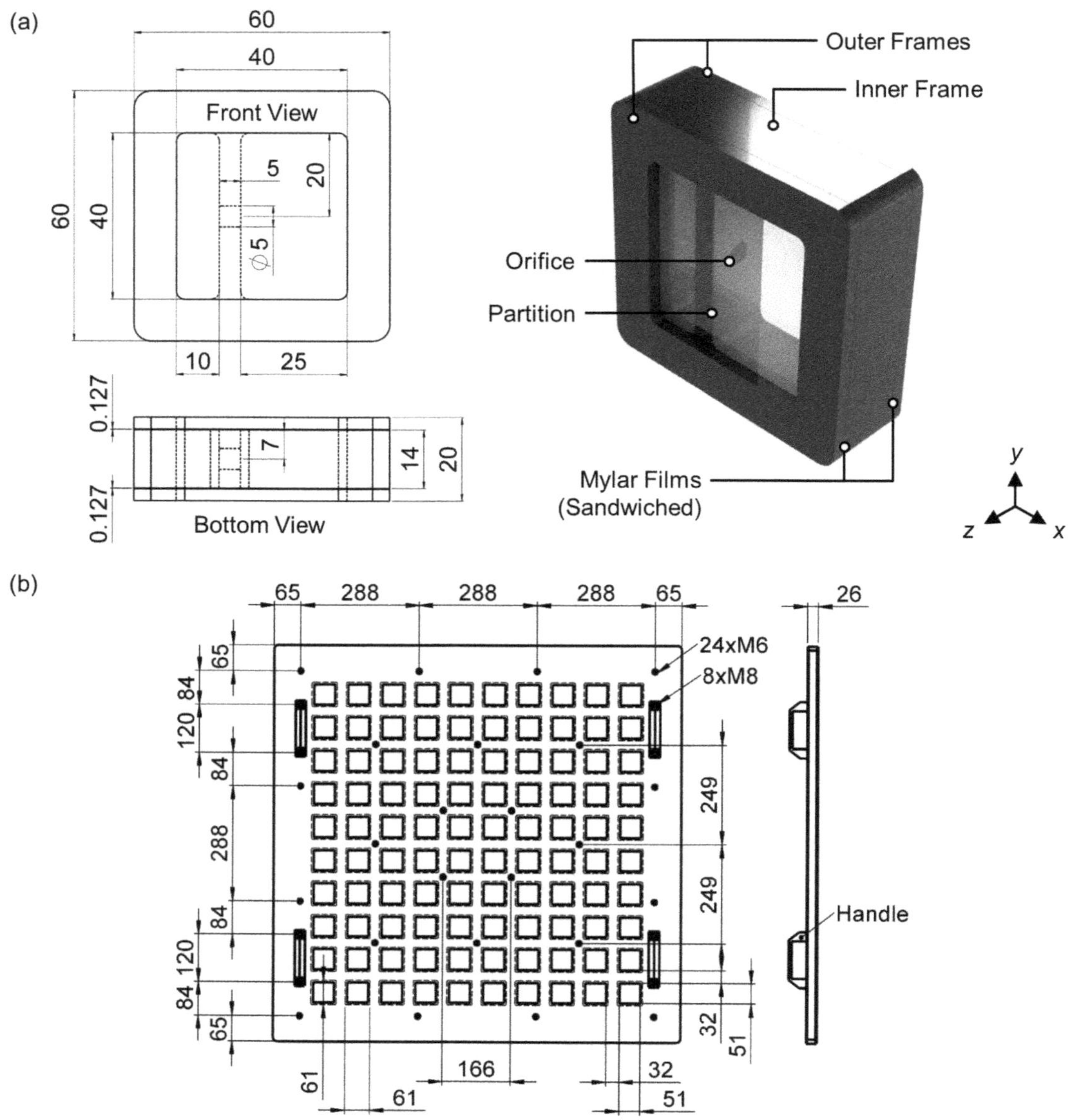

Figure 1. (**a**) Technical drawing (front and bottom views) and three-dimensional rendering of the specimen with resonator. (**b**) Technical drawing (front and side views) of the host structure. Hidden lines indicate the cutouts on the inner panel. All dimensions are in millimetres.

2.2.1. Sound Transmission Class

The sound transmission loss (STL) of the meta-panels was first determined to obtain the STC through post-processing. The measurements were conducted according to ASTM E90 [27] in two reverberation chambers. The source and the receiver chambers had a volume of 65.5 and 81.1 m^3, respectively, and a surface area of 97.2 and 109.5 m^2, respectively. The chambers were separated by a commercial acoustic wall, which was designed with an opening to mount the meta-panel from the receiver chamber. The opening had a surface area of 1 m^2 and a depth of 10 cm. Three toggle clamps were used to secure each edge of the meta-panel. As the meta-panel was 2.6 cm thick (Figure 1b), a wooden frame with a cross-sectional area of 8×2.5 cm^2 was required to transfer the clamping forces onto the edges of the meta-panel. A flange was designed around the opening to prevent further translation of the meta-panel towards the source chamber. Gasket strips were adhered along the flange to minimise sound leakages. Figure 2 shows the described set-up when viewed from each chamber.

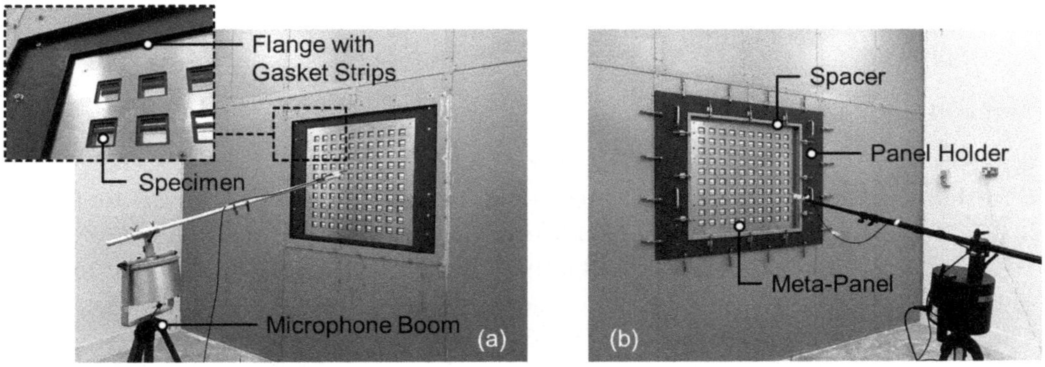

Figure 2. View of the test set-up in the (**a**) source and the (**b**) receiver chambers.

Pink noise was produced by a noise generator (Brüel & Kjær Type 2734) and transmitted into the source chamber by a pair of loudspeakers (Yamaha DXR15). Each loudspeaker was positioned at the trihedral corner furthest away from the panel holder. The noise signal was ensured to be loud enough such that the sound pressure level (SPL) at each one-third octave band frequency (80 Hz to 5 kHz) was more than 10 dB above the background noise in the receiver chamber. Consequently, background noise correction was not required. The SPL in each chamber was measured by a microphone boom (Norsonic Nor265). As recommended in ASTM E90 [27], at least six measurements (30 s each) were collected for averaging. Subsequently, the measurements were post-processed to calculate the STL by [27]

$$STL = \overline{L}_s - \overline{L}_r + 10\log_{10}\frac{S}{A_r} \tag{1}$$

where $\overline{L}_s$ and $\overline{L}_r$ denote the average SPL in the source and the receiver chambers, respectively; S denotes the surface area of the meta-panel; and A_r denotes the measured absorption area of the receiver chamber, which is given by [28]

$$A_r = 55.26\frac{V_r}{\overline{T}_{60}c_0} \tag{2}$$

where V_r denotes the volume of the receiver chamber (81.1 m^3); $\overline{T}_{60}$ denotes the average reverberation time; and c_0 denotes the speed of sound in air at 24.7 °C, which was the average temperature of the receiver chamber. In Equations (1) and (2), all variables are frequency-dependent except for S, V_r, and c_0.

From Equation (1), the STC was then determined based on the requirements and the reference curve given in ASTM E413 [29]. The requirements include (1) the deficiency at each one-third octave band frequency from 125 Hz to 4 kHz must not exceed 8 dB and (2) the sum of all deficiencies must be less than or equal to 32 dB. A deficiency is a point in which the measured STL is less than that of the reference curve. If both requirements are met, the STC will be equal to the STL of the reference curve at 500 Hz.

To include experimental uncertainty, the results are presented alongside the 95% confidence intervals ΔX calculated by [27]

$$\Delta X = Cs_X \tag{3}$$

where C denotes the constant corresponding to N measurements ($C = 1.05$ for $N = 6$ [27]) and s_X denotes the standard deviation of the variable X. For instance, X can be substituted with A_r to obtain the 95% confidence intervals for Equation (2). Only s_X is frequency-dependent. For Equation (1), the 95% confidence intervals ΔSTL were calculated by [27]

$$\Delta STL = \left[\Delta L_s^2 + \Delta L_r^2 + 18.9(\Delta A_r/A_r)^2\right]^{0.5} \tag{4}$$

where ΔL_s and ΔL_r denote the 95% confidence intervals of the SPL in the source and the receiver chambers, respectively.

2.2.2. Outdoor-Indoor Transmission Class

Since 1955, the STC has been used to classify building elements that could be treated as some sort of noise barrier with a single-value rating based on the overall acoustical performance from 125 Hz to 4 kHz [27]. Typical building elements include exterior/interior walls, doors, and windows, not to mention soundproofing canvases. Developed in 1990, the OITC classifies the same group of building elements in a similar way as the STC but places more emphasis on the low-frequency range (80 Hz to 4 kHz) [30]. The classification considers down to 80 Hz instead of 125 Hz, which is the case for the STC. The upper frequency limit remains the same at 4 kHz. According to ASTM E1332 [30], the OITC was designed to evaluate how effective the partition could mitigate outdoor ground and air transportation noises that are predominantly low-frequency (<500 Hz).

To obtain a reliable OITC, the noise source should exhibit a frequency spectrum that follows the trend of the reference curve given in ASTM E1332 [30]. The reference curve was derived by averaging the frequency spectra of aircraft takeoff noise, train passing noise, and traffic noise. In this study, the pink noise was considered as the noise source because its frequency spectrum could follow closely to the trend of the reference curve (Figure 3). As opposed to the reference curve used to determine the STC, adjustment of the curve to fit the measured STL curve based on deficiencies is not required. The OITC is mathematically expressed by [30]

$$OITC = 100.13 - 10\log_{10}\sum_{f=80}^{4000} 10^{(L_{ref,f}+L_{A,f}-STL_f)/10} \tag{5}$$

where L_{ref} denotes the SPL of the reference curve; L_A denotes the A-weighting correction factors; STL denotes the values obtained from Equation (1); and the subscript f denotes that the variable is dependent on the one-third octave band frequency from 80 Hz to 4 kHz.

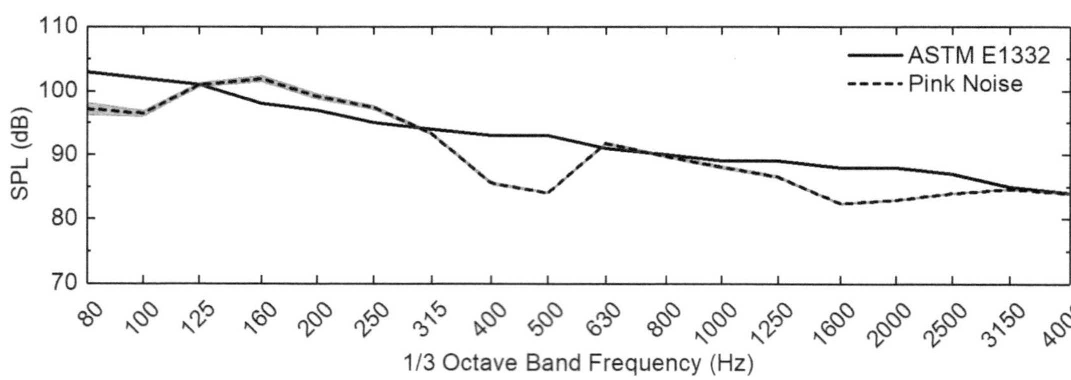

Figure 3. Comparison between the reference curve given in ASTM E1332 [30] and the measured SPL curve of the pink noise in the source chamber. Shaded region denotes the limits of the experimental uncertainty at a confidence level of 95%.

2.2.3. Noise Reduction

As mentioned at the end of Section 1, three different noise environments were considered—traffic noise, aircraft flyby noise, and construction noise. The first two noises were on-site recordings in Singapore, while the latter was downloaded from the Internet. These noises were considered because they are commonly experienced in Singapore. The test set-up was identical to that shown in Figure 2. Imagine the receiver chamber as an office room or a residential unit and the source chamber as the exterior environment polluted by each considered noise. The meta-panel can be imagined as the building element—a see-through panel, for example—separating the two domains. Based on this set-up, it is valid to determine how well the meta-panel would perform in terms of noise reduction L_{nr} as given by

$$L_{nr} = \overline{L}_s - \overline{L}_r \tag{6}$$

As defined in Equation (1), $\overline{L}_s$ and $\overline{L}_r$ denote the average SPL in the source and the receiver chambers, respectively, and are frequency-dependent.

3. Results

This section first presents the results based on the comparison between the meta-panels and the two commercially available noise barriers (Section 3.1). Next, the acoustical performance of the meta-panels in three different noise environments is discussed (Section 3.2). The detailed analysis of the specimen's characteristic was performed and can be found in the previous works [21,22,26].

3.1. Comparison with Commercially Available Noise Barriers

The scope of this comparative study is limited to the consideration of commercially available noise barriers found in Singapore—ANB 80 Steel and PEB Super Galum Steel (Figure 4). To allow a fair comparison, it is crucial to be aware of the thickness and the surface density of the meta-panels and the noise barriers (Table 1). This section first compares the measured STL curves (Section 3.1.1) before moving on to compare the STC and the OITC (Sections 3.1.2 and 3.1.3, respectively).

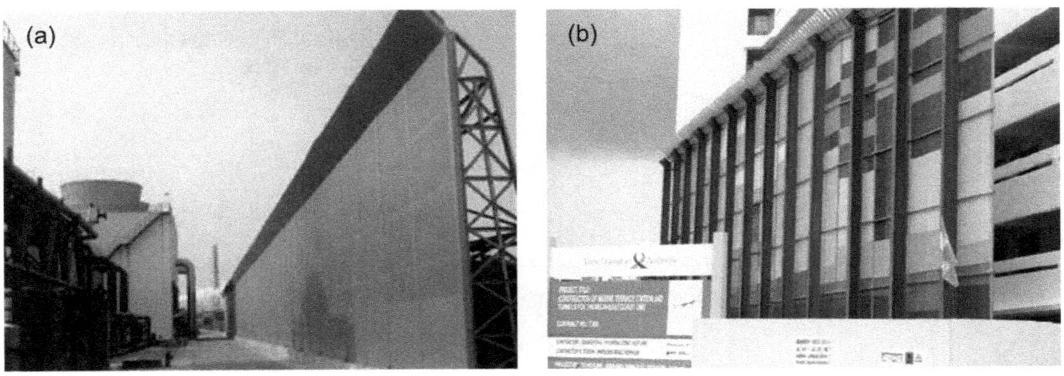

Figure 4. Commercially available noise barriers found in Singapore—(**a**) ANB 80 Steel and (**b**) PEB Super Galum Steel. Retrieved from the respective manufacturer's brochures.

Table 1. Thickness and surface density of the meta-panels—aluminium and wooden—and the noise barriers—ANB 80 Steel and PEB Super Galum Steel.

Product Name	Thickness (mm)	Surface Density (kg/m^2)
Aluminium Meta-Panel	26	51.1
Wooden Meta-Panel	26	16.4
ANB 80 Steel *	80	20.0
PEB Super Galum Steel *	30	10.1

* From manufacturer's brochure.

3.1.1. Sound Transmission Loss

To facilitate the analysis, the difference in STL was calculated by subtracting the STL curves of the noise barriers from that of each meta-panel (Figure 5). A positive difference indicates superiority of each meta-panel; a negative difference indicates inferiority of each meta-panel. The measured STL curves of the noise barriers are provided in Figure A1 in Appendix A to supplement this discussion.

At 80–500 Hz, both meta-panels were superior to the noise barriers. The aluminium meta-panel performed better than the ANB 80 Steel and the PEB Super Galum Steel by up to 18.4 dB at 200 Hz and up to 25.4 dB at 315 Hz, respectively (Figure 5a). For the wooden meta-panel, it performed better than the ANB 80 Steel and the PEB Super Galum Steel by up to 10.4 dB at 400 Hz and up to 19.1 dB at 125 Hz, respectively (Figure 5b). At this frequency range, the superiority of the meta-panels to the noise barriers was attributed to the stiffness region of the films, preventing the fundamental frequency of the entire structure from adversely affecting the acoustical performance. This phenomenon was justified in the previous work [26]. For the noise barriers, there was no additional feature to minimise the impact of the structural resonance, providing a possible explanation to the poorer acoustical performance.

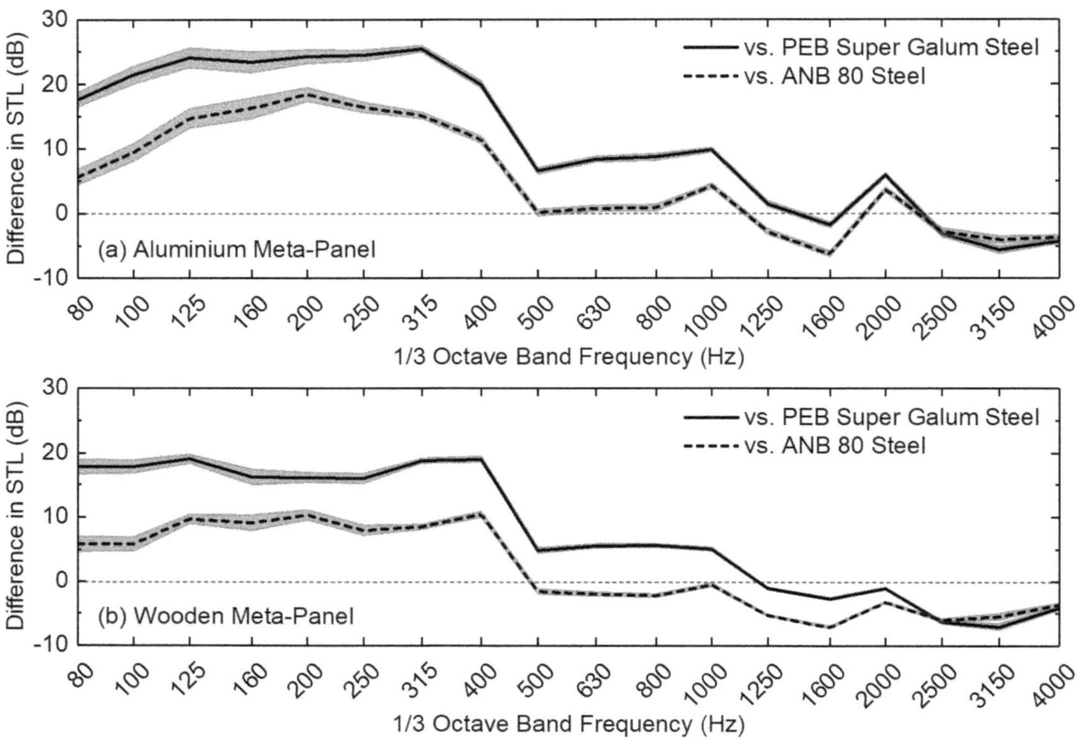

Figure 5. Difference in STL between the meta-panels—(**a**) aluminium and (**b**) wooden—and the noise barriers—PEB Super Galum Steel and ANB 80 Steel. Shaded region denotes the limits of the experimental uncertainty at a confidence level of 95%.

The superiority of both meta-panels extended further from 500 to 1250 Hz when compared against the PEB Super Galum Steel. The aluminium and the wooden meta-panels performed better by up to 9.7 dB at 1 kHz and up to 5.1 dB at 1 kHz, respectively. The maximum difference in STL at 1 kHz was contributed by one of the "noise shield" effects of the films. The "noise shield" effect refers to a phenomenon in which the displacement profile of the second film was nearly motionless relative to the first film. As such, the second film functioned like a "noise shield", minimising sound transmission. Again, this phenomenon was justified in the previous work [26]. Conversely, at the same frequency range (500–1250 Hz), the acoustical performance of each meta-panel was comparable to that of the ANB 80 Steel. The aluminium meta-panel performed better by only 0.1–0.9 dB except at 1 kHz (4.2 dB), while the wooden meta-panel fared poorer by 0.4–2.2 dB.

Unfortunately, the superiority of the meta-panels did not extend further into the high-frequency range (1.25–4 kHz), indicating that the meta-panels would not be ideal for high-frequency noise control. The wooden meta-panel performed poorer than the ANB 80 Steel and the PEB Super Galum Steel by up to 7.2 dB at 1.6 kHz and up to 7.1 dB at 3.15 kHz, respectively (Figure 5b). For the aluminium meta-panel, it performed poorer than the ANB 80 Steel and the PEB Super Galum Steel by up to

6.3 dB at 1.6 kHz and up to 5.7 dB at 3.15 kHz, respectively (Figure 5a). Despite the poorer acoustical performance at high frequencies, an improvement in STL was observed at 2 kHz due to the "noise shield" effect of the films. In particular, this effect provided the aluminium meta-panel with a positive difference in STL over the noise barriers at the same frequency. In this case, the aluminium meta-panel performed better than the ANB 80 Steel and the PEB Super Galum Steel by up to 3.7 and 5.9 dB, respectively. Despite the acoustical drawback at high frequencies, it may not pose any concern because most problematic noises fall at around and/or below 1 kHz.

From the above analysis, it is justifiable that the meta-panels may provide acoustical advantage over the noise barriers if the primary aim is to mitigate low-frequency noise (<500 Hz). Depending on the configuration, the meta-panels could also mitigate mid-frequency noise (0.5–1 kHz). Of course, it can be argued that the significant difference in STL between the aluminium meta-panel and the noise barriers could also be contributed by its high surface density (Table 1), which was 2.46 and 4.87 times higher than that of the ANB 80 Steel and the PEB Super Galum Steel, respectively. To refute this argument, it may be more convincing to compare the products that were of similar surface density. Based on the surface densities given in Table 1, the wooden meta-panel and the ANB 80 Steel—16.4 and 20.0 kg/m^2, respectively—may provide a fairer comparison. In fact, the surface density of the latter was higher than the former by 7.0%, giving it acoustical advantage over the wooden meta-panel based on the concept of the mass law [31]. This advantage is indeed the case in which the ANB 80 Steel performed better than the wooden meta-panel at 0.5–4 kHz (Figure 5b). According to the mass law [31], the higher the surface density, the higher the STL at a given frequency. However, at 80–500 Hz, the same noise barrier (ANB 80 Steel) performed poorer than the wooden meta-panel despite its higher surface density. At this frequency range, the wooden meta-panel benefited from the dynamic response of the specimens slotted within it. This comparison implies that the meta-panel, regardless of material, could still be superior to the noise barriers even with comparable surface densities.

3.1.2. Sound Transmission Class

At this juncture, the comparison of the measured STL curves showed that the meta-panels could perform better than the noise barriers especially if the primary focus is to mitigate low-frequency noise (<500 Hz). In the industry, the STC is often preferred to rate how effective the product is in noise mitigation. Therefore, to be more comprehensive, this section extends the previous section to compare the STC of the meta-panels and the noise barriers (Table 2).

Table 2. STC of the meta-panels—aluminium and wooden—and the noise barriers—ANB 80 Steel and PEB Super Galum Steel.

Product Name	STC
Aluminium Meta-Panel	36
Wooden Meta-Panel	33
ANB 80 Steel	33
PEB Super Galum Steel	25

Expectedly, the aluminium meta-panel was superior to the noise barriers with a STC of 36—3 and 11 points higher than that of the ANB 80 Steel (STC of 33) and the PEB Super Galum Steel (STC of 25), respectively. For the wooden meta-panel, it fared slightly poorer than the aluminium meta-panel by 3 points with a STC of 33. This rating was 8 points higher than that of the PEB Super Galum Steel. Recall the comparison in the previous section between the wooden meta-panel and the ANB 80 Steel. Even though the latter performed better than the former across a wider frequency range (0.5–4 kHz as opposed to 80–500 Hz), the STC showed that both products were in fact comparable in acoustical performance. The reason is because the STC emphasises more on the low-frequency range, which is the frequency content of most problematic noises [29]. Knowing that both products were identical in STC, it may be preferable to consider the wooden meta-panel over the ANB 80 Steel because of its higher

STL at 80–500 Hz, not to mention its lower surface density (Table 1). Moreover, the present design of the meta-panels is only a proof-of-concept and has yet to undergo design optimisation. Consequently, the decision-making process should be easier.

3.1.3. Outdoor-Indoor Transmission Class

To justify that the meta-panels had acoustical advantage over the noise barriers especially at low frequencies (80–500 Hz), this section compares their OITC with that of the noise barriers (Table 3).

Table 3. OITC of the meta-panels—aluminium and wooden—and the noise barriers—ANB 80 Steel and PEB Super Galum Steel.

Product Name	OITC
Aluminium Meta-Panel	34
Wooden Meta-Panel	31
ANB 80 Steel	26
PEB Super Galum Steel	18

When evaluated based on the OITC instead of the STC, the ANB 80 Steel and the PEB Super Galum Steel performed even poorer by 7 points. The former dropped from a STC of 33 to an OITC of 26; the latter dropped from a STC of 25 to an OITC of 18. This drop in rating suggests that both noise barriers were inherently poorer in acoustical performance at low frequencies (80–500 Hz)—a shortcoming that was amplified when the OITC was considered. As such, the aluminium meta-panel became even more superior to the noise barriers with an OITC of 34—12 and 16 points higher than that of the ANB 80 Steel and the PEB Super Galum Steel, respectively. The significant drop in rating difference from 3 (STC) to 12 (OITC) showed by the ANB 80 Steel suggests that it would not be ideal in low-frequency noise environments (<500 Hz). For the wooden meta-panel, it performed poorer than the aluminium meta-panel by 3 points—the same number of points when comparing between the STC.

In Section 3.1.2, the comparison between the wooden meta-panel and the ANB 80 Steel showed that they were comparable in acoustical performance based on the STC even though the latter performed better than the former across a wider frequency range (0.5–4 kHz in contrast to 80–500 Hz). However, when evaluated based on the OITC instead of the STC, the acoustical advantage of the wooden meta-panel at low frequencies (80–500 Hz) was clearly demonstrated in which it performed better than the ANB 80 Steel by 5 points with an OITC of 31. This comparison substantiates the recommendation in the last paragraph of Section 3.1.2 stating that it may be preferable to consider the wooden meta-panel over the ANB 80 Steel because of its higher STL at 80–500 Hz and its lower surface density (Table 1).

3.2. Acoustical Performance in Different Noise Environments

Having shown the acoustical advantage of the meta-panels at low frequencies (80–500 Hz), it was of interest to evaluate their acoustical performance in three different noise environments—traffic noise, aircraft flyby noise, and construction noise—as shown in Figure 6. The measured SPL curves of the noise environments are provided in Figure A2 in Appendix B to supplement this discussion.

Both meta-panels exhibited consistent acoustical performance in the three different noise environments. The noise reduction curves followed closely to the overall trend of the STL curves that were reported previously for the same configuration (i.e., solely occupied by specimens with resonator) [26]. Focusing on the low-frequency range (80–400 Hz), the aluminium meta-panel provided a noise reduction of 18.4–30.9 dB depending on the noise environment (Figure 6a). For the wooden meta-panel, it achieved a noise reduction of 18.3–26.9 dB depending on the noise environment (Figure 6b). Averaging over the same frequency range, the aluminium and the wooden meta-panels provided a noise reduction of 27.4 and 22.7 dB, respectively. This observation suggests that the meta-panel, regardless of material, could significantly reduce the noise exposure of the occupants.

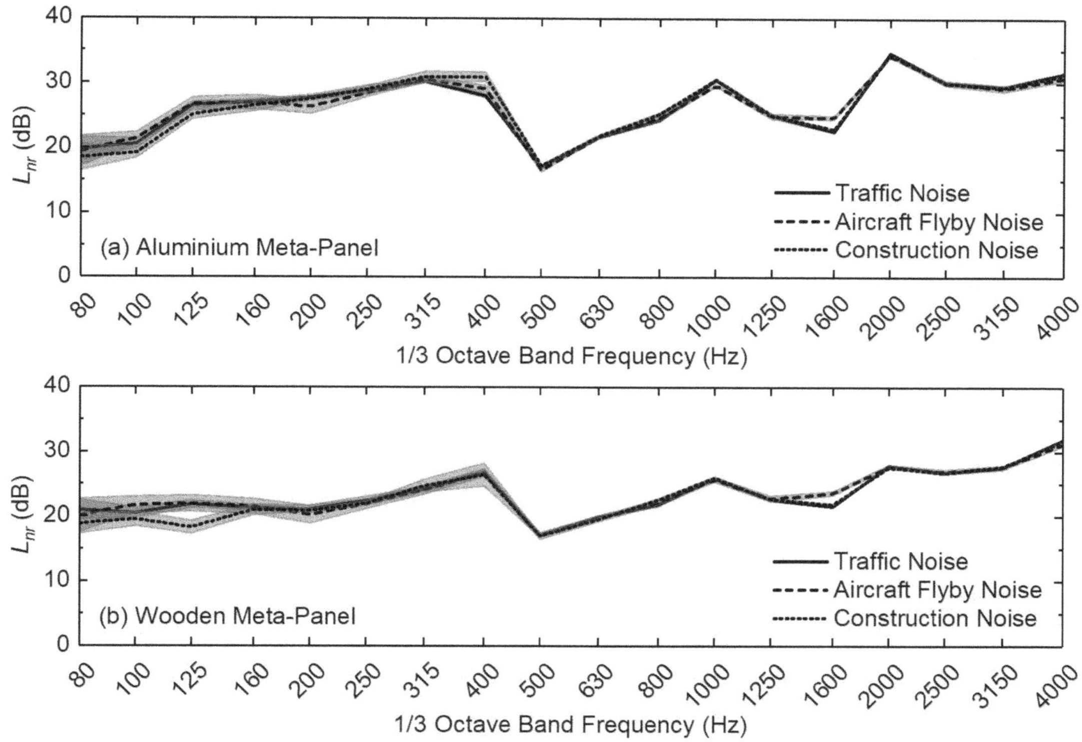

Figure 6. Noise reduction L_{nr} curves of the meta-panels—(**a**) aluminium and (**b**) wooden—due to traffic noise, aircraft flyby noise, and construction noise. Shaded region denotes the limits of the experimental uncertainty at a confidence level of 95%.

4. Discussion

Naturally, being a proof-of-concept, it is no doubt that further development and optimisation are necessary to arrive at a meta-panel design suitable for the noise environment of interest. This section discusses some of the avenues for future research.

The influence of different weather conditions on the meta-panel has yet to be considered. In Singapore, the weather conditions include heavy rain, strong wind, and hot sun. If the meta-panel is considered for outdoor applications—as a noise barrier, for example—the specimens and the host structure must be designed with adequate structural integrity. A protective feature may be required to prevent the films from excessive deflection, which could alter the resonant behaviour of the films and influence the characteristic of the meta-panel. For example, the protective feature could simply be a continuous panel fastened to each side of the meta-panel to enclose the specimens. Consequently, the overall acoustical performance of the meta-panel may also be improved based on the understanding of the mass law. The protective feature could also serve to prevent surface damage of the films and dust accumulation on the films. In actual environments, the former could be caused by foreign objects, while the latter is inevitable. If both factors are not considered, the durability of the meta-panel will be affected, not to mention its reliability in noise mitigation.

The concept of modularity could provide the meta-panel with versatility and aesthetics. By having a library of uniquely designed specimens, the meta-panel could be configured such that its acoustical performance would be optimised for the noise environment of interest. The specimens could also be colour-coded to categorise them according to their acoustical performance. Since the acoustical performance of the meta-panel was previously shown [26] to be insensitive to the distribution and the arrangement of the specimens, both factors could be varied to form a pixel art, providing aesthetics to the meta-panel.

Being a design that is highly customisable for different noise environments is no doubt advantageous especially in terms of implementation cost. As the host structure does not need to be replaced during the change in configuration, the cost incurred from the customisation should only be dependent on the cost of producing new specimens. In this study, the production cost of each specimen was approximately S$4. This amount may be reduced further if the specimens are produced in large quantities via conventional manufacturing techniques.

Lastly, as the scope of this study was limited to experimental evaluation, it is essential to develop a numerical model of the meta-panel so that the desired acoustical performance could be predicted prior to fabrication. Consequently, high computational power will be expected.

5. Conclusions

To reiterate, the purpose of this study was to justify whether the aluminium and the wooden meta-panels solely occupied by specimens with resonator could indeed be viable for noise control in the industry. By comparing the STL with that of two commercially available noise barriers, the meta-panels were shown to be superior at low frequencies (80–500 Hz). This superiority was substantiated when the STC and the OITC were compared. The meta-panels were then evaluated in three different noise environments, showing that they could provide an average noise reduction of 27.4 and 22.7 dB, respectively, at 80–400 Hz. In conclusion, the meta-panels were justified to be viable for low-frequency (80–500 Hz) noise control in the industry. Being a proof-of-concept design, there is a large room for improvement to obtain a meta-panel that is lightweight and yet has good acoustical performance at low-frequency range (<500 Hz), which is the frequency content of most problematic noises.

Author Contributions: L.Y.L.A. conceived the idea, designed the methodology, fabricated the specimens, performed the measurements/simulations, analysed the data, and wrote/revised the manuscript. Y.K.K. and H.P.L. provided feedback and edited/approved the manuscript for submission.

Funding: This research was funded by the Singapore Economic Development Board (EDB) and ST Engineering Land Systems Ltd. under the EDB Industrial Postgraduate Program (EDB-IPP) (Grant No. COY-15-IPP/120005).

Acknowledgments: The authors acknowledge the advice from Samuel Lee and the support from Kok Seng Cheng during the design and the experimental phases of this research, respectively.

Conflicts of Interest: The authors declare no conflict of interest.

Abbreviations

The following abbreviations are used in this manuscript:

STC	Sound Transmission Class
OITC	Outdoor-Indoor Transmission Class
STL	Sound Transmission Loss
SPL	Sound Pressure Level

Appendix A. Measured STL Curves of the Noise Barriers

Figure A1 compares the measured STL curves of the aluminium and the wooden meta-panels to that of the noise barriers. The measured STL curves of both noise barriers were extracted from the respective manufacturer's brochures. As the STL at 80 Hz was not available for both noise barriers, it was taken as the same STL at 100 Hz. The STL curves were used to calculate the difference in STL between the meta-panels and the noise barriers (Figure 5).

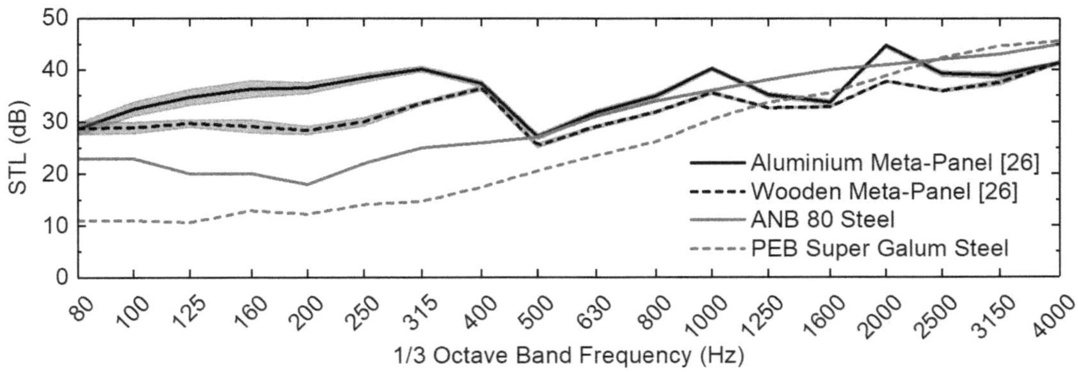

Figure A1. Measured STL curves of the aluminium and the wooden meta-panels compared to that of the noise barriers—ANB 80 Steel and PEB Super Galum Steel. Shaded region denotes the limits of the experimental uncertainty at a confidence level of 95%.

Appendix B. Measured SPL Curves of the Noise Environments

Figure A2 shows the measured SPL curves of the noise environments in the source chamber. The values of the SPL curves were used to calculate the noise reduction L_{nr} curves (Figure 6).

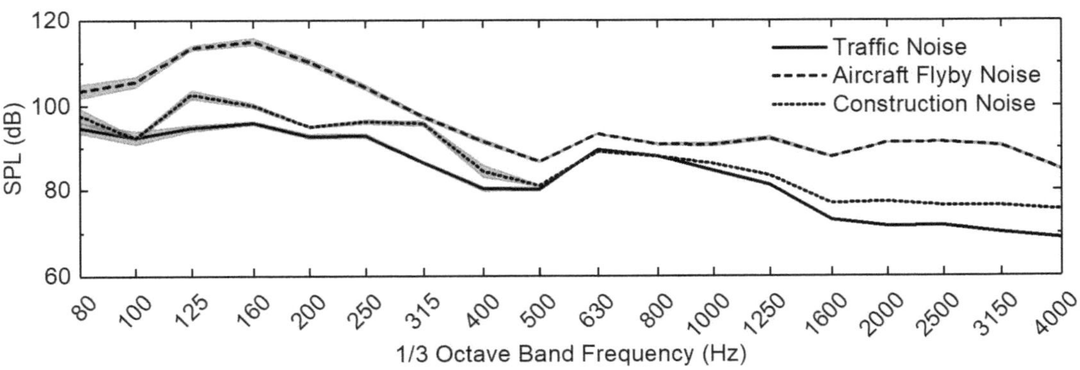

Figure A2. Measured SPL curves of the three different noise environments—traffic noise, aircraft flyby noise, and construction noise—in the source chamber. Shaded region denotes the limits of the experimental uncertainty at a confidence level of 95%.

References

1. Ang, L.Y.L.; Koh, Y.K.; Lee, H.P. Broadband sound transmission loss of a large-scale membrane-type acoustic metamaterial for low-frequency noise control. *Appl. Phys. Lett.* **2017**, *111*, 041903. [CrossRef]
2. Langfeldt, F.; Kemsies, H.; Gleine, W.; von Estorff, O. Perforated membrane-type acoustic metamaterials. *Phys. Lett. A* **2017**, *381*, 1457–1462. [CrossRef]
3. Mei, J.; Ma, G.; Yang, M.; Yang, Z.; Wen, W.; Sheng, P. Dark acoustic metamaterials as super absorbers for low-frequency sound. *Nat. Commun.* **2012**, *3*, 756. [CrossRef] [PubMed]
4. Naify, C.J.; Chang, C.M.; McKnight, G.; Nutt, S.R. Transmission loss and dynamic response of membrane-type locally resonant acoustic metamaterials. *J. Appl. Phys.* **2010**, *108*, 114905. [CrossRef]
5. Naify, C.J.; Chang, C.M.; McKnight, G.; Scheulen, F.; Nutt, S.R. Membrane-type metamaterials: Transmission loss of multi-celled arrays. *J. Appl. Phys.* **2011**, *109*, 104902. [CrossRef]
6. Naify, C.J.; Chang, C.M.; McKnight, G.; Nutt, S.R. Transmission loss of membrane-type acoustic metamaterials with coaxial ring masses. *J. Appl. Phys.* **2011**, *110*, 124903. [CrossRef]
7. Naify, C.J.; Chang, C.M.; McKnight, G.; Nutt, S.R. Scaling of membrane-type locally resonant acoustic metamaterial arrays. *J. Acoust. Soc. Am.* **2012**, *132*, 2784–2792. [CrossRef]

8. Wang, X.; Zhao, H.; Luo, X.; Huang, Z. Membrane-constrained acoustic metamaterials for low-frequency sound insulation. *Appl. Phys. Lett.* **2016**, *108*, 041905. [CrossRef]
9. Wang, X.; Luo, X.; Zhao, H.; Huang, Z. Acoustic perfect absorption and broadband insulation achieved by double-zero metamaterials. *Appl. Phys. Lett.* **2018**, *112*, 021901. [CrossRef]
10. Yang, Z.; Mei, J.; Yang, M.; Chan, N.H.; Sheng, P. Membrane-type acoustic metamaterial with negative dynamic mass. *Phys. Rev. Lett.* **2008**, *101*, 204301. [CrossRef]
11. Yang, Z.; Dai, H.M.; Chan, N.H.; Ma, G.; Sheng, P. Acoustic metamaterial panels for sound attenuation in the 50–1000 Hz regime. *Appl. Phys. Lett.* **2010**, *96*, 041906. [CrossRef]
12. Yang, M.; Ma, G.; Yang, Z.; Sheng, P. Coupled membranes with doubly negative mass density and bulk modulus. *Phys. Rev. Lett.* **2013**, *110*, 134301. [CrossRef] [PubMed]
13. Sui, N.; Yan, X.; Huang, T.Y.; Xu, J.; Yuan, F.G.; Jing, Y. A lightweight yet sound-proof honeycomb acoustic metamaterial. *Appl. Phys. Lett.* **2015**, *106*, 171905. [CrossRef]
14. Li, S.; Mao, D.; Huang, S.; Wang, X. Enhanced transmission loss in acoustic materials with micro-membranes. *Appl. Acoust.* **2018**, *130*, 92–98. [CrossRef]
15. Huang, T.Y.; Shen, C.; Jing, Y. Membrane- and plate-type acoustic metamaterials. *J. Acoust. Soc. Am.* **2016**, *139*, 3240–3250. [CrossRef] [PubMed]
16. Ma, F.; Huang, M.; Wu, J. Ultrathin lightweight plate-type acoustic metamaterials with positive lumped coupling resonant. *J. Appl. Phys.* **2017**, *121*, 015102. [CrossRef]
17. Ma, F.; Huang, M.; Wu, J. Acoustic metamaterials with synergetic coupling. *J. Appl. Phys.* **2017**, *122*, 215102. [CrossRef]
18. Ma, F.; Huang, M.; Xu, Y.; Wu, J. Bilayer synergetic coupling double negative acoustic metasurface and cloak. *Sci. Rep.* **2018**, *8*, 5906. [CrossRef]
19. Ma, F.; Huang, M.; Xu, Y.; Wu, J. Bilayer plate-type acoustic metamaterials with Willis coupling. *J. Appl. Phys.* **2018**, *123*, 035104. [CrossRef]
20. Varanasi, S.; Bolton, J.S.; Siegmund, T.H. Experiments on the low-frequency barrier characteristics of cellular metamaterial panels in a diffuse sound field. *J. Acoust. Soc. Am.* **2017**, *141*, 602–610. [CrossRef] [PubMed]
21. Ang, L.Y.L.; Koh, Y.K.; Lee, H.P. Plate-type acoustic metamaterial with cavities coupled via an orifice for enhanced sound transmission loss. *Appl. Phys. Lett.* **2018**, *112*, 051903. [CrossRef]
22. Ang, L.Y.L.; Koh, Y.K.; Lee, H.P. A note on the viscous boundary layer in plate-type acoustic metamaterials with an internal tonraum resonator. *Appl. Acoust.* **2018**, *140*, 160–166. [CrossRef]
23. Langfeldt, F.; Wolfgang, G.; von Estorff, O. Enhancing the low-frequency noise reduction of a double wall with membrane-type acoustic metamaterials. In Proceedings of the 45th International Congress and Exposition on Noise Control Engineering (INTER-NOISE 2016), Hamburg, Germany, 21–24 August 2016; Kropp, W., Ed.; Deutsche Gesellschaft Fuer Akustik: Hamburg, Germany, 2016; pp. 3413–3424.
24. Ma, G.; Fan, X.; Sheng, P.; Fink, M. Shaping reverberating sound fields with an actively tunable metasurface. *Proc. Natl. Acad. Sci. USA* **2018**, *115*, 6638–6643. [CrossRef] [PubMed]
25. Varanasi, S.; Bolton, J.S.; Siegmund, T.H.; Cipra, R.J. The low-frequency performance of metamaterial barriers based on cellular structures. *Appl. Acoust.* **2013**, *74*, 485–495. [CrossRef]
26. Ang, L.Y.L.; Koh, Y.K.; Lee, H.P. Plate-type acoustic metamaterials: Evaluation of a large-scale design adopting modularity for customisable acoustical performance. *Appl. Acoust.* **2019**, *149*, 156–170. [CrossRef]
27. ASTM. *ASTM E90-09: Standard Test Method for Laboratory Measurement of Airborne Sound Transmission Loss of Building Partitions and Elements*; ASTM International: West Conshohocken, PA, USA, 2009. [CrossRef]
28. ASTM. *ASTM E2235-04: Standard Test Method for Determination of Decay Rates for Use in Sound Insulation Test Methods*; ASTM International: West Conshohocken, PA, USA, 2012. [CrossRef]
29. ASTM. *ASTM E413-10: Classification for Rating Sound Insulation*; ASTM International: West Conshohocken, PA, USA, 2010. [CrossRef]
30. ASTM. *ASTM E1332-16: Standard Classification for Rating Outdoor-Indoor Sound Attenuation*; ASTM International: West Conshohocken, PA, USA, 2016. [CrossRef]
31. Fahy, F.J.; Gardonio, P. Transmission of Sound Through Partitions. In *Sound and Structural Vibration: Radiation, Transmission, and Response*, 2nd ed.; Academic Press: Cambridge, MA, USA, 2007; Chapter 5, pp. 277–373.

9

Sound Archaeology: A Study of the Acoustics of Three World Heritage Sites, Spanish Prehistoric Painted Caves, Stonehenge, and Paphos Theatre

Rupert Till

Department of Music and Drama, University of Huddersfield, Huddersfield HD1 3DH, UK; R.Till@hud.ac.uk

Abstract: This paper explores the acoustics of three UNESCO World Heritage Sites: five caves in Spain that feature prehistoric paintings that are up to 40,000 years old; Stonehenge stone circle in England, which is over 4000 years old; and Paphos Theatre in Cyprus, which is 2000 years old. Issues with standard acoustic methods are discussed, and a range of different possible approaches are explored for sound archaeology studies, also known as archaeoacoustics. The context of the three sites are examined followed by an analysis of their acoustic properties. Firstly, early decay time is explored, including a comparison of these sites to contemporary concert halls. Subsequently, reverberation, clarity of speech, and bass response are examined. Results show that the caves have a wide range of different naturally occurring acoustics, including reverberation, and strong bass effects. Stonehenge has acoustics that change as the design of the site develops, with some similarities to the effects in the caves. Acoustic effects vary considerably as you move further into the centre of the stone circle, and as the stone circle develops through time; these effects would be noticeable, and are a by-product of the human building of ritual sites. At Paphos Theatre, acoustics vary from the best seats on the front rows, backwards; here, the architects have considered acoustics in the design of the building. The paper illustrates the changing acoustics of ritual sites in human cultures, showing how sound contributed to giving spaces an individual character, helping to afford a sense of contextualized ritual place.

Keywords: sound; archaeology; archaeoacoustics; acoustics; reverberation; clarity; Stonehenge; cave; theatre; EDT; music

---

1. Introduction

This paper asks how one best studies the acoustic ecologies of archaeological spaces? Are standard acoustics methods appropriate in such sound archaeology or archaeoacoustic research? It explores how acoustics in sites that feature ritual use change through different stages of human culture, from Palaeolithic painted caves in Northern Spain, with a context dated from 43,000 years in the past, through the monumental 5000-year-old architecture of Neolithic Stonehenge in Britain, to the public theatres of Roman culture, in Paphos on the Mediterranean island of Cyprus. All three are UNESCO World Heritage sites, and the paper explores what the changes in the acoustic effects present suggest about how human culture changes over time. It asks what problems are encountered when examining the acoustics of ancient sites. The three sites are contextualised—prehistoric caves in Spain, Stonehenge, and Paphos Theatre—followed by a metrical analysis of their acoustics, before finishing with conclusions.

To comprehensively explore changes in acoustics over time in antiquity would take a large-scale project with many more examples, which is beyond the scope of the present study. Instead, metrical results are generated through the author's study of the acoustics of these three archaeological spaces, and are compared in order to illustrate both to what extent they differ, and the range of acoustics present in different archaeological contexts and periods. This paper provides a number of novel perspectives. While previous studies of the acoustics of these and other archaeological sites have provided a range of results, this paper provides for the first time a comparison of different archaeological eras and contexts, and of acoustic metrics, such as early decay time (EDT), clarity, speech transmission, consonant intelligibility, and bass ratio. The key measure of perceived reverberation, early decay time, is discussed in some detail, and compared to results for modern concert halls, exploring results across a wide frequency range, rather than the restricted averages often referred to within industrial acoustics. The limitations imposed by approaches, such as this averaging, are discussed, and the effects are discussed of adopting an uncritical application to archaeological contexts of the standard acoustic state of the art as used in contemporary industrial research. The comparative study of three different sites illustrates the challenges for standard acoustics methods of working within this field, and provides examples of how and why one might construct a bespoke acoustics study method for these sites. Comparison is made between the caves explored, something a previous study did not consider. Acoustics of a digital model of Stonehenge at a number of receiver positions within three different phases of development of the monument are discussed. No previous study of Stonehenge has explored the acoustics of a range of receiver positions, or of the various arrangements of stones of this iconic site. Paphos Theatre has never been studied acoustically, and in general, there have been no studies comparing Greek or Roman ancient performance venues to modern concert halls. As well as adopting empirical quantitative approaches, this project includes a qualitative comparative assessment of the sites being studied, as is appropriate when studying the experience of sound within ritual sites, and working within an interdisciplinary field that addresses archaeology, sound, and music, as well as scientific acoustics.

2. Materials and Methods

2.1. Problems with Standard Acoustic Methodology

Acoustics is usually conceived of as a purely empirical scientific field. Its purpose is to come to reliable, repeatable results, focused on certainties that can be used within applied engineering, industrial, or professional contexts. The results of acoustical studies are used to adjust the architectural design of buildings and other environments, and thus a focus is on results that can provide useful solutions. This aims to produce definitive conclusions, to find answers and solutions, although even this engineering approach has to struggle against compromises between results in different positions and frequency ranges, and the impossibility of achieving perfect acoustics for all purposes. In contrast, post-processual archaeological thought accepts that archaeological data requires interpretation, and when studying human history, the further back in time one moves, the less certain results become. In an archaeological context, it is accepted that there are many possible perspectives and interpretations, and that answers depend on the context of the archaeological situation, as well as the focus of the researcher and their own interests [1–3]. Studies of archaeoacoustics, involving as they do the use of acoustics methods to inform archaeological research, have to resolve such different approaches. This situation is further problematised in music and sound archaeologies, which are often centred on even more creative and artistic perspectives [4] and the phenomenological experience of sound [5,6]. Sound is a time-based medium, and thus needs an approach that includes action and experience, which addresses embodiment and emotion as much as mechanistic rationalism [7]. Sound archaeology requires a non-representational approach (NRT), as described by Thrift [8] and Dewsbury [9], as it attempts to interpret human meanings from antiquity, where the involvement of physical activities and direct experience necessitates a fusing together of separate methodologies. In order to address these concepts, and integrate elements of experiential acoustic phenomenology, the experience of the author is deliberately discussed within this paper. Although this is perhaps unusual in acoustics publications, it is more commonplace in archaeological, musical, and sound-focused research. Indeed, field studies are common elements both of archaeology and ethnomusicology, the two subjects from which archaeoacoustics and sound archaeology have emerged. It is proposed that discussion of the author's experiences during fieldwork is appropriate content for this study.

Architectural acoustics has traditionally focused on issues related to industrial practice, such as concert hall design, traffic noise, and health and safety issues. It has as a result a range of standards, leading to a number of assumptions and set practices in standard acoustics methods. These can be problematic when using acoustics to study archaeological sites. For example, Long [10] suggests that "in a room whose dimensions are large enough that there is a sufficient density of modes, it is customary to describe the space in terms of a statistical model known as a diffuse field." (p. 327). In a diffuse field it is expected that a similar acoustic result will occur, independent of the positioning of the source of sound, or receiver, as no matter where you make the sound or hear it, reverberation is considered as consistent across the space. Such simplification is appropriate for standard acoustics purposes, in modern buildings with regular shapes.

Archaeological sites often have a complex or irregular shape, and as in any space, an experience of acoustics can vary considerably depending on listening (receiver) position, and where sound is being made (source position). Although it is customary in acoustics to assume a diffuse field, this assumption overlooks the range of sonic experiences possible for an individual within such a field. Modal effects can mean that sound near a wall is different from that in the centre, and unusually shaped rooms can have odd and varied frequency and amplitude responses. In archaeological sites, a large number of measurements may have to be taken if one wishes to comprehensively understand the acoustics of the space, for example focusing on positions of particular interest, perhaps with important archaeological features, or that seem to be a particular focus of activity.

Standard acoustics approaches often act to remove measurement of variation, aiming in many cases for an averaged result. BS EN ISO 3382-1:2009 is the British Standard for Acoustics–Measurement of Room Acoustic Parameters Part 1: Performing Spaces [11]. This describes the state of the art of engineering standards for acoustic measurement practice, which is well-known and thus not described in detail in this study, other than where sound archaeology necessitates an alternative approach. It is a useful document for commercial purposes, to enable comparable standardised results, or to describe a contemporary performance space, but can be problematic for a forensic analysis of the acoustics of an archaeological site. A cave or temple is not a performance space per se, but this standard is referred to as such sites are closer in use than to the other available options for such standards, such as a room, office, or laboratory.

The standard raises the issue of "whether single spatial averages will adequately describe the room" (p. 5), and suggests that if "the room is likely to show areas with differing reverberation times, these shall be investigated and measured separately" (p. 6). The standard suggests this option for spaces that do not feature a diffuse field, but the averaging of results is the more common practice, and no definition is provided of how different reverberation results have to be, to require individual consideration of a range of source and receiver positions. The standard requires that "a minimum of two source positions shall be used" (p. 5). It suggests (p. 20) that 6 different microphone/receiver positions are needed for a venue with 500 seats, and 10 positions are required for a venue with 2000 seats. These suggestions are for a performance space, and an archaeological site with no seats requires different consideration. The multiple source and receiver results are usually averaged; "it is necessary to average over a number of measurements at each position in order to achieve an acceptable measurement uncertainty" (p. 7). In a sound archaeology project, this study suggests that both individual and averaged results should be considered, as uncertainty is often present in any case, and certainty is not the only focus. The intention of the standard is to establish a result that describes an overall impression of a space, whereas sound archaeology may be interested in a detailed identification of the variation and distribution of different acoustic effects present, and in results in individual positions of archaeological significance.

The standard recommends a minimum distance of 1.5 m between an acoustic test signal source (such as a sine signal played through a loudspeaker) and a receiver (microphone) when carrying out an acoustic measurement (p. 3). In a sound archaeology study, one might be additionally interested in results where the source and receiver are in the same place (to mimic a performer making sound and hearing it themselves), are far away from one another, or where one or the other are moving. Proximity to a wall or floor can significantly affect the frequency response recorded, as there might be more early reflections present as a result; near a curved wall, the sound may be different than next to a straight wall. In the standard, the "distance from any microphone position to the nearest reflecting surface, including the floor, shall be at least a quarter of a wavelength, i.e., normally around 1 m" (p. 5). If one is interested in the sound next to a cave painting, sound made by scraping a prehistoric engraving into the wall, the sound made by a drum placed on the floor, or by the strange effects created by singing next to a particular surface, one might need to change this standard practice. An assessment method based on sources and receivers at set distances assumes a standard separation between a performer and an audience, a contemporary Western paradigm that is not shared in all archaeological contexts, where participation may be the norm. Although it is acceptable in acoustics to deal with a venue as though it were empty, if one wants to understand people's experience of acoustics, it may be more relevant to explore a venue's acoustics with an audience present, something that is problematic due to audience noise.

Reverberation is described with a standard metric, such as T30 or early decay time (EDT), as a set value, an invariant amount that never changes. Consideration of early reflections allows EDT to provide a better indication of perceived reverberance than T30 [12] and may be more useful as a result in some sound archaeology contexts. Reverberation varies at different frequencies and may be longer or shorter at lower bass frequencies. Acoustic studies usually state the reverberation time at 1 kHz, or as an average of results at 500 Hz and 1 kHz [11] (p. 10). In a workplace, these frequencies are significant because mid-range frequencies are most likely to damage hearing, whereas low frequencies are less problematic. The standard recommends that the range from "at least 250 to 2000 Hz" is examined, and for "engineering and precision methods, the frequency range should cover at least 125 to 4000 Hz in octave bands or 100 to 5000 Hz in one-third octave bands". At an archaeological site, one might be interested in, for example, whether reverberation at low frequencies is unusually long, or whether or not mid-range reverberation is significant; as we will see in a later section, at Stonehenge, there are frequencies of interest well below 100 Hz.

Digital acoustic modelling software is sometimes used within acoustical studies to predict acoustic effects. It can be used to analyse the original effects present where an archaeological site is no longer present or is damaged or altered. Murphy et al. [13] discusses auralisation, using digital tools to represent the acoustics of spaces using digital acoustic models. Weitze et al. [14] discusses acoustic modelling of Hagia Sofia in Istanbul. Details of a site are often simplified in such modelling to provide an approximation, but when exploring the acoustics of an archaeological site, one may be interested in the effect of the detail of the architecture. Such modelling programmes have certain deficiencies. For example, the Odeon acoustic modelling programme used in this project is excellent for high frequencies, but it is ineffective at some low frequencies and as a ray-tracing method becomes problematic when the length of the rays is close in magnitude to the wavelength of the source sound. Odeon's digital acoustic modelling software competitor CATT Acoustic, uses the same modelling method, and exhibits similar issues.

The standard acoustic measurement equipment used to carry out field tests can also be problematic. Precision measurement microphone capsules often require separate power supplies, connected by banana plugs to specialist cables designed for use in the laboratory or an office. The author used these during a study of the acoustics of caves featuring Palaeolithic paintings, with a methodology described in detail in previous project publications [4,15]. Despite taking care to look after them, due to the humidity these highly expensive but fragile pieces of equipment stopped working. During the same project, we used a dodecahedron loudspeaker. These are large sound sources designed to be omnidirectional, and to meet set IEEE standards, such as ISO 10140-5 (laboratory measurements), ISO 16283-1 (field measurements), and ISO 3382 Annex A (reverberation time measurements). Optimisation for such standards inevitably means compromises. Dodecahedron loudspeakers are designed to be omnidirectional, to radiate sound equally in all directions. This directionality is adequate for testing for compliance to standards, but because the speakers need to be small so that a number can be fitted into one housing, as the enclosure's design is optimised for directionality and in terms of meeting the ISO standard, the frequency response is not flat across a wide frequency range, often dropping off at low and high frequencies. For example, the NTi Audio DS3 dodecahedron loudspeaker, though an excellent sound source in many ways, has a frequency response that drops off above 10,000 Hz and below 100 Hz [16].

For measurement of archaeological acoustic spaces, Murphy [17] proposes the use as a source of a recording studio reference monitor loudspeaker that has a flat frequency response, accompanied by a sub-woofer speaker, in addition to a dodecahedron. This is intended to balance the accuracy of directional and frequency responses; however, this approach creates other difficulties. Murphy's setup was designed to systematically capture the acoustic character of an individual site. This method uses a microphone placed on a revolving turntable to capture impulse responses from various directions. This generates a large amount of data about a single source and receiver position. Like any method, it has advantages and disadvantages. It only records a single pair of source and receiver positions, and results could be different in other positions. Murphy goes on to describe (p. 224–225) how his approach captured an accurate representation of the centre of Scottish prehistoric monument Maes Howe, identifying that in the 125 Hz octave band, reverberation is significantly higher. Though a very useful approach, the study did not address the acoustic behaviour in the side chambers of the monument or in the entrance passage. It identifies modal resonant frequencies but does not discuss a range of comparative levels, nor does it present a comprehensive frequency response of all of the space. Murphy's method takes a considerable length of time to carry out one set of measurements, and such a large a complex set up is time consuming to move. At some archaeological sites, there may be very limited time available. In some cases, archaeologists restrict access in order to protect the site. In other cases, limited time is available where no other people (and resultant noise) are present, often due to the site being open to the public. Time may also be restricted due to the battery life of equipment where other power sources are not available, or because there is limited staff time available. An acoustic study in Malta by the author [18], for example, presented many of these issues, in part because the study explored the acoustics present in different parts of the monument.

Murphy discusses the work of Jahn et al. [19], which examined a number of UK passage grave or chamber tomb monuments similar in type to Maes Howe, such as Wayland Smithy. This earlier study swept the frequency of a sound source to identify modal frequencies with a hand-held soundmeter. It also attempted to calculate resonant frequencies of the spaces theoretically, with some limited success; although results were sometimes within 10% of the frequencies measured, they were never the same. This older method illustrates the advantages of the use of modern portable digital equipment, and techniques from the ISO standard. Jahn et al. do not provide the frequency response of the loudspeaker, and one cannot thus assess whether it accurately stimulated all frequencies; in addition, subjective opinion is used to identify the most powerful resonances present. Although various receiver positions are used, only one source position is used, and in trying to calculate theoretically the various parameters, approximations are made of the size of the spaces, with curved three-dimensional surfaces theorised as cubes. On the other hand, the study does effectively report a sense of human perception of the acoustics of a number of spaces.

The dodecahedron used in the author's initial field tests in caves in Spain was used with a subwoofer [4,15]. These speakers were large and heavy; the first cave being explored was at the top of a long climb up a muddy hill, and transporting the equipment was difficult. The dodecahedron needed a separate amplifier, and both it and the subwoofer required us to access a power supply in the depths of a cave. When we arrived at the entrance to the first site, the large size of the speakers meant that it was not possible to get the dodecahedron into the cave. Had we managed to manipulate the speakers through the entrance, transporting them down ladders deep underground would have proven similarly challenging. In other studies by the author, the size of equipment and access have also been problematic: a generator used to power equipment was noisy in one situation and affected recordings and testing; in another study, the generator broke down, meaning no further measurements could be taken; power was not available in a number of archaeological sites; moving a bulky loudspeaker set-up between measurements has proven time-consuming in archaeological sites, where available time is often restricted; and setting up and moving large equipment has proven difficult where there are fragile archaeological remains present.

Any single archaeoacoustic approach does not always work for all situations, no matter how rigorously it is created. When trying to understand the individual behaviour of the acoustics of an archaeological site, an approach is needed that is designed around specific research questions, and the individual conditions of the site, with qualitative as well as quantitative analysis where necessary. This is especially the case where one is exploring sites from the ancient past, for example, in prehistoric sites, where much is unknown or uncertain. In such cases, we must move beyond the fixity of standard methods, and design a methodology that suits the particular context and captures a range of experiences of acoustics in the space, while maintaining scientific rigour and using as much of the state of the art as possible. Sound archaeology is a young field, and approaches to methods used in research studies continue to need further exploration in order to balance these issues successfully.

2.2. *The Sites*

This project focuses on three archaeological sites: Caves in northern Spain, Stonehenge, and Paphos Theatre in Cyprus. The caves are part of the Cave of Altamira and Paleolithic Cave Art of Northern Spain UNESCO World Heritage Site, a large group of caves that feature Palaeolithic paintings and engravings. The caves explored are El Castillo, Las Chimeneas, La Pasiega, and La Garma in Cantabria, and Tito Bustillo in Asturias. They were selected by Dr. Roberto Ontanon (Director of the History and Archaeology Museum, Santander) to represent a range of different contexts. Stonehenge is a well-known Neolithic stone circle in Southern England, and also a World Heritage Site. Various phases of this site's development are examined. Paphos Theatre in Cyprus, or the Hellenistic-Roman theatre of Nea Paphos, was built around 300 BC, and was used until around 365 AD. It had seating for 8000 spectators and is again a World Heritage Site.

2.2.1. Caves

The caves studied feature paintings from as long as 43,000 years ago, a tradition of decoration with visual motifs that continues until the end of the Magdalenian period, 12,000 years ago. The five caves included have various features. La Garma was discovered comparatively recently in 1995, and is in pristine condition, as it was left when the entrance collapsed in prehistory. A point at the far end of the caves system from the original entrance featuring hand-shaped markings was selected for the present paper, next to a deep recess leading down to an underground river. Las Chimeneas features a small side chamber, with distinctive black paintings. La Pasiega was sampled in two positions, in a small turret-like feature filled with paintings, and at the end of a corridor that features many more paintings; overall, the cave features long narrow tunnels. These three caves are little changed and are closed to the public. El Castillo and Tito Bustillo are large publicly accessible show caves. El Castillo has a series of connected chambers and side sections, one of which was chosen for this study; this has the oldest dated cave painting, a red dot that is more than 40,000 years old [20]. Tito Bustillo has large chambers connected by corridors, and features small secluded side chambers; a position is featured here next to polychrome paintings of horses, again near a drop to an underground river. The Songs of the Caves multi-disciplinary project (SOTC) set out to explore relationships between the acoustics of the caves and positioning of paintings. Two existing publications [4,15] provide further contextual information about the caves as well as relevant archaeological references.

SOTC was inspired by the work of Reznikoff and Dauvois [21], who hypothesised that the positioning of cave art was related to the acoustics present at a particular position. Reznikoff [22] explored a number of caves in France and elsewhere, suggesting that paintings were placed in response to acoustic effects, including reverberation, echo, and low frequency resonance, which he stimulated with his low bass singing voice. His methodology was not published in detail; he did not include any kind of statistical study, providing indicative rather than conclusive or empirically evidenced results. While this is a somewhat speculative assertion, the evidence found by SOTC in Spain supports this hypothesis to some extent, establishing a statistical relationship between cave art and acoustics, and suggesting that in some of these examples, it is feasible that modal effects present at least contributed to the selection of these particular sites. In order to understand the relationship between acoustic effects and the placement of visual motifs, a wider scale comprehensive assessment of caves in Northern Spain and Southern France would be necessary, which would need to account for other notable features or causes that could have inspired decoration.

Diaz Andreu and Garcia Benito also explored the relationships between rock art and acoustics, and used a whistle and observation of what was heard in a range of source and receiver positions, to map acoustic effects present in an outdoor site featuring rock art. This approach lacked the ability to generate acoustic metrics, but identified evidence of a link between echoes and the presence of rock art in a Spanish valley [23]. More recently, Diaz Andreu led the Artsoundscapes project [24], and explored technical approaches to studying the relationships between rock art and sound [25]. In Scandinavia, Rainio et al. [26] have explored the relationships between rock art and acoustics.

SOTC captured more than 200 impulse responses in the caves, either next to Palaeolithic paintings or carvings, or in control positions with no visual motifs. Details of archaeological context were recorded in each position, and logistic regression analysis was used to examine whether there were statistical relationships between acoustical and archaeological features. The study found a statistical association: "between acoustic response and the positions of Palaeolithic visual motifs (...) in these caves. Our primary conclusion is that there is statistical, although weak, evidence, for an association between acoustic responses measured within these caves and the placement of motifs. We found a statistical association between the position of motifs, particularly dots and lines, and places with low frequency resonances and moderate reverberation" [15]. SOTC indicated that acoustics were in some way linked to the caves better-known tradition of visual motif-making. The earliest known human musical instruments were also found in Palaeolithic caves [27], such as the Hohle Fels and Isturitz vulture bone flute-like pipes that date from c.40 to 30,000 BC, and a range of evidence suggests that sound played an important role in the ritual life evidenced by the cave art in this period.

A detailed account of the acoustic methodology used for collecting impulse responses has been published [4,15], and these collected responses were used for generating the acoustic metrics for caves discussed in this study. Sine sweep test signals were generated through a portable powered loudspeaker that had been characterised in an anechoic chamber, using WinMLS software on a laptop, results captured with a soundcard and small diaphragm condenser microphones. A dodecahedron was not used as it would have been too large, a smaller Bang and Olufsen Beolit speaker provided a source with an appropriate balance of portability, fidelity, and directionality. Two high quality DPA microphones captured two impulse responses at a time. These models are designed for recording studio rather than acoustics use and were used as they are more robust than traditional acoustics capsules, while being appropriately flat in frequency response. As the system was characterised in an acoustics lab before use, the measurement system could account for any irregularities. Further information is available on the project website [28]. In the present paper, one illustrative position that features interesting acoustics as well as significant cave paintings was selected from each cave. This is intended to provide a qualitative comparison between the acoustics in these caves and in other sites, in order to indicate some general trends, and illustrate the range of acoustical effects present at archaeological sites.

2.2.2. Stonehenge

Stonehenge is a Neolithic monument, much of which was built between approximately 3000 and 2200 BC. Parker Pearson [29,30] describes the site as a place of, or for, the dead, and the site contains a number of significant burials. It was used in the past for ritual activities that are related to the winter solstice, and is to this day the site of solstice celebrations in summer and winter. Darvill [31] suggests that Stonehenge was related to healing; in many cultures, rituals relating to healing, astronomy, ancestor worship, or the dead are related, and many such rituals use sound and music as a key element [32]. Research on the acoustics of Stonehenge was inspired by a pilot study by Aaron Watson [33,34], which showed that there were acoustic effects present at the present-day monument. Further study by the author [35,36] analysed the acoustics present in the final arrangement of the original complete monument, pointing out low frequency resonances, echoes, and reverberation. Field studies by the author with the acoustics researcher Bruno Fazenda [37,38], analysed the Maryhill Monument, a full-size concrete Stonehenge model in the United States, which has an approximately similar design. That research identified higher reverberation than in the remaining monument, and a powerful low frequency resonance at 47-48 Hz; similar low frequency modal effects are discussed further below. The small amount of reverberation and echo present is something that was described long ago by Thomas Hardy in his novel Tess of the D'Urbervilles [39]. Fazenda confirmed the presence of weak echoes and a small amount of reverberation in a further study [37]. It focused on ISO standard approaches, and did not discuss in detail low frequency reverberation time (although this is reported in a table), nor did it discuss the differences between results from different source positions or phases of the monument. The present study focuses on Stonehenge itself, rather than the concrete part-replica in Washington, DC, USA.

Stonehenge was examined in three different arrangements, using constructed digital models that draw upon a range of archaeological sources, in particular those of Parker Pearson [29,30] and Darvill [31]. A joint article involving both these researchers [40] laid out the complex sequence of Stonehenge's construction. The phases selected in the present study demonstrate the difference between an early version of the space with a large circle of bluestones, an arrangement of Sarsen stones, and a representation of the final state of the monument. Rather than following one of the various existing archaeological numberings of phases, in the present paper, the models presented are referred to simply as Stonehenge digital models A–C. The first phase modelled by the team, dated c.3000 BC, is not discussed. Only a single large stone, the Heel Stone, may have been present at this time, along with a ditch and bank. This phase has few acoustic effects, other than that the ditch and bank isolate the space acoustically to some extent from what is outside. Three subsequent phases of building were digitally modelled graphically and acoustically. The first is dated c.2900 BC (Stonehenge A), and sees the introduction of Bluestones, mostly spotted dolerite from Wales, which were erected in a large circle in what are now called the Aubrey holes, inside the Stonehenge circular ditch and bank (a reverse arrangement of a bank enclosing a ditch, which is what characterises a henge). This was a single wide circle of stones. Each stone reflects sound, and the combined result creates unusual effects in the centre, and varied acoustics across the space. A second digital model based on the arrangement of stones at around 2500 BC (Stonehenge B) sees the Bluestones removed, and replaced with large local Sarsen stones, including huge pairs of upright stones called Trilithons with lintels across the top, and an outer ring capped by a ring of lintels. A third and final phase is modelled, dated at 2200 BC (Stonehenge C), which sees the Bluestones returned in circular patterns amongst the Sarsen stones.

As many of Stonehenge's stones are fallen or missing today, digital modelling was used to create impulse responses representing what the space sounded like in the past. A model of Stonehenge was created by this author and colleagues [41] using point clouds of individual stones provided by English Heritage [42]. DXF (a standard digital graphical modelling format) graphical files of this new model were imported into Odeon acoustics software, which then generated impulse responses for pairs of source and receiver positions. These impulse responses were originally generated for use in convolution-based auralisation, within an interactive graphical model called the EMAP Soundgate [43], offering a phenomenological experiential approach to exploring the space. This was carried out as part of the European Music Archaeology Project (EMAP), a five year co-operation project carried out with the support of the Culture Program of the European Union. All of the sites in this paper were interactively modelled audio-visually within this app, allowing the user to see and hear what it might have been like to be in these sites in the past. The impulse responses were imported into Odeon software, which was used to generate a range of acoustic metrics describing the site quantitatively. These metrical results are discussed within this article.

Impulse responses of a number of different receiver positions were generated to gain a basic understanding of the differing acoustics present. A source position in the centre of the circle was used for each Stonehenge model. In the Stonehenge A model, a number of receiver positions were chosen running in a line from just inside the stone circle to the centre, to understand how sound changed as one moved through the space. In the Stonehenge B and C models, a receiver position was chosen next to the Heel Stone outside the stone circle, because the author had observed echoes at this position onsite at the moment. Two further receiver positions were chosen, Centre Source at the centre of the circle, and Inside Source inside the monument, towards the Heelstone but still inside the stone circle (see Figure 1).

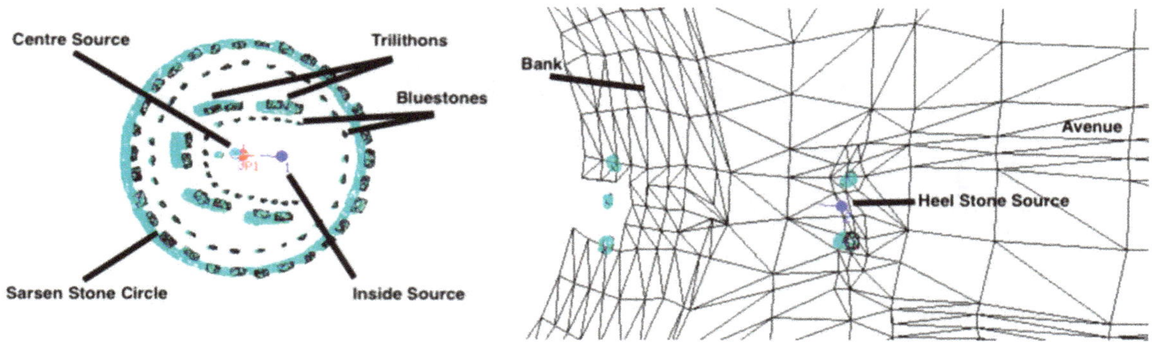

Figure 1. Source and receiver positions in the model of Stonehenge C.

2.2.3. Paphos Theatre

A monochrome archaeological digital model reconstruction of the Paphos Theatre, based on archaeological field work, was provided by the Cyprus Institute [44], and features research carried out by them with the University of Sydney [45]. Paphos was the capital of Cyprus under the Ptolemaic and then Roman administrations, the theatre was used for performance and entertainment for over six and a half centuries (c. 300 BC to the late fourth century AD). At its maximum extent during the reign of the Antonine Emperors of the second century AD, the theatre could seat over 8500 spectators. The acoustics of Greek and Roman theatres have been explored as part of the Identification, Evaluation, and Revival of the Acoustical Heritage of Ancient Theatres and Odea (ERATO) project, and in a range of publications [46–49]. The author of the present paper worked with ERATO researcher Jian Kang as part of the Acoustics and Music of British Prehistory Research Network [50], and the work of ERATO, in particular the use of raytracing-based acoustic modelling software ODEON, informed the present study. Similar to the ERATO Project, the Conservation of the Acoustical Heritage y the Revival and Identification of the Sinan's Mosques' Acoustics (CAHRISMA) project focuses on the acoustics of mosques [51].

As was the case for Stonehenge, ODEON software was used to generate a number of impulse responses and acoustic results using the digital model of Paphos Theatre. A source position on the stage was used, along with four receiver positions. Four receiver positions were chosen, in the centre of the space in front of the stage, on the third row, on the fourteenth row, and on the back row (see Figure 2). Acoustic metrics were generated for each source and receiver pair in Stonehenge and Paphos models, and for each impulse response collected in each cave space.

By the Roman period, the writings of Vitruvius [52] and the development of theatres, Odea, and other venues show that there is in Roman culture by this period an understanding that architectural design can influence the acoustics of a space. Not only do Romans appreciate acoustics, but they attempt to intentionally control them. At Stonehenge, there is no evidence that architectural design of the site is selected with the intent of controlling acoustics, but acoustics in the space change as the monument develops. Acoustic effects are noticeably present and would impact upon ritual activities in the space. In Palaeolithic caves, there is some relationship between acoustics and the paintings that are present, which suggests an appreciation of acoustics and sound in the past [15]. The acoustic ecologies of human ritual culture developed over time, from an appreciation of the acoustics of natural spaces, such as caves, to human-generated acoustics in buildings, such as Stonehenge, that were created for ritual purposes, perhaps taking advantage or notice of the acoustics present. Eventually, ritual sites feature deliberate attempts to manipulate acoustics in order to support speech and music. These sites illustrate the change of human ritual acoustic ecologies over time, as well as their variety. How the acoustics of these sites change is discussed below.

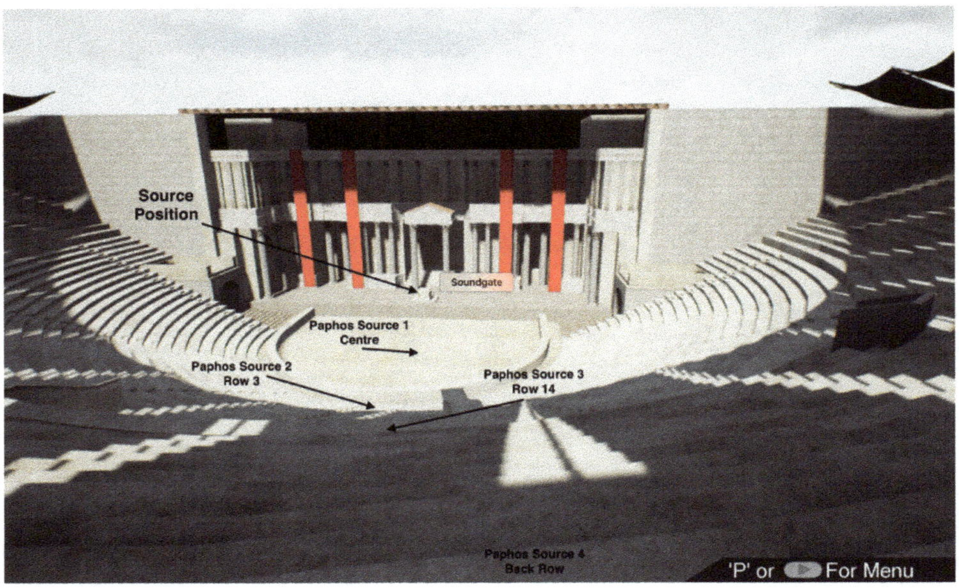

Figure 2. Source and receiver positions in the Paphos Theatre model.

2.3. Acoustic Metrics and Normal Values

Ahnert and Schmidt [53] provide a useful overview of acoustic metrics and normal values. They discuss overall parameters, including reverberation time (T30), early decay time (EDT), and bass ratio (BR); metrics related to speech, such as definition (D50), clarity (C50), articulation loss (ALcons), speech transmission index (STI or RASTI), and echo criterion for the perception of annoying reflections (EKspeech); and metrics related to music, such as the direct sound measure of the sensation of directness and nearness of the sound source (C (7)), and clarity of transparency of musical time and register (C (80)). Table 1 describes typical values of these metrics. For example, C (50) should be greater than or equal to −2 dB to avoid syllable intelligibility decreasing below 80%; this is the bottom admissible limit value for good speech intelligibility. Overall, long reverberation reduces articulation of consonants, and thus intelligibility.

In order to assess how the acoustics of ritual sites in human culture have changed over time, results for a number of these metrics in a number of sites are compared. Initially, early decay time (EDT) is discussed. Barron [54] (p. 330) compares EDT at a number of concert auditoria, and these results are used to provide contrast and a point of reference relevant to contemporary readers. As discussed above, EDT is representative of perceived reverberation [12] (p. 2), related to specific source and receiver positions. Reverb time (T30) in comparison represents the overall reverberation of the space. Comparing Barron's EDT results to those from the archaeological sites being studied is illuminating, in order to illustrate in what ways ancient and modern sites are similar and different. These results are used as a benchmark against the caves, Stonehenge, and Paphos. EDT is compared at five octaves, from 125 to 2000 Hz, as well as in two averaged results that are used in standard acoustics approaches (125–1000 Hz and 500–1000 Hz). The two cave spaces that are the closest to the concert halls are selected for comparison. The percentage difference between EDT in a selection of concert halls and in each ancient site are compared with the just noticeable difference of EDT, to provide an objective standard acoustic comparison. The variety of EDT in concert halls and caves is also illustrated. This method provides an interesting reference point and satisfactory results.

Table 1. Typical ranges [53] p. 12.

Clarity Metric	Type	Typical Range
C (7)	Directness/nearness of musical sources	>−10 to −15 dB
C (50)	Clarity (speech)	>−2 dB (>3 dB is good)
C (80)	Clarity (music)—classical music	>−1.6 dB
C (80)	Clarity (music)—romantic music	>−4.6 dB
C (80)	Clarity (music)—sacral music	>5 dB
Metric	**JND**	**Typical Range**
G	1 dB	−2 to +10 dB
EDT/T20/T30	5%	1 to 3 s
C80	1 dB	−5 to +5 dB
D50	0.05	0.3 to 0.7
Articulation Loss of Consonants (ALcons)		
ideal intelligibility		≤3%
very good intelligibility		=3% to 8%
good intelligibility		=8% to 11%
poor intelligibility		>11% to 20%
worthless intelligibility (limit value 15%)		>20%
Syllable Intelligibility		**RASTI Value**
Poor		0 to 0.3
Satisfactory		0.3 to 0.45
Good		0.45 to 0.6
Very Good		0.6 to 0.75
Excellent		0.75 to 1

Subsequently, EDT results at Paphos are compared. Results at four receiver positions are examined, as well as averaged results from these positions. EDT is then analysed in some detail at Stonehenge, the results illustrating the difficulties of standard methodology in non-standard spaces. Barron examines frequency octaves at 125, 250, 500, 1000, and 2000 Hz, whereas results for the ancient spaces are presented here an octave lower at 63 Hz and two octaves higher at 4000 and 8000 Hz. Individual octave results are explored, as are individual receiver positions, rather than a more standard approach of averaging a number of results. Frequency responses, reflection geometries, and impulse response waveforms are also examined. Finally, results at all sites are presented together, in order to address differences between EDT in different sites and eras.

A number of other metrics are then addressed. Reverberation time (T30) is discussed, showing how non-standardised distances between source and receiver positions can provide varied results. Clarity (C50) is presented at octave bands between 63 and 8000 Hz, in order to appreciate the variety at different sites and positions. Individual results for loss of consonants (ALcons), bass response (BR), and speech intelligibility (RASTI) are also presented for all sites, in order to compare a number of acoustic features. These results are then used to reveal changes in acoustics over time.

3. Results

3.1. *Early Decay Time*

Mostly, the caves have lower EDT times than modern concert halls, as can be seen in Table 2, but in some cases, they are remarkably similar. La Pasiega Turret has EDT that is similar to that at Royal Festival Hall, Wigmore Hall, Assembly Hall, and Free Trade Hall, and that is larger than EDT at Wembley Conference Centre.

Table 2. Difference in EDT between a position in La Pasiega cave, and a number of British concert halls.

Space	EDT 125	EDT 250	EDT 500	EDT 1000	EDT 2000	Average 125–1000 Hz	Average 500–1000 Hz
La Pasiega Turret	1.92	1.62	1.58	1.4	1.16	1.54	1.49
Royal Festival Hall, London	1.33	1.37	1.43	1.57	1.66	1.47	1.5
% difference	−31	−15	−9	12	43	−4	1
Wigmore Hall, London	1.49	1.61	1.55	1.57	1.48	1.54	1.56
% difference	−22	−1	−2	12	28	0	5
Assembly Hall Watford	1.53	1.74	1.58	1.47	1.25	1.51	1.525
% difference	−20	7	0	5	8	−1	2
Free Trade Hall, Manchester	1.47	1.47	1.65	1.77	1.62	1.60	1.71
% difference	−23	−9	4	26	40	4	15

The just noticeable difference of EDT is 5% (see Table 1), and the differences between EDT in the caves and concert halls are listed in Table 2. EDT at a number of frequencies in a number of modern venues is less than 5% different to that measured in La Pasiega, and thus not noticeably different. The average of EDT between 125 and 2000 Hz in Royal Festival Hall and Free Trade Hall, two well-known UK concert venues, is only 4% different to that measured in the La Pasiega Turret. EDT in Assembly Hall is only 1% different, and in Wigmore Hall is even more similar, only 0.26% different. If one averages EDT at only 500 and 1000 Hz, which is a standard approach in acoustics, then Royal Festival Hall is only 1% different, whereas Free Trade Hall is, in this case, 15% different, which is noticeably different. With the latter approach, Wigmore Hall is 5% different, and Assembly Hall is 2% different (Table 2). It is somewhat surprising that a cave made up of a corridor with a diameter and height of approximately 2 m, can produce reverberation which is not perceived as different to a number of modern concert halls.

Table 3 shows the equivalent results for Tito Bustillo cave. When averaged across 125 to 1000 Hz, the differences in EDT between the cave and the halls are similar to those observed at La Pasiega. Across this frequency range, EDT values in three of the halls are within 1% to 2% of that in Tito Bustillo, below the just noticeable difference. Free Trade Hall has 6% higher EDT. The difference is greater when averaged only across 500 to 1000 Hz; in this case, the differences are noticeable, between 11% to 26% different.

Table 3. Difference in EDT between a position in Tito Bustillo cave, and a number of British concert halls.

Space	EDT 125	EDT 250	EDT 500	EDT 1000	EDT 2000	Average 125–1000 Hz	Average 500–1000 Hz
Tito Bustillo	2.24	1.11	1.36	1.35	1.46	1.50	1.355
Royal Festival Hall, London	1.33	1.37	1.43	1.57	1.66	1.47	1.5
% difference	−41	23	5	16	14	−2	11
Wigmore Hall, London	1.49	1.61	1.55	1.57	1.48	1.54	1.56
% difference	−33	45	14	16	1	2	15
Assembly Hall Watford	1.53	1.74	1.58	1.47	1.25	1.51	1.525
% difference	−32	57	16	9	−14	1	13
Free Trade Hall, Manchester	1.47	1.47	1.65	1.77	1.62	1.60	1.71
% difference	−34	32	21	31	11	6	26

In Figure 3, one can see that many of the concert halls have frequency responses that are flat to different degrees. Some other halls have less flat responses, as shown in the right-hand graph. Of these, Wessex Hall and Butterworth Hall have since undergone changes that have improved their acoustics, and Royal Concert Hall now has adjustable acoustics. Before these changes, these spaces exhibited stronger low frequency EDT than at higher frequencies; this is usually a sign of modal behaviour, which means that the acoustic effects present vary in different parts of the space. For audiences, this means a different experience depending on where one sits, something that concert halls aim to minimise. Wembley Conference Centre is not designed solely as a concert hall but is used regularly as a performance space. Tito Bustillo has reverberation that is not perceptibly different to that of some of these concert halls; this time, the similarity is less surprising, as the cave is the largest studied, a cavernous open stone area. The variety present in the caves was very noticeable in situ, ranging from very dry side chambers with very little reverberation (as low as 0.11 s EDT 1000 Hz), through to large open spaces with large cathedral-like acoustic sustain (above 2.5 s EDT 1000 Hz) [15]. In La Pasiega in particular, source–receiver separation made a significant difference to perceived reverberation, and the extremely low level of background noise present (caves are very quiet) meant that reverberation seemed longer, as it could be heard for longer before being masked by environmental noise, and quite subtle effects that would be unheard outside were audible.

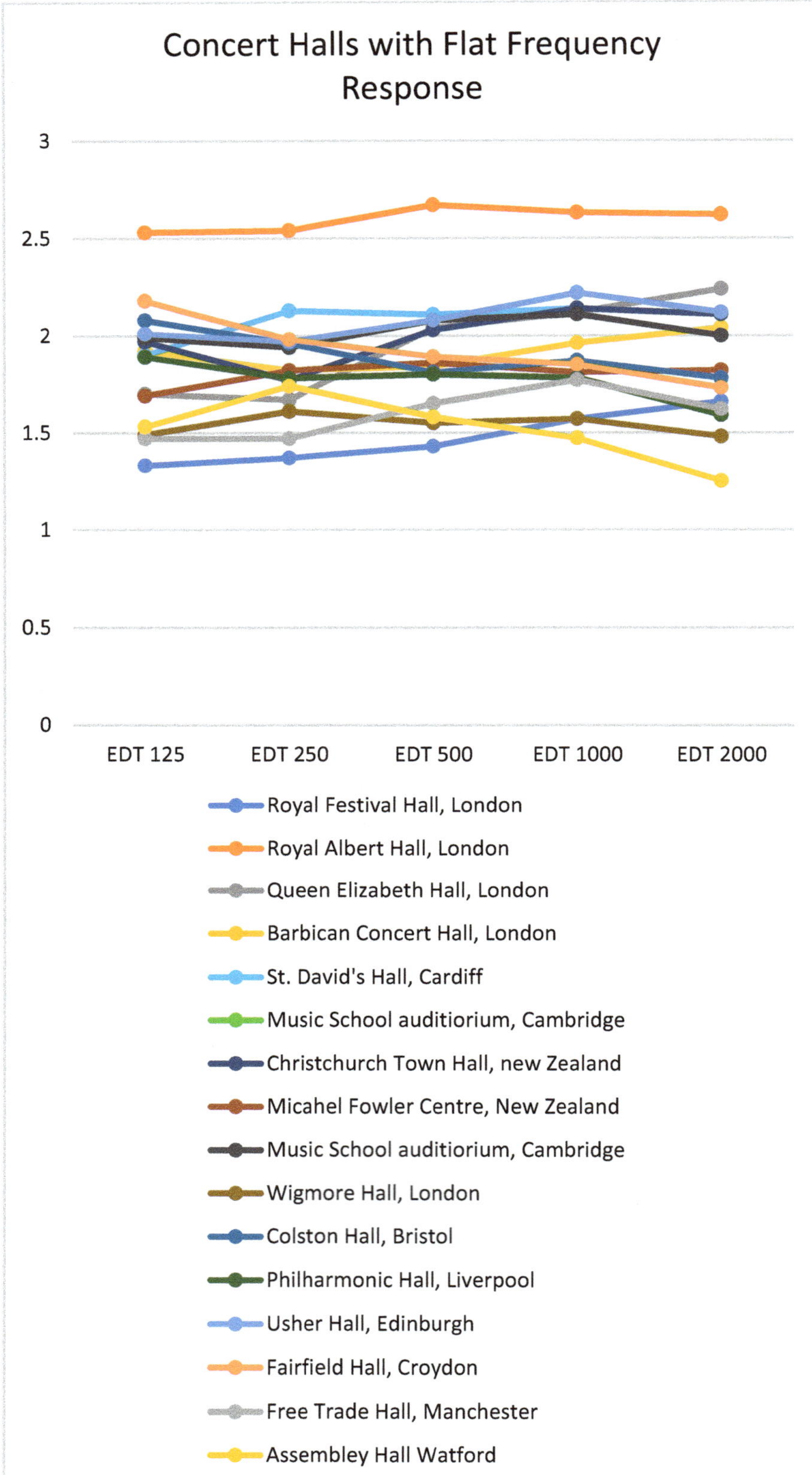

Figure 3. *Cont.*

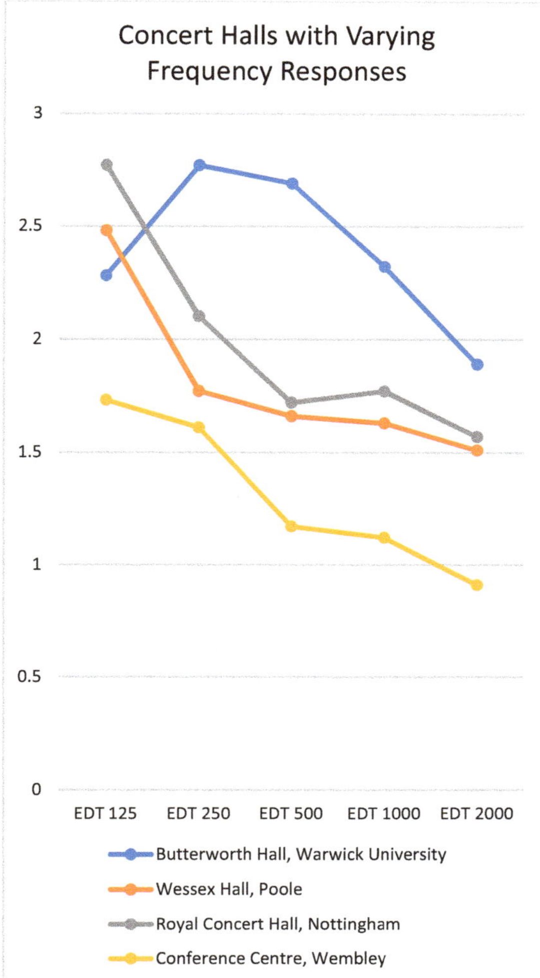

Figure 3. EDT at different frequencies in a number of British concert halls.

As can be seen in Figure 4, the caves have a range of frequency responses, but many show signs of modal resonance, evidenced by higher EDT at low frequencies. This is the case at Tito Bustillo, where EDT is noticeably different to the four concert halls discussed when one considers lower frequencies, rather than just 500 to 1000 Hz. Interestingly, the strong low frequency resonances in Tito Bustillo and La Pasiega would be missed if this study had drawn its method from the ISO standard, and only focused on 125 Hz and above. Further work is needed to explore the sound archaeology of the caves to understand the individual behaviours of the acoustics in these spaces in more detail. The existing paper by the SOTC team [15] principally studies their statistical relationships to motif positioning, further research is needed to examine the nature of the acoustic ecology of the various spaces qualitatively. The small chamber in Las Chimeneas was observed by the author when in the space to have had a noticeably powerful low frequency response. Indeed, this is the reason for its selection for use in the app discussed above [43]. Figure 4 shows high values for EDT in the 63 Hz octave, and that at higher frequencies, EDT values tail off. Las Chimeneas has some of the most dramatic images the author experienced in the caves studied (Figure 5). El Castillo has a very similar value for EDT at a low frequency of 63 Hz. El Castillo is a large cave, and the impulse response discussed here was captured in a side chamber next to a panel featuring paintings. It is likely that it is the volume of this side chamber that produces its corresponding low frequency resonance. The panel features a red dot that is over 43,000 years old and is a good example of the statistical association found by SOTC between the position of dots and lines, and low frequency resonances in particular [15] (p. 1347).

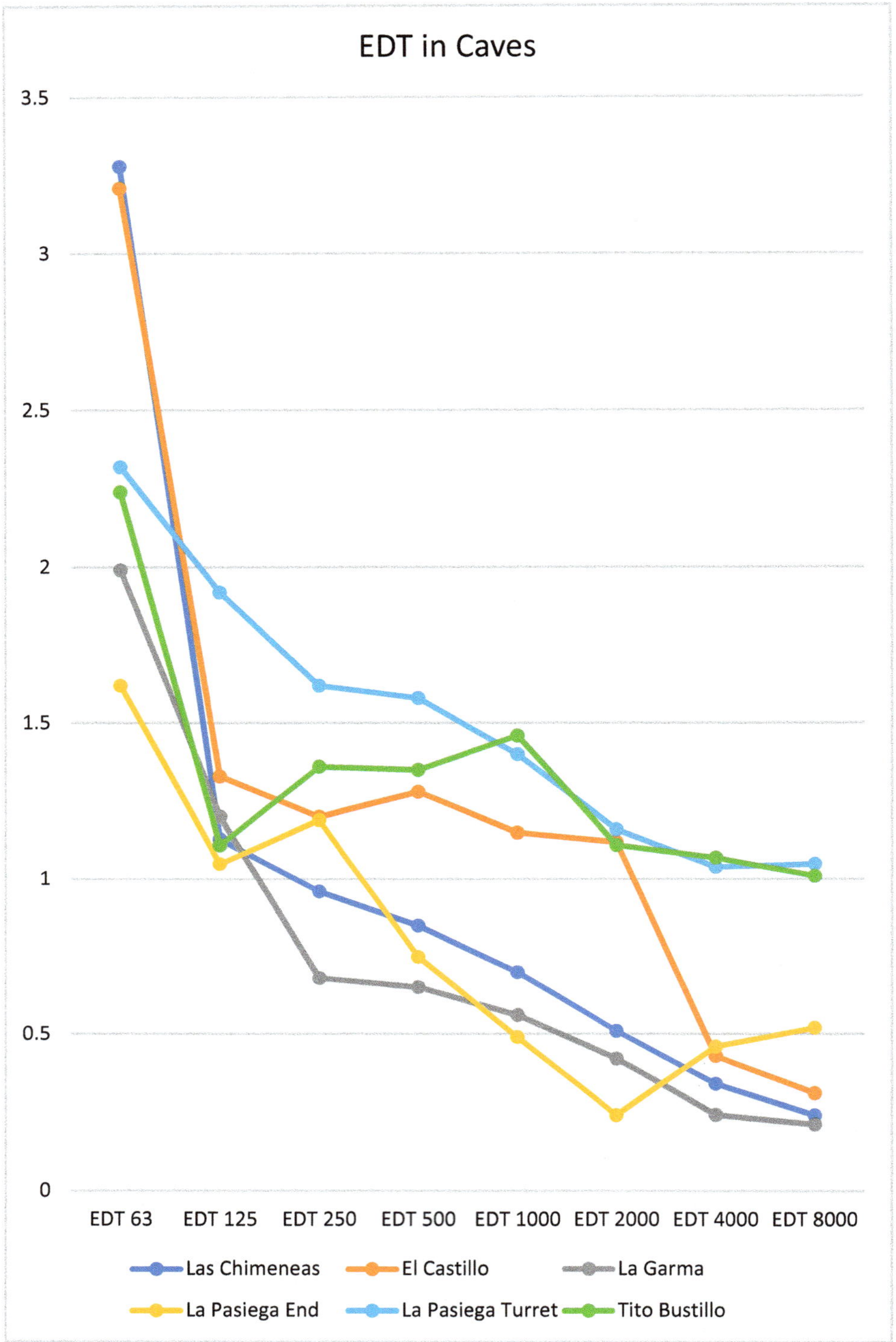

Figure 4. EDT in caves at various octave frequencies.

Figure 5. Painting of a stag from Las Chimeneas cave, image taken from the Soundgate app.

Paphos Theatre has values of EDT that are acceptably flat across a range of frequencies (Figure 6) to an extent that would make it acoustically acceptable for use as a modern concert hall. Indeed, a number of Roman amphitheatres are still in use as concert halls. Most concert halls are enclosed spaces, have a roof, but use wood or carpet for flooring. As an open roofed space, one might expect Paphos Theatre to have lower reverberation, but the stone construction and marble cladding increases reflection strength, extending reverberation time, and the stone seating contributes a stepped range of temporal reflections [47,48]. When compared across the 125 to 2000 Hz range (Figure 7), Paphos has slightly less high frequency EDT than three concert halls, but overall, it has very similar EDT values.

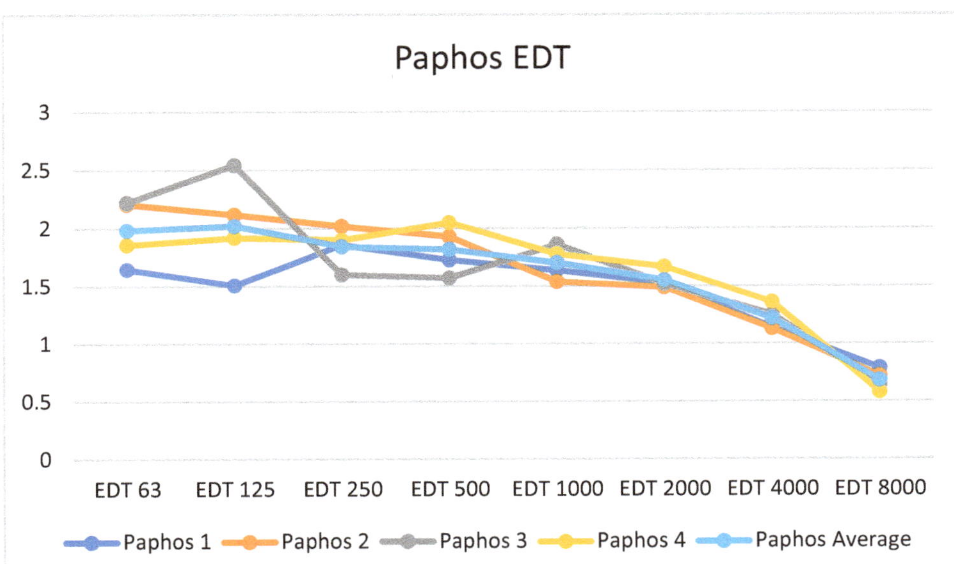

Figure 6. Early decay time (EDT) at Paphos Theatre in different source positions and at a range of frequencies.

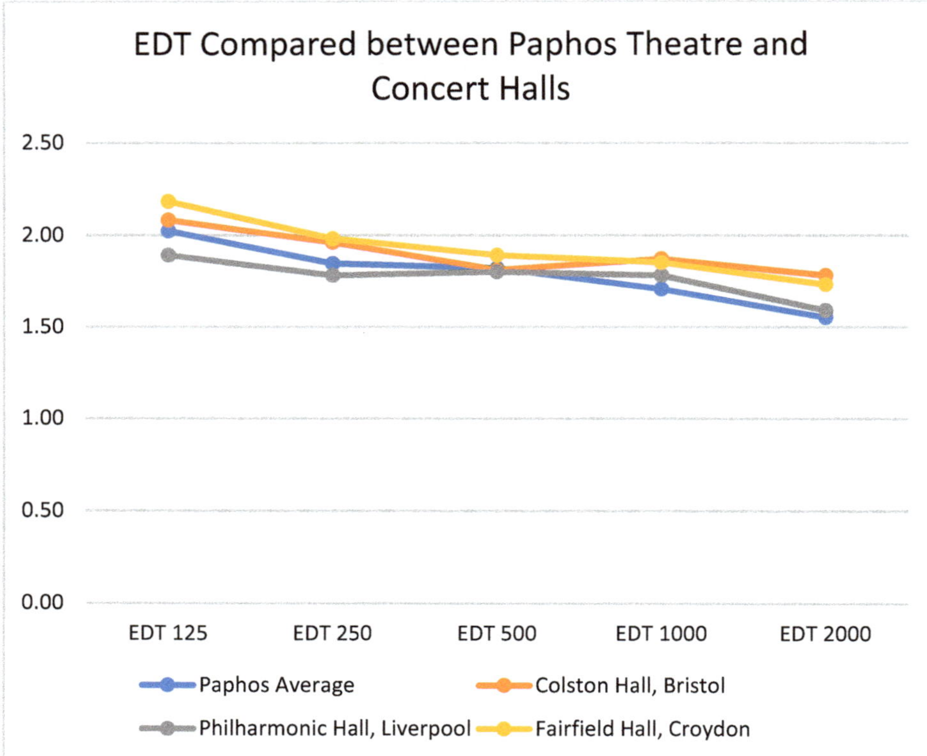

Figure 7. Comparison of EDT at Paphos Theatre with modern concert halls.

Philharmonic Hall in Liverpool is the closest match to Paphos (Table 4); the only noticeable difference between the two is in the 125 Hz octave band, Paphos has more lower frequency support, and perhaps slightly more modal behaviour. The average EDT across 125 to 2000 Hz, and the average for the standard 500–1000 Hz range is not noticeably different (2%, 1%). An audience would not notice the difference between the perceived reverberance in the two spaces according to these measures, but as we will see later, there are other differences between the acoustics of the spaces.

Table 4. Comparison of EDT at Paphos and at Philharmonic Hall in Liverpool.

	EDT 125	EDT 250	EDT 500	EDT 1000	EDT 2000	Average 125–2 kHz	Average 500–1 kHz
Paphos Average	2.02	1.85	1.82	1.71	1.55	1.79	1.76
Philharmonic Hall	1.89	1.78	1.8	1.78	1.59	1.77	1.79
% difference	−7	−4	−1	4	2	−1	2

EDT is quite different in Stonehenge (Figure 8). In Stonehenge A, a large circle of Bluestones from Wales dated to c.2900 BC [40] and with a diameter of c.87 m, there are low results for EDT both at the centre and inside the circle near the edge. The circular shape seems to cause a particular focusing of sound, with circular modes of resonance appearing. At the edge of the circle, there is a strong low frequency resonance, with an EDT of 1.31 s at 63 Hz and 1.36 s at 125 Hz, which drops off above this frequency to a very low result of 0.02 s at 2000 and 8000 Hz. This would result in a booming sound when stimulated by low frequency audio content, such as drums, thunder, or a strong wind.

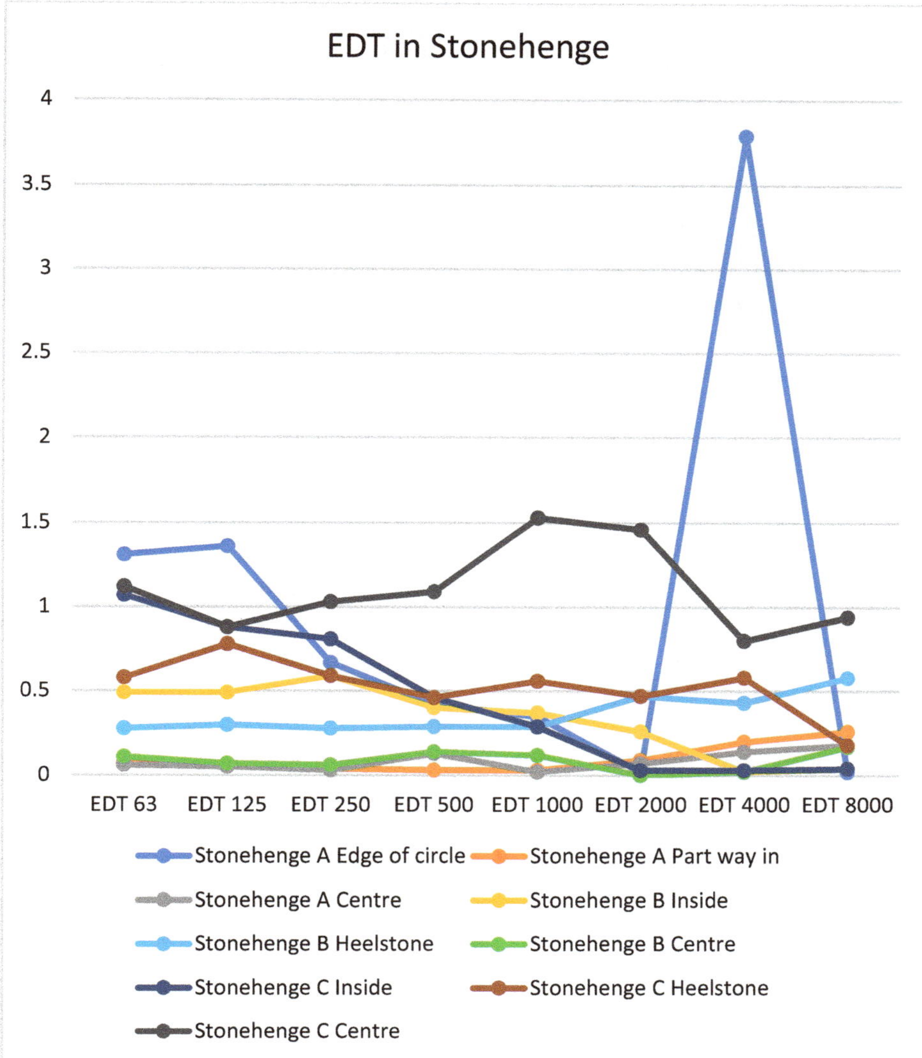

Figure 8. EDT in Stonehenge.

In the 4000 Hz octave band, there is a very high EDT result of 3.79 s. Digital modelling can cause strange results, and this figure is so extreme it seems possible this result could be a modelling error. Scattering and diffusion effects might reduce this resonance, but it seems from these results that there is a particular boost of frequencies in the 4000 Hz range. It is worth noting that 4000 Hz is where human hearing has a peak of sensitivity, and this frequency would particularly affect how one hears voices. Circular spaces can generate odd acoustic focuses and resultant modes, so this could be the result of a specific acoustic focusing. The space certainly has unusual acoustic effects that are present. This paper presents an outline of the acoustic features present, but a more in-depth investigation is warranted than is possible here to explore this and other features, and this is a suggestion for future research. Figure 9 shows the frequency response of Stonehenge A at the edge of the circle. Between 2 and 5 kHz, the space amplifies those frequencies by as much as 15 dB, whereas between 6 and 8 kHz, the level is 30 dB lower. Higher frequencies of 12 and 18 to 19 kHz are also boosted.

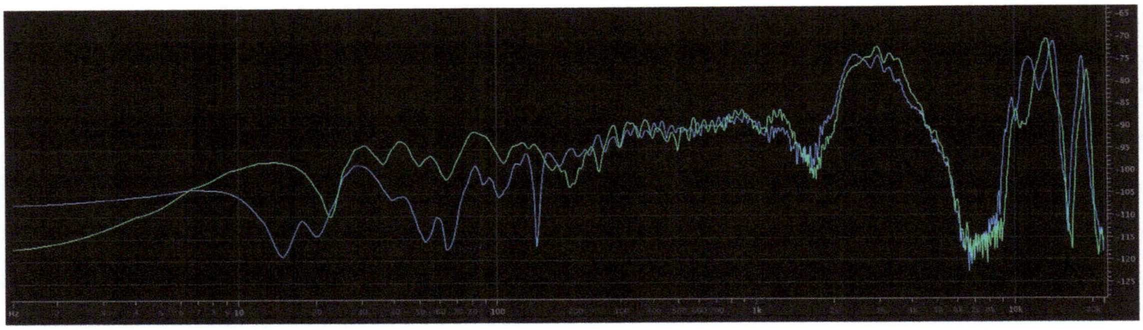

Figure 9. Stonehenge A frequency response at the edge of the circle.

Moving onwards in time to around 2500 BC, in the Stonehenge B model, the large circle of bluestones set in the Aubrey Holes is removed, and a smaller ring with a diameter of about 30 m made up of larger local Sarsen stones replaces them in the centre of the space. This arrangement is taller, with a number of stones also set inside an outer circle. It is a more complex monument, and it has less dramatic acoustic features. Low frequency resonances are present to a smaller extent, with a stronger low frequency response at the edge of the circle than at the centre (see Figure 8). The centre has very low results for EDT, and the edge of the space has EDT averaging only 0.3 to 0.4 s. This space has only subtle acoustic effects present.

The acoustics of the later (final) modelled arrangement of the site, Stonehenge C, dated c.2200 BC, has been examined in earlier publications by this author and Fazenda [35–38]. Those papers used various theoretical approaches to explore the acoustics of the site, as well as acoustic field tests at a full-size concrete replica that has a design based on Stonehenge, the Maryhill Monument in Washington State, USA. It found a T30 of 1.5 s and EDT of 1.8 s, as well as observing echoes. Two neighbouring circular/cylindrical modal resonances at approximately 47–48 Hz were identified at the Maryhill Monument by using a sine wave generator with variable frequency, as well as being calculated theoretically based on the shape of Stonehenge. In the corresponding modelled acoustics in this present study, a low level of EDT is found outside the circle, alongside little low frequency energy (see Table 5). At the centre of the circle, the EDT is over 1 s at 63 Hz, and then rises, with focused reflections from the surrounding circular stone walls generating reverberation. With an average EDT of 1.3 s (500–1000 Hz) or 1.4 s (500–2000 Hz), this suggests there would be clearly noticeable reverberation at the centre of the circle, similar results to those measured in Maryhill. A presentation of the octave band EDT does not, however, tell the whole story.

From the plotted frequency responses (Figures 10 and 11), one can see the specific resonant frequencies present. At the edge of the circle, individual low frequencies are strong, suggesting modal low frequency resonances below 200 Hz. This frequency response becomes flatter above this point, although with individual frequencies having strong resonance, for example, at around 400 Hz. A standard EDT reverberation measure based on average values at 500 and 1000 Hz would be 0.38 s, and an average of 500 to 2000 Hz is even lower at 0.26 s. EDT at 63 Hz has a value of 1.07 s when measured near the edge of the circle in the modelled acoustics, and the value would most likely be higher still at 47–48 Hz. The frequency response plot of the impulse response shows various frequencies having strong dips in EDT, perhaps caused by nodal points, which are as strong an indication of modal resonance as high values. Above 10 kHz, the EDT is very strong, another result missed by standard averages. This illustrates an issue with standard acoustic methods, in a space such as this that has odd acoustic behaviour. The space has unusual resonances at low and high frequencies, as well as less prominent but idiosyncratic mid-range responses.

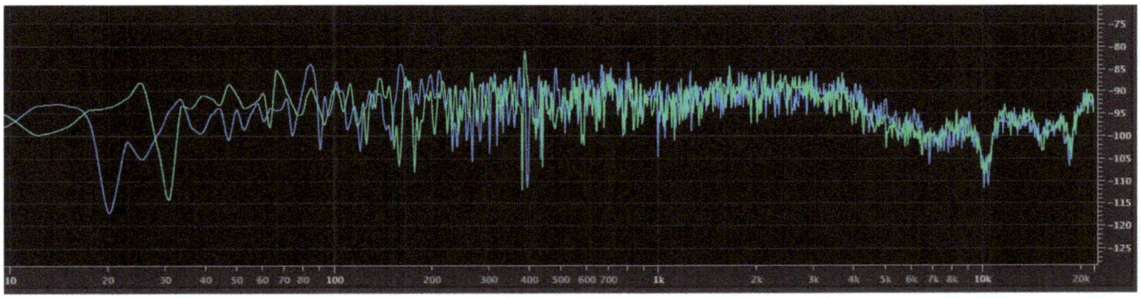

Figure 10. Frequency response of Stonehenge C digital model at the edge of the stone circle.

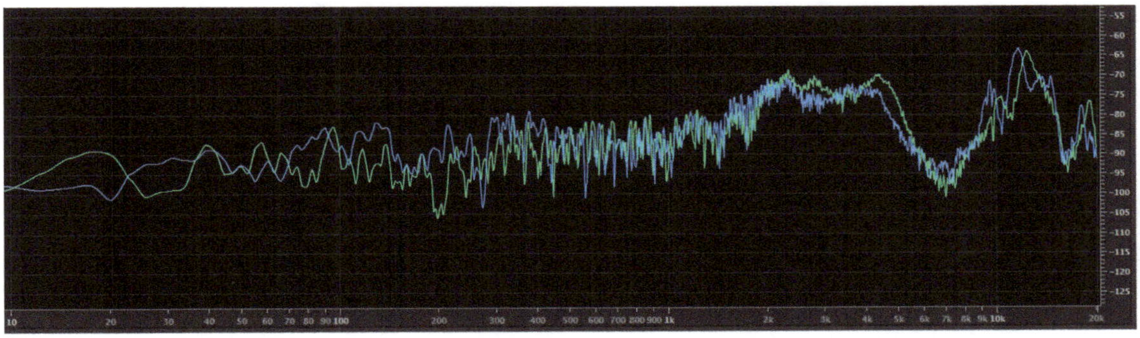

Figure 11. Frequency response of Stonehenge C digital model at the centre of the stone circle.

In both Stonehenge itself and the Maryhill Monument, the author found that speech was subjectively particularly clear, and was remarkably so at specific source and receiver positions. When walking around the space, suddenly one would hear a voice from another part of the circle with great clarity, despite coming from some distance. It is suggested that this effect is due to circular focusing, which may take place in elliptic, circular, parabolic, or hyperbolic patterns (Figure 12) [55], depending on the source position.

Figure 13 overlays an image of Stonehenge model C over these focusing effects. Elliptic focusing would mean that the voice of someone standing just inside the circle, between the arms of the horseshoe shape of Bluestones, might find their voice focused between the stones of the largest stone trilithon, just past the altar stone. Archaeologists now suggest that the horseshoe shape illustrated here was actually an complete oval shape, and the source position illustrated would be just inside this oval, which conveniently marks out where one might stand in the space to achieve elliptic focusing effects. The central position in Stonehenge could produce a particularly strong focusing of sound for someone standing there. Making sound just in front of the largest trilithon would produce parabolic focusing, adding support to a voice or other sound source. The focusing could also work in reverse, and sound from far off may be easier to hear. A source just inside the Bluestone circle, the second circle inside the outer sarsen circle but outside the horseshoe shape of Sarsen trilithon pairs, may have created hyperbolic focusing, which could make a sound made just inside the stone circle, or when hidden behind the largest trilithon, appear to come from outside the stone circle. When one explores these focusing effects on the larger Stonehenge A model, this position of focus outside the circle is at the position of the Heel Stone. The two circles of stones would produce combined focusing effects, and with a number of people present in the circle, acoustic focusing effects could have been complex, unpredictable and confusing, subtle but certainly noticeable, and possibly considered by Neolithic people to be supernatural in a monument to ancestors within an animist culture.

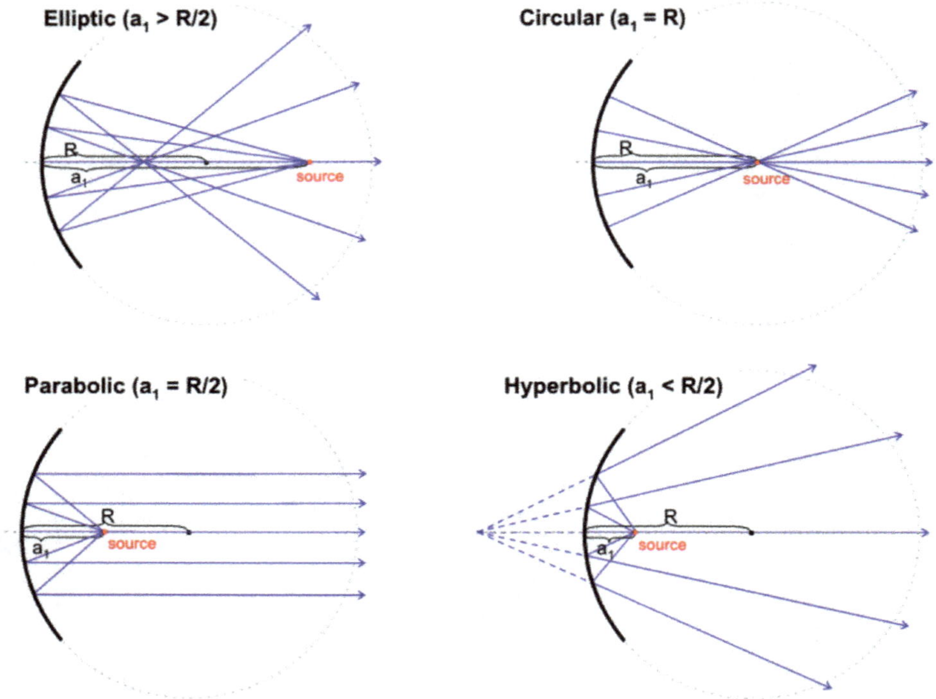

Figure 12. Focusing effects in a circle.

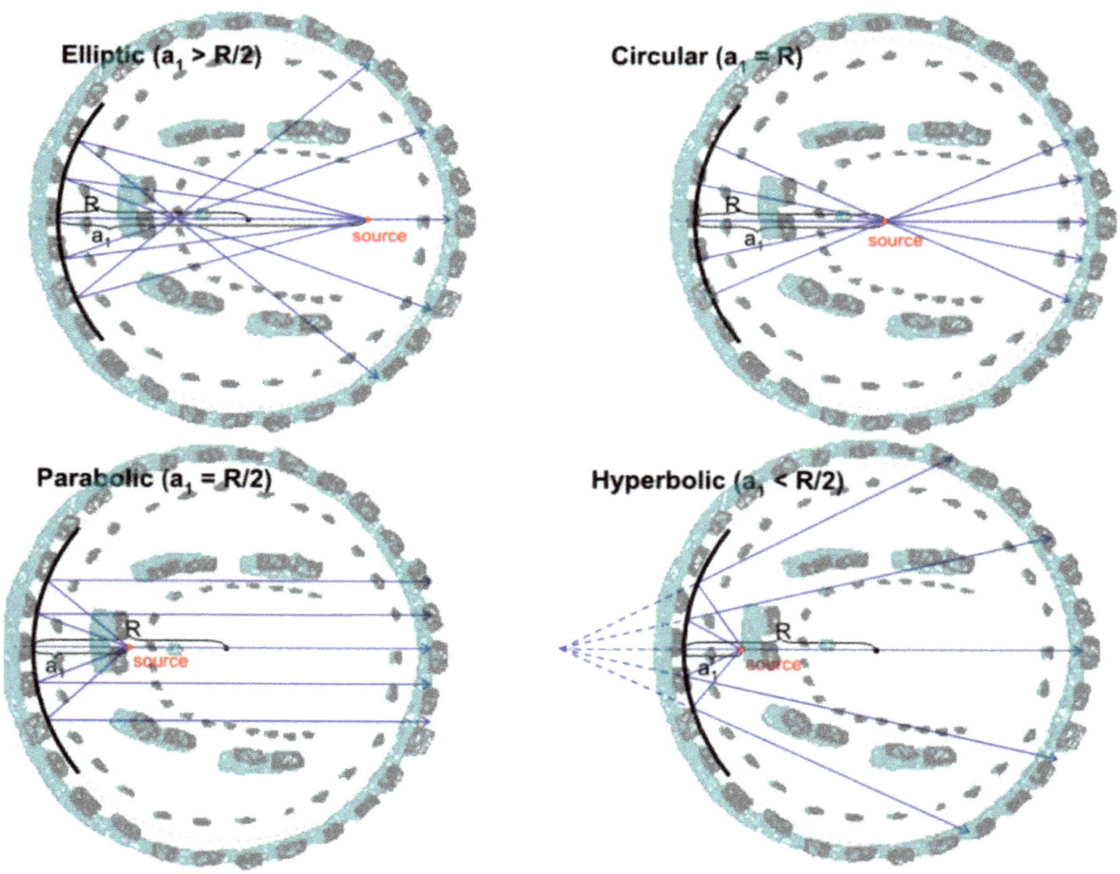

Figure 13. Possible circular acoustic focusing effects at Stonehenge.

Figure 13 also illustrates how sound bounces around the space, creating echoes. These echoes can be seen in the representations of impulse responses in the space that can be seen in Figure 14. Figure 15 and Table 5 illustrate EDT in various spaces. Overall, Stonehenge has acoustics that are irregular, and are very different to those at Paphos. They are closer in nature to some of the caves discussed, in that both have strong low frequency effects, as well as a non-linear acoustic response. The caves have varied EDT results, some of the which are closer to those in Paphos.

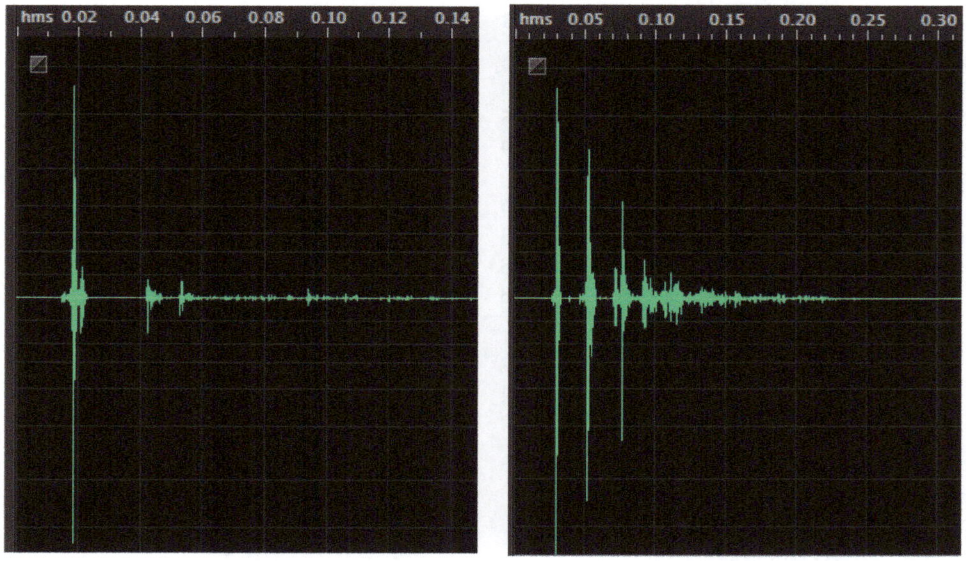

Figure 14. Waveform of Stonehenge C digital model at the edge (**left**) and centre (**right**) of the stone circle.

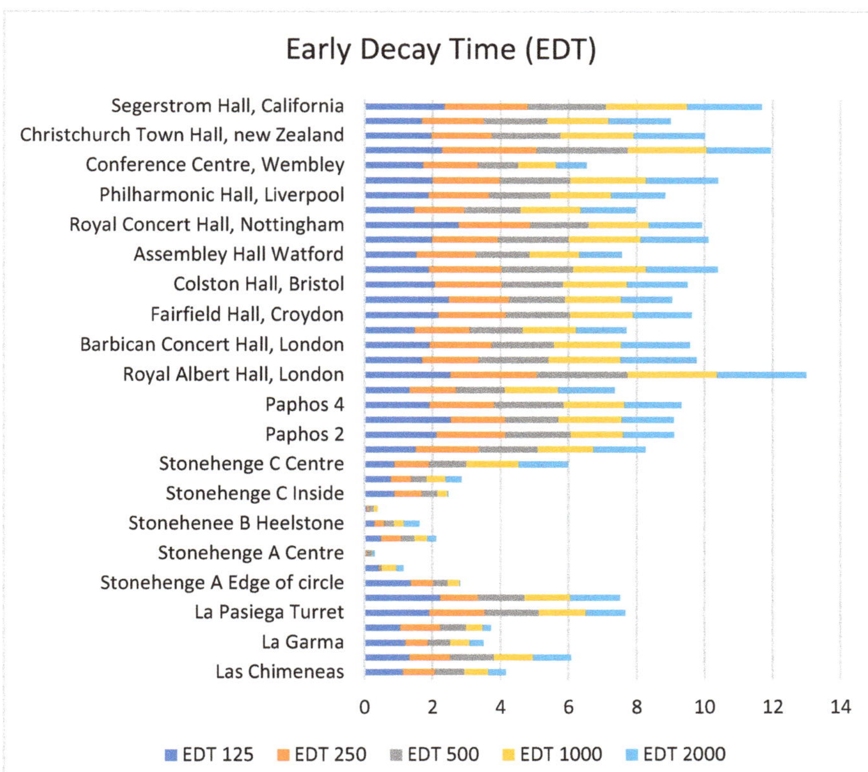

Figure 15. EDT in various sites.

Table 5. EDT in various spaces.

Space	EDT 125	EDT 250	EDT 500	EDT 1000	EDT 2000
Las Chimeneas	1.13	0.96	0.85	0.7	0.51
El Castillo	1.33	1.2	1.28	1.15	1.12
La Garma	1.2	0.68	0.65	0.56	0.42
La Pasiega End	1.05	1.19	0.75	0.49	0.24
La Pasiega Turret	1.92	1.62	1.58	1.4	1.16
Tito Bustillo	2.24	1.11	1.36	1.35	1.46
Stonehenge A Edge of circle	1.36	0.67	0.42	0.35	0.02
Stonehenge A Part way in	0.43	0.05	0.03	0.43	0.2
Stonehenge A Centre	0.05	0.03	0.13	0.02	0.07
Stonehenge B Inside	0.49	0.59	0.4	0.37	0.26
Stonehenge B Heelstone	0.3	0.28	0.29	0.29	0.47
Stonehenge B Centre	0.07	0.06	0.14	0.12	-
Stonehenge C Inside	0.88	0.81	0.46	0.29	0.03
Stonehenge C Heelstone	0.78	0.59	0.46	0.56	0.47
Stonehenge C Centre	0.88	1.03	1.09	1.53	1.46
Paphos 1	1.51	1.86	1.73	1.64	1.53
Paphos 2	2.12	2.02	1.93	1.54	1.49
Paphos 3	2.54	1.6	1.57	1.86	1.52
Paphos 4	1.92	1.9	2.05	1.78	1.67
Royal Festival Hall, London	1.33	1.37	1.43	1.57	1.66
Royal Albert Hall, London	2.53	2.54	2.67	2.63	2.62
Queen Elizabeth Hall, London	1.7	1.67	2.04	2.12	2.24
Barbican Concert Hall, London	1.92	1.82	1.84	1.96	2.04
Wigmore Hall, London	1.49	1.61	1.55	1.57	1.48
Fairfield Hall, Croydon	2.18	1.98	1.89	1.85	1.73
Wessex Hall, Poole	2.48	1.77	1.66	1.63	1.51
Colston Hall, Bristol	2.08	1.96	1.81	1.87	1.78
St. David's Hall, Cardiff	1.9	2.13	2.11	2.14	2.11
Assembly Hall Watford	1.53	1.74	1.58	1.47	1.25
Music School auditorium, Cambridge	1.99	1.94	2.08	2.11	2
Royal Concert Hall, Nottingham	2.77	2.1	1.72	1.77	1.57
Free Trade Hall, Manchester	1.47	1.47	1.65	1.77	1.62
Philharmonic Hall, Liverpool	1.89	1.78	1.8	1.78	1.59
Usher Hall, Edinburgh	2.01	1.97	2.08	2.22	2.12
Conference Centre, Wembley	1.73	1.61	1.17	1.12	0.91
Butterworth Hall, Warwick University	2.28	2.77	2.69	2.32	1.89
Christchurch Town Hall, New Zealand	1.97	1.77	2.03	2.14	2.11
Michael Fowler Centre, New Zealand	1.69	1.82	1.86	1.81	1.82
Segerstrom Hall, California	2.36	2.44	2.3	2.38	2.2

3.2. Reverberation, Clarity, and Bass Response

T30 is a measure of reverberation that relates to the overall reverberation of the space, rather than the relationship between source and receiver positions. In the caves, the results for T30 are mostly similar to those for EDT, other than in La Pasiega End (Table 6). In this space, T30 is very long, 18.4 s at 125 Hz, reducing gradually at higher frequencies. The space is very long and narrow, a corridor or tunnel. While in the space, the research team heard little reverberation when the source and receiver were close together, whereas there was long complex reverberation when the source and receiver were further apart. At Stonehenge A, T30 was higher than EDT. At Stonehenge B, T30 was a little higher inside the circle, while at Stonehenge C, T30 was a little longer than EDT, but results were less extreme. At Paphos, results for T30 were comparable to EDT results, but with less high frequency drop off.

Table 6. T30 Reverberation in La Pasiega Cave.

	T30 63	T30 125	T30 250	T30 500	T30 1000	T30 2000	T30 4000	T30 8000
La Pasiega End	7.2	18.4	13.7	8.8	7.2	4.3	2.5	0.9

Speech clarity C (50) should be greater than or equal to −2 dB to avoid syllable intelligibility decreasing below 80%, this is the bottom admissible limit value for good speech intelligibility; greater than −3 dB is a good value. In Table 7, results for C (50) that indicate poor speech clarity are marked in red; typical ranges are all taken from British Standards [11] (p. 12).

Table 7. C (50) Clarity of Speech results.

	C (50) 63	C (50) 125	C (50) 250	C (50) 500	C (50) 1000	C (50) 2000	C (50) 4000	C (50) 8000
Las Chimeneas	−8.3	−3.7	0	2	2	4.9	8.2	12
El Castillo	−22.2	−26.4	−32.2	−30.2	−31.8	−37.3	−37.5	−39.5
La Garma	−2.9	−1.8	3.6	3.3	4.2	7.3	11.1	12.3
La Pasiega End	−9.5	−7.7	−6.7	−4.2	−4.6	−2.3	−1.7	−5.5
La Pasiega Turret	−15.3	−16.9	−22.5	−26.7	−32	−35.7	−35.3	−36.4
Tito Bustillo	−13.7	−18.9	−19.4	−24.4	−28.5	−34.5	−32.4	−32.6
Stonehenge A 1m in circle	1.6	0.2	5	7.8	11	26.4	34.9	35.4
Stonehenge A Inside	24.8	19.1	21.4	17.3	18.5	26	29.5	36.3
Stonehenge A Centre	25.1	27.4	26.3	25.3	24.6	32.5	35.2	39.2
Stonehenge B Inside	5.1	6.2	5.7	8.5	10.7	17.1	18.9	21.4
Stonehenge B Heelstone	10.5	8.2	9.5	11.3	10.6	10.5	11.3	17.2
Stonehenge B Centre	24.5	26.9	28.2	29.6	29.3	-	41.5	44.6
Stonehenge C Inside	−0.2	0.4	2.7	6	10.8	19	21.3	21.8
Stonehenge C Heelstone	−2.8	−2.4	−1.4	0.3	−0.4	4.8	4.5	12.9
Stonehenge C Centre	−10.5	−12	−13.7	−16.9	−14.9	−7.7	−7.8	−7.9
Paphos 1	−26.5	−19.3	−16.4	−14.2	−10.2	−3.2	−1.5	4.4
Paphos 2	−8.9	−7.7	−5.6	−2.3	1	1.1	2.3	2.6
Paphos 3	−3.7	−5.3	−3.4	−4.6	−6.7	−4	−0.7	3.1
Paphos 4	−11.7	−6.6	−6.8	−6.9	−4.8	−2.7	−0.5	2.1

In the caves, there are very mixed C (50) results, with Las Chimeneas and La Garma having very high values, while in the other caves clarity is poor. Other measures of intelligibility produce differing results. ALcons measures loss of consonants due to acoustic effects. These results were averaged across the frequency spectrum, and all the caves have results between 3 and 8, which indicates very good intelligibility. RASTI values for the caves indicate that speech intelligibility would be very good or excellent, according to the typical values in Table 1.

At Stonehenge, C (50) speech intelligibility is very good (Table 7) except in the final stage of development at the centre of the circle. At this position, C (50) results suggest that at each octave speech clarity is poor. However, this is a measurement in which the source and receiver position are both near the centre, and as we have seen, focusing could produce very different results at specific source and receiver positions. ALcons shows little loss of consonants, and RASTI shows mostly excellent results with some very good speech intelligibility (Table 8). At Paphos, the measurements represent a source on the stage, and receivers placed in the centre, in the third row, near the front, in the middle of the seating in row 14, and right at the back. C (50) results suggest that the first position is not good in terms of clarity. Indeed, this is a position for performers, rather than the audience. All positions have poor clarity in low frequencies, but position 2, the third row, has the best speech clarity, with good results at 1000 and 2000 Hz, while on row 14 and at the back, results suggest intelligibility is not good. From 4000 Hz and above, all positions have good speech clarity. ALcons and RASTI results for intelligibility are all good.

Table 8. ALcons, RASTI, and bass ratio results.

	ALcons	RASTI	BR (SPL)
Las Chimeneas	5.01	0.68	2.2
El Castillo	4.51	0.68	11
La Garma	3.74	0.73	5.3
La Pasiega End	3.88	0.77	2
La Pasiega Turret	5.88	0.62	3.2
Tito Bustillo	6.98	0.64	5.5
Stonehenge A 1m in circle	2.07	0.91	−1.5
Stonehenge A Inside	1.22	0.97	−2.8
Stonehenge A Centre	1.91	1	−7.8
Stonehenge B Inside	1.74	0.87	−4.6
Stonehenge B Heelstone	1.91	0.88	0.9
Stonehenge B Centre	1.14	1	−2.9
Stonehenge C Inside	2.03	0.86	−5.3
Stonehenge C Heelstone	5.33	0.67	−2.2
Stonehenge C Centre	4.52	0.71	−2.8
Paphos 1	9.65	0.5	1.8
Paphos 2	9.05	0.55	−2.1
Paphos 3	10.02	0.52	−0.6
Paphos 4	10.66	0.5	0.3

Speech intelligibility in the spaces varies. In Las Chimeneas and La Garma caves, voices with low frequency content, men with particularly low voices for example, would have an effect added to their voices, which would blur consonants, and yet allow for speech to be intelligible as sibilant higher frequencies, for example, were unaffected. At the end of La Pasiega, low frequencies would be most affected, but all frequencies would be modulated. In El Castillo, the La Pasiega Turret, and Tito Bustillo, men with higher voices and women would find that their voices would be most affected, as they contain greater high frequency content. In these caves, speech would be more intelligible at some frequencies than others, but an effect would be added to all voices. At Stonehenge, speech intelligibility is good. In the final phase of development, someone standing in the centre would find the lower range of their own speech somewhat unintelligible; others listening to it from another position would still be able to hear it, underlining again the significance of this specific position. At Paphos, speech intelligibility results tell us that the third row would be a far preferable seating position to the 14th row, or back row; one would still understand speech at the back, but it would be less clear. All of these results would be affected by people filling the space, as people act as high frequency absorbers, further increasing clarity.

Musical clarity C (80) is very low in El Castillo and La Pasiega Turret, as well as in the centre of Stonehenge C, and at position Paphos 1, the stage position. Directness of musical source C (7) results are typical in Las Chimeneas and La Garma, and very low in other caves. Musical sources directness is typical in all Stonehenge models at the inside receiver positions for low frequencies, but music with higher frequencies would be highly direct. At the Heelstone positions of models B and C, musical sources do not seem near or direct, especially at low frequencies. In model A, musical sounds would be the most direct, whereas in model B, results are typical, and in model C, directness is very low. Once more, this illustrates the unusual acoustic nature of the central of Stonehenge. Low frequency musical instruments, such as large drums, could have had an interesting effect at Stonehenge.

Bass ratio, BR(SPL), describes reverberation time at low frequencies. For music, the modern standard desirable BR is 1.0 to 1.3; for speech, it should be 0.9 to 1.0 [53]. In the caves, all results are higher than what one might desire for speech. Other metrical results at low frequency octaves also suggest that bass effects are present in some positions. BR measures the ratio between reverberation energy in the 125 and 250 Hz octaves and in the range between 500 and 1000 Hz. This gives no information about the 63 Hz octave range; it is a measurement designed to assess the warmth of a concert hall acoustic, and most concert halls have results close to 1. In all of the caves, the results are considerably higher than one would expect in a concert hall, and there is far more bass than is normal or desirable. At Stonehenge, all but one of the results are negative, and the ratios are much higher than one would wish. Again, the bass response is unusual and high. At the Heel Stone receiver position of Stonehenge B, the bass response is more acceptable, but this is quite some distance from the source. In the Paphos model, the bass ratio is very high, and at the back it is very low; it is again negative in some positions. There may be a particular position in the space where the bass ratio is perfect, but it seems that the bass response is irregular and uncontrolled at Paphos.

4. Conclusions

The study of the acoustics of these sites involved working with digital models, and as mentioned above, these were later integrated into an audio-visual museum exhibit and app for PC, Mac, tablets, and smartphones, the EMAP Soundgate [43], which provided an easily appreciable understanding for a diverse audience. The impulse responses discussed in this paper are all auralised, integrated into that app, and one can explore and compare the acoustics intersubjectively in a kind of experimental phenomenology. Further research will integrate a greater range of impulse responses into the app to more accurately represent the acoustic ecologies of the sites. As these sites do not have diffuse reverberant fields, and because the variations in acoustics in the sites are interesting, a large number of impulse responses are needed to adequately represent the acoustic effects present. Although the mixed methodology in this paper should be relevant to a range of readers, the technical terminology means that such use of apps assists with dissemination to a wide audience. This paper explored a very small number of impulse responses, and further research is needed to understand and appreciate the acoustic effects present in each of these sites, and what we might learn from them, including an in-depth individual assessment of each of the three types of site.

Further work might include: An exploration of lateral fraction and envelopment metrics; exploration of low frequency resonances; mapping of patterns of modal resonance; exploring echoes; further identification and exploration of positions of particular sonic interest; and gaining a better understanding of why some of these effects are present. Many other archaeological sites have had no study of their acoustic properties, and a more complete survey of a range of sites will help to provide a better understanding of how acoustic ecologies have changed over time. Further experimental work will explore the use of virtual reality headsets in order to immersively experience the auralised acoustics of ancient sites. While this study provides examples of how acoustic effects present can differ in archaeological sites, such as Palaeolithic caves, Neolithic stone circles, and classical Hellenistic antiquity, much further work is required to comprehensively survey how human experiences of acoustics changed across the millennia. Comparison of the acoustics in a wider range of painted French and Spanish Palaeolithic painted caves would help to better characterise their acoustic ecology. Study of a larger number of source and receiver positions in the various phases of Stonehenge would help us to understand the effect of acoustics on ritual activities related to this famous stone circle. Exploration of the different phases of development of Paphos Theatre, as well as the effect of different surface materials, would assist in knowing more about audiences' experiences in the space.

This research illustrated the importance of constructing a specific set of methods suited to sound archaeology rather than uncritically adopting standard methods from contemporary building or industrial acoustics, which often has quite different research problems and questions. In this case, the methods adopted differ from standard approaches in that they: include comparison of different spaces; use a range of different source and receiver positions; explore a wider frequency range, including 63 and 8000 Hz octaves; evaluate individual octave results for metrics such as EDT, T30, and C (50); explore metrics related to musical use, such as C (80), C (7), and BR, as well as speech-related metrics, such as RASTI, C (50), and ALcons; explore low frequency effects in detail; compare results to those for contemporary performance venues, such as concert halls; use a consumer B&O Beolit loudspeaker as a source instead of, for example, a dodecahedron, because of its portability; and use recording studio/live professional audio DPA microphones as receivers, rather than acoustic testing capsules and power conditioners, due to their robustness.

The acoustic metrics examined produced a number of findings. The caves and Stonehenge have varied EDT and T30 reverberation, with modal responses at low frequencies in particular, meaning that there are acoustic effects present in particular positions, and specific frequency ranges. Speech clarity metrics are different depending on whether you use C (50), ALcons, or RASTI values. In some cases, speech is very clear, in others, effects are added to voices or speech is incomprehensible. In the caves, speech is clear when judged using ALcons or RASTI, but at different octaves identified by C (50), clarity of speech can be either extremely high or low, indicating speech is understandable, but may be changed by support or transformation of one or other frequency range. At Stonehenge, speech transmission is good, but the voices of people standing at the centre of the monument are particularly transformed. Stonehenge A has a particular boost of frequencies between 3 and 6 kHz, which would aid the hearing of human voices, and has a further boost of very high frequencies, affecting the perception of sibilants or other high frequency consonants. At Paphos, speech is transmitted clearly in general, but clarity is better when sitting near the front of the seating, with sound becoming less good as one moves towards the back of the venue; as one might expect, the best seats in the house are at the front. Overall, the musical clarity results similarly imply acoustics effects are present, rather than the acoustic supporting or detracting from musical performance.

The bass response metric shows that the caves have strong low frequency support, Stonehenge has strong and unusual low frequency effects, and that at Paphos, the bass response is variable and uncontrolled. Low frequencies would not be common in antiquity, and it is notable that all three sites feature low frequency effects, something that may have been a particular marker of special experiences. Low frequencies are often enhanced in these spaces, sometimes in an extreme manner. Sounds are transformed as one enters or moves through a space, with effects added that contribute to a perception of its special or sacred nature. Each cave is inhabited by an individual sonic character that plays a part in how one perceives the space, providing a contextualisation, a sense of place, and of identity. In a prehistoric culture dominated by animism, in which all of nature was alive with spirits, the acoustics of each cave helped to bring the space to life, providing intersubjective agency that transformed the experience of anyone present.

Acoustics at Stonehenge change during its developing phases of construction, and change as one moves towards its centre, suggesting the site is more dynamic than might be expected of a contemporary home for the dead. The centre of the final phase of development of the site provides a particular focus of sonic affect. Focusing effects in the space are complex and unpredictable in the outdoor, natural, and windswept environment, and the changes in sound present, depending as they do on the specific position where sound is made and heard, would have contributed a sense that this was a place that was alive, as well as making individual positions in the space of particular importance, or adding a sacred sonic effect to the person standing there making sound or listening. The human architectural choices of Stonehenge's construction create an interaction between space, place, and acoustic ecology, allowing human agency to creatively participate in the generation of the voice of a home for the ancestors. Echoes, low frequency effects, sound focusing, amplification of particular frequencies, and reverberation provide acoustic effects that change and shift in different circumstances, changing voices and the sound of other ritual activities in a variety of ways. Specific positions, occupied by ritual specialists, afford human culture the opportunity to structure communication with sacred otherworldly powers. Paphos Theatre has a more refined sound, comparable to that of a concert hall, human designers overtly working to create an acoustic ecology controlled by them, and that supports the secular-if-ritualised performances occurring in the building. Such human rituals develop social relationships, with the most wealthy or powerful occupying prime positions, visible to those seated behind, and lower down the social stratigraphy.

The deeper parts of caves provided a very particular and powerful acoustic for humans in Palaeolithic times. In an animist cosmology, the lack of background and environmental noise differentiated caves from outdoor spaces. With no experience of stone buildings, these were alien natural spaces that featured variable reverberation, low frequency effects, and transformation of sounds made by human speech and movement. These were natural formations, humans entered into them, becoming enveloped by these other worlds' acoustics, going into an environment over which they had no control, leaving as a record of their presence visual motifs charged with spiritual meanings, and engaging with altered states and what they probably regarded as powerful supernatural forces. Stonehenge is human-built, representing or a home for ancestral spirits or a place of healing, with acoustics that its builders could observe changing as they redesigned the site in a number of phases of development.

Over time, the acoustics of the site become more sophisticated, much as ritual culture similarly developed, the landscape being imbued with sonic character by human agency. Humanity began to exert control over its environment, clearing forests and moving from a hunter-gatherer lifestyle towards agriculture, beginning to construct an architectural tradition to monumentalise these cultural changes, with acoustics an artefact and evidence of that process. At Stonehenge, acoustic ecology is human-made, but a by-product of wider cultural factors. At Paphos, design choices are made to overtly manipulate the acoustics present, in order to facilitate the support of speech and music. Humans began by this point to attempt to control their acoustic ecologies, to consciously design their surrounding soundscapes. Over time, acoustics become more predictable, as ritual acoustics move from a Palaeolithic variety of selected natural spaces in which the acoustics present are appreciated by early humans; to the construction of Neolithic monuments, which resulted in human-made sites with a range of unusual acoustic effects present; and finally, a literate Hellenistic culture, which consciously manipulated acoustics designed to flatter speech and music. It is perhaps these changes in acoustic cultures that are a more significant development than the individual variations in the acoustic metrics present.

The earliest (cave) sites have natural sonic effects, which seem to have played some part in attracting Palaeolithic humans to carry out ritual acts within them. These effects are replicated in the human built stone circle at Stonehenge, and over time, the development of the architecture of the site made these effects more focused on specific positions, affecting low frequencies while allowing higher frequencies associated with speech to retain clarity. Although we cannot know to what extent acoustic concerns were in the minds of those designing the arrangement of stones, the change in acoustic ecology would have been noticeable and may well have become associated with the rearrangement of the site in different phases; specific acoustic effects may well have become associated with particular architectural shapes. The reverberation and echoes present would have diminished or disappeared when stones were removed, and reappeared when new stones were erected. Such acoustic changes and effects may have been attributed to supernatural forces. One possible interpretation could be, for example, that they were thought of as the voices of spiritual forces associated with individual or groups of stones.

It should be stressed that there is no evidence that the design of Stonehenge was based on acoustics, but the acoustic effects present in each stage are prominent and different enough to suggest that acoustics may well have begun to be associated with specific architecture, much as concert hall architects draw upon the designs of existing spaces with good acoustics, when planning new buildings. It is at least possible that Stonehenge's designers believed that animist or ancestral spirits were pleased by the circular arrangements of the stones when they were erected, when they heard the voices of these spirits seem to appear through acoustic effects. This may even have provided confirmation that these spirits had successfully inhabited the monument, validating the design choices made. This might have perhaps transferred a tradition of spiritual power from sites with similar acoustics, such as at the Welsh source of the bluestones, or through a vestigial memory of acoustics in sacred caves. With no written sources for confirmation, such suggested interpretations remain inevitably speculative. We do know that by the Roman period these processes of human architecture had developed enough to allow for the conscious design of performance venues with supportive acoustics, as is evidenced in the theatre at Paphos. Caves present a naturally existing acoustic, a place occupied by spirits, which humans could visit to attempt to interact with these acoustically manifesting spiritual powers; Stonehenge has a human-made acoustic in which a complex ritual culture collaborates with audible ancestral spirits in order to exert human control over natural forces; Paphos illustrates a sophisticated rational culture, in which a human logic has established how acoustic effects can be manipulated, and interact with social stratification, free from the influence of gods.

Changes in cosmology are reflected and represented in the acoustic ecologies present in these sites, from a strong relationship as part of nature in hunter-gatherer prehistoric cultures in caves, through the monumental efforts of early farming communities to entreat spiritual powers to control the land, to the larger societies and complex performative social interaction of Rome. The acoustic ecology of these archaeological sites contributed to their ritual significance and afforded particular effects that made the activities within seem alive, or even larger than life, something different to the natural spaces outside, places where meanings were manipulated and transformed. These spaces have different acoustic characters, but each has noticeable acoustic effects present, which would have been integrated into any activities in the space, contributing to the generation of meaning, and playing an important part in creating a sense of set and setting. Over time, those sets and settings became increasingly controlled and manipulated by their users, enhancing the experiences of those within, turning space into place, position into context. This process of change at these three types of sites illustrates the emergence of architectural acoustics, and outlines its significance to human culture.

Funding: This research was funded by the UK Arts and Humanities Research Council and Engineering and Physical Sciences Research Council, (grant number AH/K00607X/1) as part of the Science and Heritage Programme; and by the Culture Program of the European Union (EACEA reference number 536370).

Acknowledgments: Access to the caves was only possible because of the support and commitment of the Gobierno de Cantabria and Gobierno Del Principado de Asturias, the regional government bodies in the area. Research on Stonehenge was assisted by the kind advice and support of English Heritage and Historic England, as well as Mike Parker Pearson and Ben Chan. Paphos Theatre research was with the support of the Cyprus Institute and University of Sydney.

Conflicts of Interest: The authors declare no conflict of interest. The funders had no role in the design of the study; in the collection, analyses, or interpretation of data; in the writing of the manuscript, or in the decision to publish the results.

References

1. Hodder, I. *Symbolic and Structural Archaeology*; Cambridge University Press: Cambridge, UK, 1982.
2. Hodder, I.; Hutson, S. *Reading the Past: Current Approaches to Interpretation in Archaeology*, 3rd ed.; Cambridge University Press: Cambridge, UK, 2003.
3. Renfrew, C.; Bahn, P. *Archaeology: Theories, Method and Practice*, 4th ed.; Thames & Hudson: London, UK, 2004.
4. Till, R. Sound Archaeology: Terminology, Palaeolithic Cave Art and the Soundscape. *World Archaeol.* **2014**, *46*, 292–304. [CrossRef]
5. Solomos, M. A Phenomenological Experience of Sound: Notes on Francisco Lopz. *Contemp. Music Rev.* **2019**, *38*, 1–2. [CrossRef]
6. LaBelle, B. *Acoustic Territories: Sound Culture and Everyday Life*; Continuum: New York, NY, USA, 2010.
7. Ansell-Pearson, K.; Schrift, A.D. *The New Century: Bergsonism, Phenomenology and Responses to Modern Science*; Taylor and Francis: Abingdon, UK, 2014.
8. Thrift, N. *Non-Representational Theory: Space, Politics, Affect*; Taylor and Francis: Abingdon, UK, 2008.
9. Dewsbury, J. Witnessing space: Knowledge without contemplation. *Environ. Plan.* **2003**, *35*, 1907–1932. [CrossRef]
10. Long, M. *Architectural Acoustics*, 2nd ed.; Academic Press: Waltham, MA, USA, 2014.
11. British Standards Institute. *Acoustics: Measurement of Room Acoustic Parameters. Part 1: Performance Spaces (ISO 3382-1:2009)*; International Organization for Standardization: Brussels, Belgium, 2009.
12. Bradley, J.S. Review of objective room acoustics measures and future needs. *Appl. Acoust.* **2010**, *72*, 713–720. [CrossRef]

13. Murphy, D.; Shelley, S.; Foteinou, A.; Brereton, J.; Daffern, H. Acoustic Heritage and Audio Creativity: The Creative Application of Sound in the Representation, Understanding and Experience of Past Environments. *Internet Archaeol.* **2017**, *44*. [CrossRef]
14. Weitze, C.A.; Rindel, J.H.; Christensen, C.L.; Gade, A.C. The Acoustical History of Hagia Sophia revived through Computer Simulation. In Proceedings of the Forum Acusticum, Sevilla, Spain, 16–20 September 2002.
15. Fazenda, B.; Scarre, C.; Till, R.; Pasalodos, R.J.; Guerra, M.R.; Tejedor, C.; Peredo, R.O.; Watson, A.; Wyatt, S.; Benito, C.G.; et al. Cave Acoustics in Prehistory: Exploring the Association of Palaeolithic Visual Motifs and Acoustic Response. *J. Acoust. Soc. Am.* **2017**, *142*, 1332–1349. [CrossRef] [PubMed]
16. NTi-Audio Online, Dodecahedron Speaker Set Product Data. Available online: https://www.nti-audio.com/Portals/0/data/en/NTi-Audio-Dodecahedron-Speaker-Set-Product-Data.pdf (accessed on 2 March 2019).
17. Murphy, D. Archaeological Acoustic Space Measurement for Convolution Reverberation and Auralization Applications. In Proceedings of the 9th International Conference on Digital Audio Effects (DAFx-06), Montreal, Canada, 18–20 September 2006.
18. Till, R. An Archaeoacoustic Study of the Ħal Saflieni Hypogeum in Malta. *Antiq. J.* **2017**, *91*, 74–89. [CrossRef]
19. Jahn, R.; Devereux, P.; Ibison, M. Acoustical resonances of Assorted Ancient Structures. *J. Acoust. Soc. Am.* **1996**, *99*, 649–658. [CrossRef]
20. Pike, A.W.G.; Hoffman, D.L.; García Diez, M.; Pettitt, P.B.; Alcolea González, J.; Balbín Behrmann, R.; de González, S.C.; de las Heras, C.; Lasheras, J.A.; Montes, R.; et al. U-series dating of Palaeolithic art in 11 caves in Spain. *Science* **2012**, *336*, 1409–1413. [CrossRef]
21. Reznikoff, I.; Dauvois, M. The sound dimension of painted caves (original in French). *B. Soc. Prehist. Fr.* **1988**, *85*, 238–246. [CrossRef]
22. Reznikoff, I. The evidence of the use of sound resonance from Palaeolithic to Mediaeval times. In *Archaeoacoustics*; Scarre, C., Lawson, G., Eds.; McDonald Institute Monographs: Cambridge, UK, 2006; pp. 77–84.
23. Diaz-Andreu, M.; Garcia-Bnito, C. The sound of rock art: The acoustics of the rock art of Southern Andalusia (Spain). *Oxf. J. Archaeol.* **2014**, *33*, 1–18. [CrossRef]
24. Available online: http://www.archeoacustica.net/home-page/ (accessed on 1 April 2019).
25. Diaz-Andreu, M.; Mattioli, T. Archaeoacoustics of Rock Art: Quantitative Approaches to the Acoustics and Soundscape of Rock Art. In *CAA 2015, Keep the Revolution Going: Proceedings of the 43rd Annual Conference on Computer Applications and Quantitative Methods in Archaeology*; Campana, S., Scopigno, R., Carpentiero, G., Cirillo, M., Eds.; Archaeopress: London, UK, 2015.
26. Rainio, R.; Lahelma, A.; Aikas, K.; Okkonen, J. Acoustic Measurements and Digital Image Processing Suggest a Link Between Sound Rituals and Sacred Sites in Northern Finland. *J. Archaeol. Method Theory* **2018**, *25*, 453–474. [CrossRef]
27. Conard, N.; Malina, M.; Munzel, S.C. New Flutes Document the Earliest Musical tradition in Southwestern Germany. *Nature* **2009**, *460*, 737–740. [CrossRef] [PubMed]
28. Songs of the Caves. Available online: http://SongsoftheCaves.Wordpress.com (accessed on 2 April 2019).
29. Parker, P.M. *Stonehenge: Exploring the Greatest Stone Age Mystery*; Simon and Schuster: London, UK, 2012.
30. Willis, C.; Marshall, P.; McKinley, J.; Pitts, M.; Pollard, J.; Richards, C.; Richards, J.; Thomas, J.; Waldron, T.; Welham, K.; et al. The Dead of Stonehenge. *Antiquity* **2016**, *90*, 337–356. [CrossRef]
31. Darvill, T. *Stonehenge: The Biography of a Landscape*; History Press: Stroud, UK, 2006.
32. Gilbert, R. *Music and Trance: A Theory of the Relations Between Music and Possession*; University of Chicago Press: Chicago, IL, USA, 1985.
33. Watson, A.; Keating, D. Architecture and Sound: An acoustic analysis of megalithic monuments in prehistoric Britain. *Antiquity* **1999**, *73*, 325–336. [CrossRef]
34. Watson, A. (Un)intentional sound? Acoustics and Neolithic monuments. In *Archaeoacoustics*; Scarre, C., Lawson, G., Eds.; McDonald Institute for Archaeological Research: Cambridge, UK, 2006; pp. 11–22.
35. Till, R. Songs of The Stones: The Acoustics of Stonehenge. In *BAR 504 2009: The Sounds of Stonehenge, Centre for the History of Music in Britain, the Empire and the Commonwealth*; CHOMBEC Working Papers No. 1; Banfield, S., Ed.; Archaeopress: Oxford, UK, 2009; pp. 17–42.
36. Till, R. Songs of the Stones: An Investigation into the Musical History and Culture of Stonehenge. *IASPM J.* **2011**, *1*, 1–18. [CrossRef]
37. Fazenda, B. The Acoustics of Stonehenge. *Acoust. Bull.* **2013**, *38*, 32–37.

38. Fazenda, B.; Drumm, I. Recreating the Sound of Stonehenge. *Acta Acust. United Acust.* **2013**, *99*, 110–118. [CrossRef]
39. Hardy, T. *Tess of the d'Urbervilles: A Pure Woman Faithfully Presented*; Harper and Brothers: New York, NY, USA, 1892.
40. Darvill, T.; Marshall, P.; Parker, P.M.; Wainwright, G. Stonehenge Remodelled. *Antiquity* **2012**, *86*, 1021–1040. [CrossRef]
41. Unver, E.; Taylor, A. Virtual Stonehenge Reconstruction. In *Progress in Cultural Heritage Preservation, Proceedings of the 4th International Conference, EuroMed 2012, Lemessos, Cyprus, 29 October–12 November 2012*; Lecture Notes in Computer Science Subseries: Information Systems and Applications, Incl. Internet/Web, and HCI, 7616 (XXV); Springer: Berlin, Germany, 2012; pp. 449–460.
42. Bryan, P.G.; Clowes, M. CAA96, Computer Applications and Quantitative Methods in Archaeology, BAR International Series 845. In *Stonehenge—Mapping the Stones*; Lockyear, K., Sly, T.J.T., Mihăilescu-Bîrliba, V., Eds.; Archaeopress: Oxford, UK, 2012; pp. 41–48.
43. EMAP Soundgate App. Available online: http://www.emaproject.eu/content/soundgate-app.html (accessed on 28 March 2019).
44. 3D Model of the Antonine Final Alteration of the Paphos Theatre (Phase 5). Available online: http://www.omnia.ie/index.php?navigation_function=2&navigation_item=%2F2020720%2FDR_5d8cc774e23c67c16ffba8653d4ee24d&repid=1 (accessed on 2 April 2019).
45. Paphos Theatre Archaeological Project. Available online: http://www.paphostheatre.org (accessed on 4 April 2019).
46. Kang, J.; Chourmouzuadou, K. Acoustic Evolution of Greek and Roman Theatres. *Appl. Acoust.* **2008**, *69*, 514–529.
47. Lynge Christensen, C.; Rindel, J.H. A new scattering method that combines roughness and diffraction effects. In Proceedings of the Forum Acusticum, Budapest, Hungary, 29 August–2 Septemer 2005.
48. Gade, A.C.; Lisa, M.; Lynge Christensen, C.; Rindel, J.H. Roman Theatre acoustics: Comparison of acoustic measurement results from the Aspendos Theatre, Turkey. In Proceedings of the 17th ICA, Kyoto, Japan, 4–9 April 2004.
49. Farnetani, A.; Prodi, N.; Fausti, P.; Pompol, R. Acoustical measurements in ancient Roman theatres. *J. Acoust. Soc. Am.* **2004**, *115*, 2477. [CrossRef]
50. Acoustics and Music of British Prehistory Research Network. Available online: http://AMBPNetwork.wordpress.com (accessed on 28 March 2019).
51. Karabiber, Z. The conservation of acoustical heritage. In Proceedings of the First International Workshop on 3D Virtual Heritage, Geneva, Switzerland, 26–27 September 2002; pp. 286–290.
52. Maconie, R. Musical acoustics in the Age of Vitruvius. *Musical Times* **2005**, *146*, 75–82. [CrossRef]
53. Ahnert, W.; Schmidt, W. Fundamentals to Perform Acoustical Measurements. Available online: http://renkusheinz-sound.ru/easera/EASERAAppendixUSPV.pdf (accessed on 19 April 2019).
54. Barron, M. Interpretation of Early Decay Times in Concert Auditoria. *Acta Acust. United Acust.* **1995**, *81*, 320–331.
55. Ottley, M. Designing for Speech in a Circular Room. In Proceedings of the ACOUSTICS, Sydney, Australia, 6–9 November 2018; Available online: https://au.marshallday.com/media/2802/speech-in-circular-rooms-aas2018-mott.pdf (accessed on 3 March 2019).

10

Towards Italian Opera Houses: A Review of Acoustic Design in Pre-Sabine Scholars

Dario D'Orazio [1,*,†] **and Sofia Nannini** [2]

1 Department of Industrial Engineering, DIN University of Bologna, 40126 Bologna, Italy
2 Department of Architecture and Design, DAD Politecnico di Torino, 10125 Torino, Italy; sofia.nannini@polito.it
* Correspondence: dario.dorazio@unibo.it;
† Current address: Viale Risorgimento 2, 40126 Bologna, Italy.

Abstract: The foundation of architectural acoustics as an independent science is generally referred to Sabine's early studies and their application. Nevertheless, since the 16th Century, a great number of authors wrote essays and treatises on the design of acoustic spaces, with a growing attention to the newborn typology of the Opera house, whose evolution is strongly connected to the cultural background of the Italian peninsula. With roots in the Renaissance rediscovery of Vitruvius's treatise and his acoustic theory, 16th- to 19th-Century Italian authors tackled several issues concerning the construction of theatres—among them, architectural and structural features, the choice of the materials, the social meanings of performances. Thanks to this literature, the consolidation of this body of knowledge led to a standardisation of the forms of the Italian Opera house throughout the 19th Century. Therefore, the scope of this review paper is to focus on the treatises, essays and publications regarding theatre design, written by pre-Sabinian Italian scholars. The analysis of such literature aims at highlighting the consistencies in some 19th-Century minor Italian Opera houses, in order to understand to what extent this scientific and experimental background was part of the building tradition during the golden age of the Italian Opera.

Keywords: opera house; cultural heritage; shape optimisation; room acoustics

---

1. Introduction

Almost all Italian Opera houses with outstanding room acoustical reputations have been built before what is considered the beginning of architectural acoustics as an independent science, which dates back to 1898 [1,2]. The acoustics of historic Opera houses can be considered the result of a Darwinian type evolution, with architects designing halls inspired by the success of former "good" halls, frequently destroyed by fires. According to previous works [3–5], it is possible to identify three main room acoustic design approaches which were the starting point of this evolutionary process: the theory of "circulation of sound", the proto-geometrical acoustics and the "echo theory".

The theory of "circulation of sound" was based on the Renaissance rediscovery of Vitruvius's writings [6]. The Latin author argued that the voices of actors should be unobstructed in order to create favorable room-acoustics [5]. Such method thus influenced the building of auditoria with round shapes, rounded proscenium arches, without obstructions which "slowed the circulation" or "broke the voice" [3–5].

The proto-geometrical acoustics—from the 17th Century—was based on the assumption that the trajectory of sound was analogous to the sound rays reflected from a surface. Therefore, sound could be conveyed by modifying the plan shape. The similarity between light and sound was first assumed by Giuseppe Biancani (1566–1624), who defined the so-called "echometria" [7]. Assuming that rays spread evenly over a plan after being reflected from parabolic or elliptically shaped walls [5], scholars discussed the best plan for the cavea: elliptical, bell- and horseshoe-shaped—the latter becoming the standard shape for 19th-Century theatres. Moreover the raising of proscenium arches, directly bound both to scenography and acoustics, increased the early reflections—or, better, the first reflection; lastly, the shape of the ceilings was never designed as concave in order not to concentrate reflections.

The "echo-theory" is a recent guideline based on geometrical acoustics which dates back to the end of the 18th Century [5,8]. This theory focused on a "quantification of the perception threshold between direct and reflected sounds: when a first order reflection exceeded this threshold, an echo would be perceived, which was considered detrimental for the acoustics of the hall" [5]. Instead of uniform radiation pattern, a directional one was assumed and the "echo-theory" guideline was grounded on simple measurements, which employed a speaking person and a human observer judging audibility or intelligibility as a function of distance and direction. Because of such voice directivity and propagation guideline, the size of audience areas in Opera houses was limited.

Analyses through virtual reconstructions [9–13] provided further information on the acoustic quality of historical theatres, also taking into account the occupied conditions. The results generally highlighted situations that are more reverberating than the current standards. If we compare these conditions to the limited available literature on the current state of such theatres, following the ordinary measuring procedures, it is possible to define local building characteristics that are deeply connected to the politic fragmentation of Italy [14–20]. A collation of such data was suggested by Prodi et al. [20]: this work sought the correlation between acoustic and geometric descriptors of the environment (volume, seats, volume/audience ratios, etc.). Moreover, some authors [21,22] suggested the adjustment of some descriptors used in international theatres to the small-medium typology of Italian Opera houses. Lastly, Garai et al. [23] examined the coupling effects between fly tower and cavea.

Thanks to the reading of several original texts, spanning between the 16th and the 19th Century, this paper aims at adding another contribution to the scientific discussion on the pre-Sabine theories. Indeed, these texts show a particular attention on materials and other features that can be interpreted by means of architectural acoustics.

2. The Late-Renaissance Theatre

Until the end of the 17th-Century, the acoustic spaces were treated using the Vitruvian categories, taken from Vitruvius's well known treatise De Architectura (On Architecture). The treatise was written around the 1st Century B.C. and rediscovered in the 15th Century, becoming one of the bases of Renaissance humanist culture. Vitruvius's theory on theatre design was based on the propagation of spherical sound waves—the so called "circulation" [4,5]:

- Dis-sonantes, when the circulation is dissipated by destructive interference provided by the architectural elements (hard and sharp corners, etc ...);
- Con-sonantes, when the environment increases the circulation;
- Circum-sonantes, when curved surfaces create a reverberation effect;
- Re-sonantes, when echo phenomena occur.

One of the first music rooms of the modern age was the Odeo Cornaro in Padua, designed by Giovanni Maria Falconetto (1468–1534) around 1524–1530 for the nobleman and patron Alvise Cornaro (1484–1566). This project was included by the architect and theorist Sebastiano Serlio (1475–1554) in his Sette libri dell'architettura (Seven Books of Architecture) [24]. Serlio's drawing of the Odeo Cornaro shows a central symmetric plan with four niches on the sides and it is covered by a dome (see Figure 1). This small room fully agreed with the features of Vitruvian circum-sonantes spaces, as the round niches increased the diffusivity (As Serlio writes, "Et li quattro niccij per la sua ritonditá concava ricevono le voci, & le ritengono" (And the four niches, due to their round concavity, receive the voices and retain them) [24].). Indeed, this space was oriented to music listening rather than mere speech.

Serlio's great interest in theatre design can be particularly understood by reading the last pages of his second book on perspective, where a specific paragraph is dedicated to theatres and scenographies. By blending his readings of Vitruvius's acoustic theory and the laws of geometrical perspective, Serlio included a section and a plan of what might have been the model of his timber theatre designed for the courtyard of the palazzo Porto in Vicenza in 1539. On one hand, the first terraces of the cavea are semicircular and then acquire an almost elliptical shape [25]; on the other hand, the stage shows a sloped surface for a better perspective view from the audience (see Figure 2).

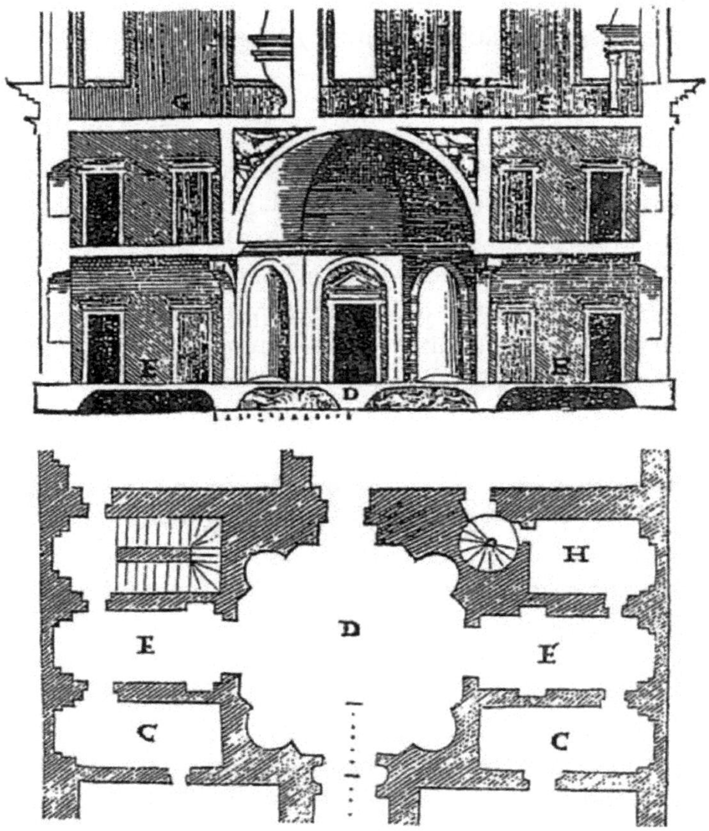

Figure 1. Plan of Odeo Cornaro in Padua (Hall D in the figure) (after [4]).

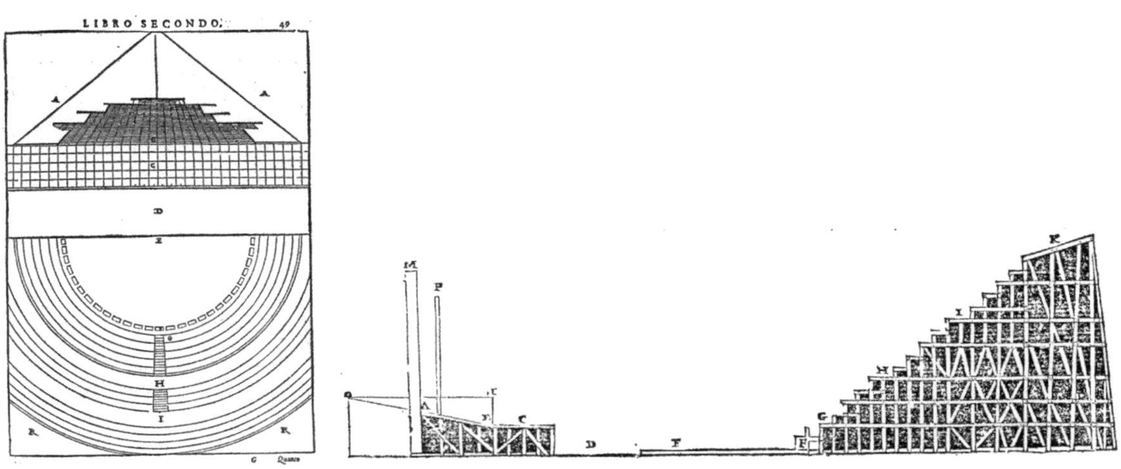

Figure 2. Plan and section of Sebastiano Serlio's timber theatre, in the book dedicated to Perspective (after [24]).

However, it is important to keep in mind that Vitruvius's acoustic and musical theory was based on considerations that mainly regarded the propagation of the human voice, as this was the key element of ancient Graeco-Roman theatre. Therefore, the acoustics of Greek and Roman theatres was generally revolving around the speech intelligibility of the actors [26–31]. Clearly, the evidence of these theatres' exceptional acoustics is due to their amphitheatrical shapes, and thus to the knowledge of the propagation of circular sound waves. However, it is not possible to affirm that the theory of the Latin author—despite his extensively quoting of Greek sources such as Aristotle's pupil Aristoxenus—was also applied to former Greek and Hellenistic theatres. Nevertheless, Vitruvius's rediscovered writings became the core of the Renaissance architectural theory, thus necessarily influencing also its acoustic outcomes [32,33].

A few decades after, with the aim of increasing the intelligibility of the speech and still following the models of Graeco-Roman theatres, late–Renaissance architects started experimenting with con-sonantes spaces. Such is the case of the Teatro Olimpico in Vicenza, by Andrea Palladio (1508–1580), inaugurated in 1585 with Sophocle's Oedipux Rex, with choruses by Andrea Gabrieli (1533–1585) [4,13]. The theatres of this period were characterised by steps and showed a reverberation higher to the present ones. In fact, the Odeo Cornaro shows a measured reverberation time at mid frequencies of 2.4 s in an unoccupied condition and a simulated reverberation time in an occupied condition of 1.3 s [34]; the Teatro Olimpico shows a measured reverberation time at mid frequencies of 3.3 s in unoccupied conditions and a simulated reverberation time of 2.2 s in occupied conditions [13]. Such acoustic conditions allowed a good intelligibility of the early writing styles of melodrama, as the Gabrieli's choruses in the Teatro Olimpico, as confirmed by historical sources [34,35] (see Figure 3).

At the same time, the development of acoustics was strongly connected to political and religious issues prompted by the event of the Protestant Reformation and the Council of Trent (1545–1563), which opened to the Counter-Reformation. On one hand, the Lutheran culture was oriented to the act of reading, due to the availability of printed Bibles thanks to Gutenberg's innovation. On the other hand, the reformed Catholic culture gave more importance to the act of listening, due to the changes in ritual offices. For example, coffered wooden ceilings in churches were preferred by Jesuits due to the decrease in reverberation time and increase in speech intelligibility [4]. An increasing interest in the propagation of sound was collected by the Jesuit Giuseppe Biancani (1566–1624), who published in 1620 the essay Sphaera Mundi [7], whose third paragraph is titled Echometria, idest Geometrica tractatio de Echo (Echometry, i.e., the Geometric Treatise of Echo). Biancani's Echometria opened a new branch of scientific knowledge blending sound and optics, which will be further called Acustica (Acoustics).

In this cultural environment, the building of the theatres corresponded to the birth of a new form of representation: the so-called Melodrama, which integrated various types of performance: mélos meaning singing, drama meaning acting. Indeed, the word Opera in Latin is the plural of opus, which means "act, performance". In the last decades of the 16th Century, the performance acquired a semi-public dimension—open to the Ruler and the court—and its own independent architecture. In the spaces designed for melodrama, one of the most significant aspects is the structural and typological background, deriving from the form of the Roman Basilica [36]. In fact, they were composed of a rectangular body ending with an apse, becoming, respectively, the audience and the stage. The theatre plan was typically divided into areas with different functions: the cavea for the audience, the stage and the rooms behind it reserved for the staff and for the artists. This is the case of Teatro all'antica (Ancient-Style Theatre) in Sabbioneta (1588–1590), by Vincenzo Scamozzi (1548–1616), which became the first building fully designed to be a free-standing theatre (see Figure 4).

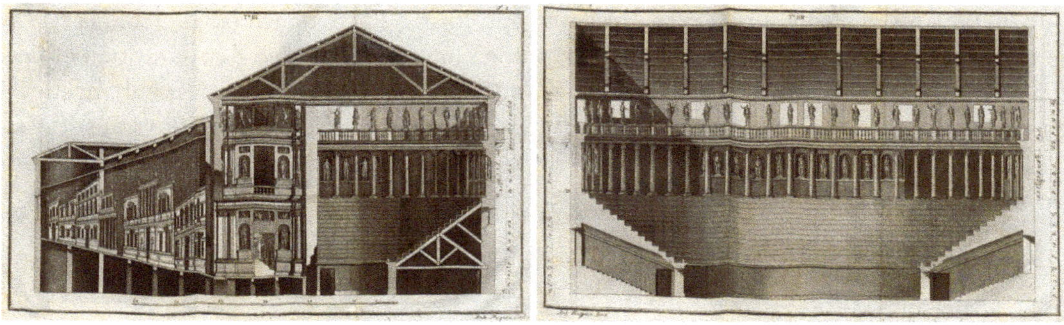

Figure 3. Section and view of Andrea Palladio's Teatro Olimpico in Vicenza (1585–1590) [37].

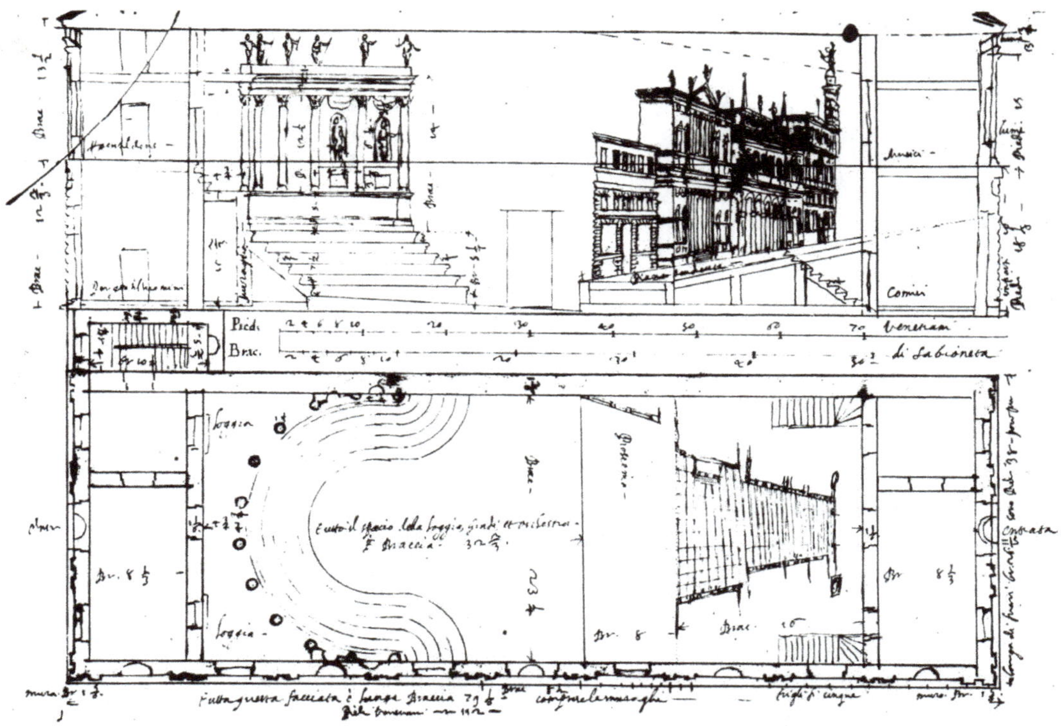

Figure 4. Vincenzo Scamozzi's draft of Teatro all'antica in Sabbioneta (1588–1590) [38].

3. The Baroque Theatre

At the beginning of the 17th Century, a new element entered the theatre design, which would influence both scenography and acoustics: the proscenium arch. This element, placed between the auditorium and the stage, was a direct consequence of the late-Renaissance studies on perspective—especially Serlio's aforementioned paragraph on perspective and scenography [39]. When it comes to its architectural use, the first and most famous architect constantly experimenting with proscenium arches was Inigo Jones (1573–1652). Inspired by his Italian architectural visits, he started introducing proscenium arches in his scenographies already in the first decade of the 17th Century, thus bringing the Italian perspective studies to Great Britain [40,41].

At the same time, the Italian architect Giovan Battista Aleotti (1546–1636) introduced such new spatial disposition in two of his works: the Teatro degli Intrepidi in Ferrara (1605) [9], now demolished, and Farnese Theatre in Parma (1610) [11] (see Figure 5). In both structures, he built a proscenium arch which materialised the difference between what is real and what is a theatrical fiction. The dimensions of the scenic arch increased with the progressive standardisation of its form, together with the increasing complexity of Baroque scenery as in the works of Giacomo Torelli

(1604–1678) [42], Nicola Sabbattini (1574–1654) [43] and Joseph Furttenbach (1591–1667) [44]. The role of this element was both visual and acoustical: on one hand, it provided a separation between the space of the representation and the space of the audience; on the other hand, it enhanced the first reflections from the stage and the orchestra, thus increasing the intelligibility in the stalls.

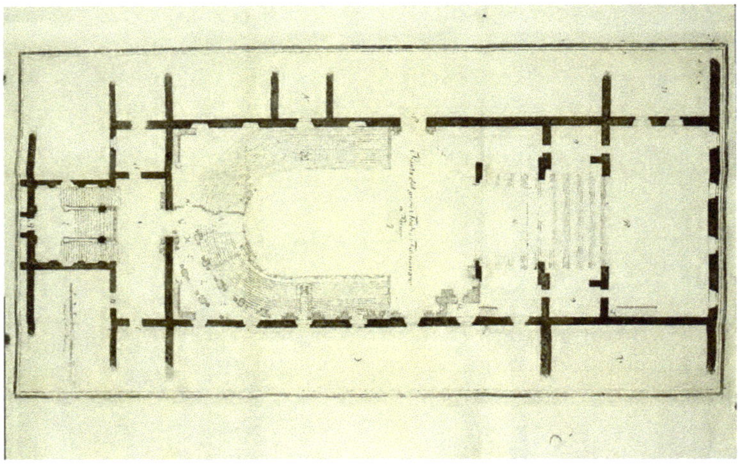

Figure 5. Plan of the Teatro Farnese in Parma (1610) (after [45]).

During the 17th Century, the most significant evolution of the theatre shape took place. It was due to the transformation of the acting performances, the appearance of the masques, and the renovation of the subjects. An important date, which represented a turning point in the theatre history, was 6th March 1637, when all social classes attended the inauguration of the theatre in campo San Cassian in Venice, by paying a ticket [46]. The paying audience and the new theatrical genre, the melodrama, consequently led to a redefinition of the theatre shape. In fact, the payment of a ticket for the performance made it possible to plan the theatrical seasons and the related investments, thus allowing for building permanent structures and to buy theatrical machineries [47]. The space for the audience became a field of architectural experimentations, with the main purpose of incrementing the seating capacity and improving the visual conditions of the attending public. At the same time, the upper classes began to claim for independent and private spaces, leading to the building of wooden partitions on the different tiers, creating the so-called "boxes". This spatial division reflected the social subdivision into classes. Moreover, such spatial configuration introduced a new kind of performance, where the main actors were the spectators themselves. Not only were the boxes rich and sophisticated showcases for the upper class, but they were also a window from which the richest could observe the common people sitting on the uncomfortable benches of the audience [46].

The first essay dealing with the so-called "Italian theatre" was written by the architect Fabrizio Carini Motta (16??–1699) in 1676 [48]. The author summarised all the architectural characteristics of the theatre and it recognised its different typologies: the 16th-Century typology with steps, with boxes or galleries, with boxes joined or not joined with the proscenium. He identified four main parts called "theatre square", "scene square", "boxes area" and "stage area". Concerning the so-called horseshoe typology, Carini Motta proposed two different models changing the ratio between the width and the length of the scene square, as shown in the author's drawing of Figure 6. The architect also dealt with the design of the partition between boxes: if the partitions were built perpendicular to the audience perimeter, the people sitting behind were not able to see the stage. Considering the possibility of choosing a focal point in the middle of the proscenium arch and building diagonal partitions by pointing to that focus, he noticed that there was unused space due to the presence of acute angles following Vitruvius's recommendation, allowing the public sitting behind to enjoy a better view of the stage area. Influenced by Carini Motta's observations, the Swedish architect Nicodemus Tessin The Younger (1654–1728) introduced some of these elements—rounded off back walls and a royal box—in the design for a theatre at the court of Denmark in the last years of the 17th-Century [5,49]. Furthermore, Carini Motta theorised a new kind of partition, shown in the last drawing of Figure 6. This was a mix between the aforementioned solutions, but it was unfortunately very difficult to realise. If a longer 'scene square' was needed, it was possible to set the wanted length with a line parallel to EF and to draw the biggest circle tangential to that line (see the last draw in Figure 6). According to Carini Motta's theory, all theatre dimensions had to be multiples or fractions of the theatre width. For instance, the height of the audience ceiling should have been higher than 2/3 of the theatre width but smaller than the theatre width itself. On the other hand, he did not give specific indications about the height of the ceiling above the stage area: he only stated that it should have been higher than the audience vault.

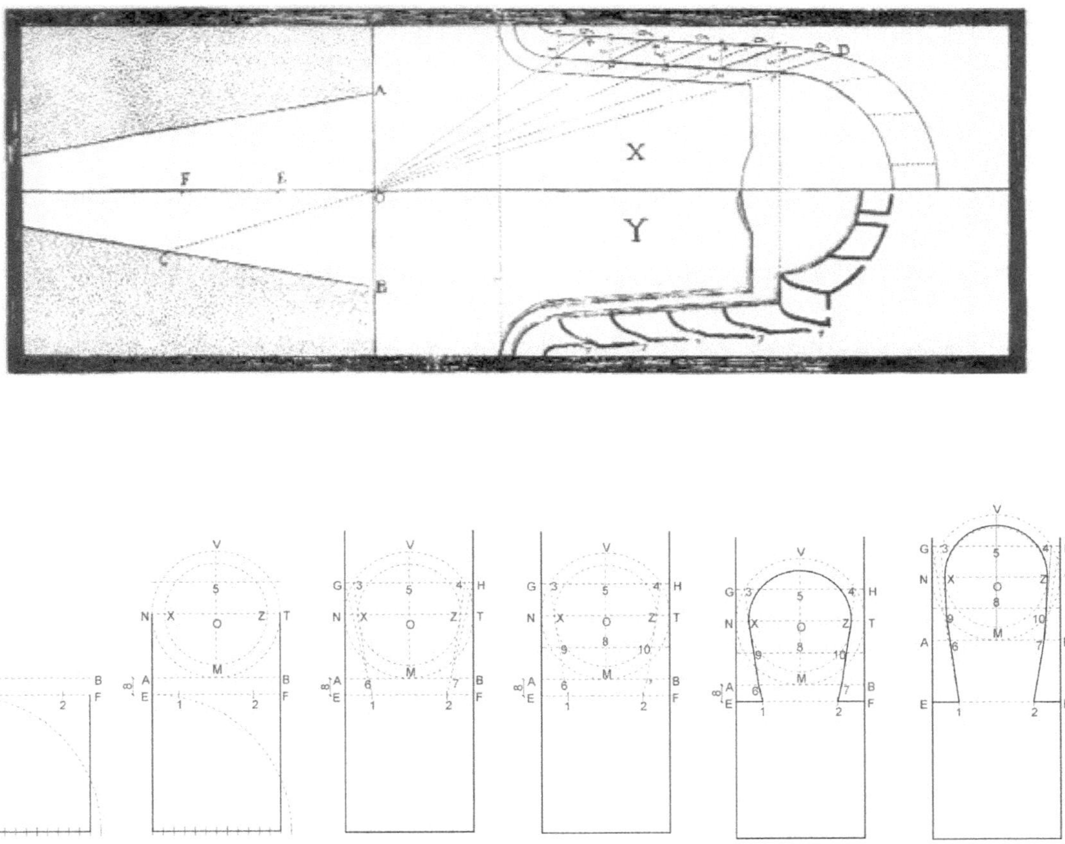

Figure 6. Fabrizio Carini Motta's original drawing (1676) [48] and redrawings [50]. The last figure refers to the possibility of obtaining a longer "scene square" if needed—there are more details in the text.

4. The Evolution of Theatrical Architecture during the 18th Century

During the 18th Century, the form and the functions of the theatre were optimised to host the melodrama: the stage was extended towards the audience—through the proscenium—in order to enhance the intelligibility of the soloists, the orchestra was placed in front of the stage; the boxes' arrangement was adapted for reasons of visibility. The mixtilinear shape of the Italian theatre was developed by many architects, who experimented with new solutions in order to improve the acoustical conditions and the possibility to have exact visuals (see Figure 7).

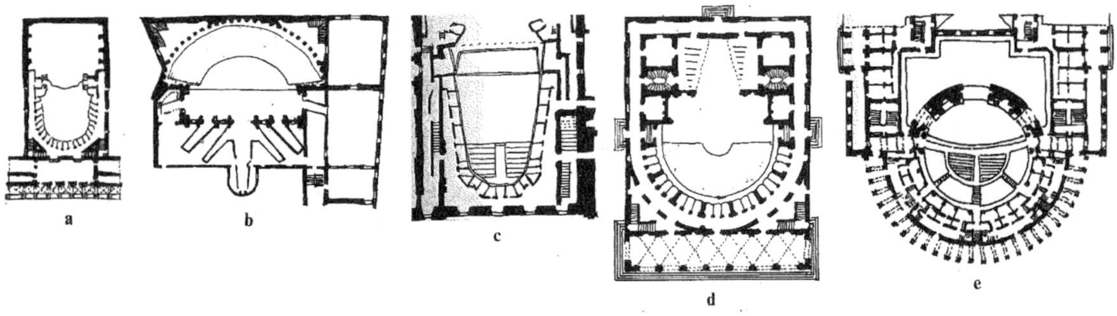

Figure 7. Theatre plan comparison: (**a**) bell shape: Teatro Comunale of Bologna (1763); (**b**) semielliptical shape: Teatro Olimpico of Vicenza (1585); (**c**) open U-shape: Molière hall in the Palais Royal of Paris (1640); (**d**) semi-circular shape: a design by Enea Arnaldi (1762); (**e**) circular shape: project by Vincenzo Ferrarese (1771) (after [36]).

Despite being extremely active in theatre and scenography design [51], the members of the famous Galli Bibbiena family did not write acoustic treatises with which we could directly retrace their thoughts. Nevertheless, some authors referred in their writings to the works by the Galli Bibbiena architects. For example, Francesco Algarotti (1712–1764) ground his study Saggio sopra l'opera in musica (Essay on the Opera, 1764) [52] on the works of Ferdinando Galli Bibbiena (1657–1743) and his sons Giuseppe (1696–1757) and Antonio (1697–1774). The former designed the Margravial Opera House in Bayreuth (1750) and several theatres in Europe; the latter designed the Teatro Comunale in Bologna (1763), the Teatro Scientifico in Mantua (now Teatro Bibiena,1769) and theTeatro dei Quattro Nobili Cavalieri in Pavia (now Teatro Fraschini, 1773) (see the plans in Figure 8).

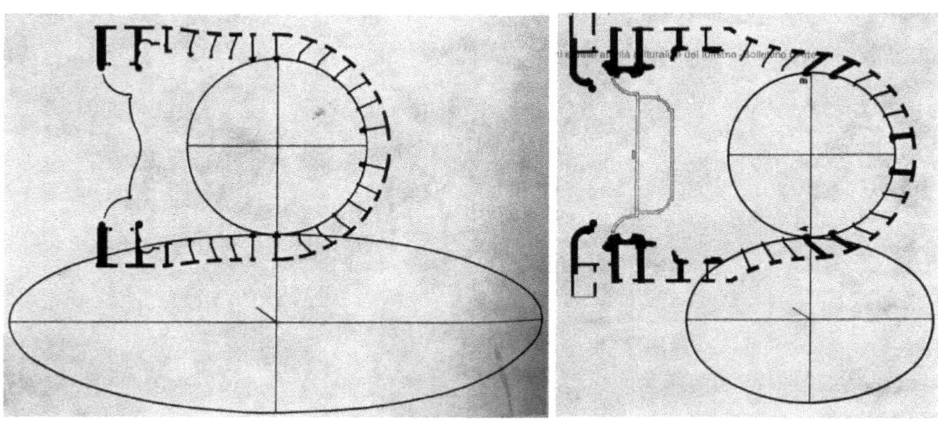

(**a**) Teatro Comunale in Bologna (1763) (**b**) Teatro Fraschini in Pavia (1773)

Figure 8. Plans of two bell-shaped theatres by Antonio Galli Bibbiena [53].

Algarotti agreed with some building techniques used by Galli Bibbiena, e.g., the mixed use of masonry and wood in the theatre: masonry for the bearing structure in order to prevent fire and timber for indoor cladding in order to "equalise" the sound. He proposed the use of timber as cladding, analysed the case of a room—considered as a finite volume—with different boundary conditions (walls, tapestries, timber). Moreover, he linked their effects both on the vocal timbre and on the sound level spatial decay. He also noted that a room with flat walls made the voice too "sharp", while a room covered by tapestry made the voice too "silent". Instead of this, a room cladded by timber made the voice loud and pleasant. Therefore, he stated that the all the wooden parts inside the theatre had to be aged. Building solutions of the Teatro Comunale in Bologna, built by Antonio Galli Bibiena in the same years of Algarotti's essay, reflected this approach. Moreover, Algarotti discussed about the maximum dimensions of a theatre, which should be limited taking into account the sound strength of the voice, quoting Vitrivius's fifth book [6]. Concerning the shape of the audience, Algarotti denied the bell-shape for sake of visibility, although it is ideal for the strength of the voice. He proposed the use of the elliptical shape, being the semicircular shape an ideal solution but unachievable due to the too large width of proscenium. Furthermore, he avoided the decorations with sharp elements, referring again to Vitruvius's dissonantes category. In order to optimise the hall volume, he also proposed that all the tiers of boxes had the same height: the architectural praxis was to increase the width of boxes in the higher tiers in order to emphasise the cavea perspective. Algarotti's essay was extensively quoted in the work by Antonio Planelli (1737–1803), titled Dell'opera in musica. Trattato del cavalier Antonio Planelli dell'ordine gerosolimitano (On the Musical Opera. Treaty by Knight Antonio Planelli of the Jerosolimitan Order, 1772), in the section related to the construction of theatres [54].

Another late 18th-Century author, who analysed the effects of boxes in Italian theatres, was Francesco Riccati (1718–1790) [55–58]. He compared the Italian theatre with the one developed in the European courts, characterised by open galleries, but he also pointed out the benefits, in terms of social connections, that the boxes ensured to their owners. In particular, he did not agree with the presence of not proportioned openings in theatre cavea, which was in contrast with the architectural rules and with an idea of harmony and elegance of the structure. According to Riccati, a theatre should satisfy at least two main requirements: it should ensure the possibility to see and to hear properly. In his book Della costruzione de' teatri secondo il costume d'Italia vale a dire divisi in piccole logge (About the Construction of Theatres According to the Italian Style, i.e., with Small Boxes, 1790) [55], he dealt with both such themes (The essay opens with the following words: "Fra tutte le produzioni dell'Architettura Civile la più disastrosa, la più difficile, la più contumace ad assoggettarsi alle teoriche della scienza, ed ai precetti dell'arte, ella, non v'ha dubbio, è la costruzione de' Teatri giusta il costume d'Italia" (Among all the outcomes of Civil Architecture, the most disastrous, the most difficult, the most insubordinate towards the theories of science, and the norms of art, it is, without doubt, the Italian construction of theatres).). First of all, he identified the curved shape as the best solution for the cavea because it allows for aligning a high number of boxes along the cavea itself, ensuring a sufficient visibility to all of them. Secondly, among the different kinds of curved shapes, he suggested the use of the "divergent" shapes. In fact, the adoption of a concave curve ensured the visibility of the stage to the major part of the public, especially if the part of the curve overlooking the proscenium was made as large as possible. According to the author, the horseshoe shape originated from the degeneration of this attempt. In particular, he disagreed with this particular shape as it enlarged the part of the curve facing the stage way too much, leading to the necessity of narrowing the part of the curve closer to the stage itself. In this way, the boxes located at the sides of the cavea were necessarily disadvantaged. An appropriate shape for the audience was, in Riccati's opinion, the bell shape, which was often adopted by the architects Galli Bibbiena in their theatres. Indeed, this shape could ensure a good visibility to the whole public. Riccati strongly criticised the use of the elliptical shape, which was instead proposed by Luigi Rizzetti (17??–18??), in a piece titled Memoria intorno alla più perfetta costruzione di un Teatro (Essay on the Most Perfect Construction of a Theatre) [59], quoted by Riccati in his treatise and later on published as the small pamphlet Risposta del sig. Conte Luigi Rizzetti alle accuse date al Teatro da lui proposto (Response to the Allegations Made to the Theatre He Proposed, 1792) [60]. Rizzetti's choice of this shape was probably due to a property of the ellipse: the rays coming from a focus reflect one another. For this reason, Rizzetti placed the stage border on a focus of the ellipse, thinking that it would support the sound diffusion. Riccati observed that the public sitting in the boxes at the stage sides would not be able to have comfortable visuals to it because it would directly face the other side of the cavea. Probably influenced by the political and social issues that revolved around the concept of an "ideal" theatre (see next paragraph), Riccati claimed that the best shape was the circular one, although it was possible to adopt it only in small theatres, otherwise the stage—which would have been placed on the diameter of the circle—would become too wide.

Riccati thus proposed an elliptical shape obtained by different circle sectors. These sectors were drawn fixing the circle centre on the diameter of the wanted ellipse. In Figure 9, an ellipse is shown built up with three circular sectors (CF, FG, GD). The sectors CF and GD are drawn centering, respectively, the compasses on the points N and O, located on the shorter diameter out of the ellipse perimeter. The sector FG is drawn centering the compasses on the longer diameter, on the point E, located on the cross point of the segment NF and OG. The whole construction is ruled by the dimension of the segment BE. Increasing this dimension, it is possible to obtain an ellipse with a larger apex, which can contain a consistent amount of boxes facing the stage.

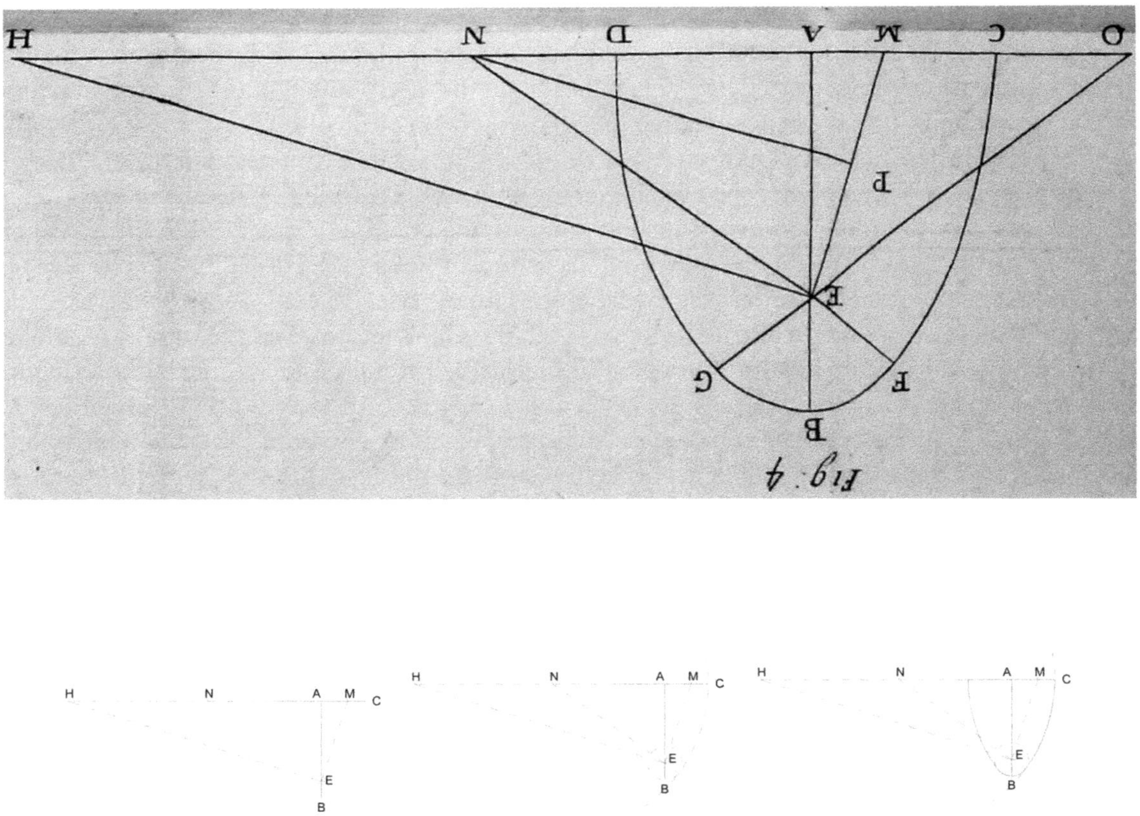

Figure 9. Francesco Riccati's guidelines to determine the elliptical shape of a theatre with circular sectors. Riccati's original drawing (1790) and re-drawings—more details are in the text.

The elliptic shape was used also by Cosimo Morelli (1732–1812) [61]; at the time, one of the most influential architects of the Papal States [62]. Due to visibility properties of elliptical shape, Morelli designed a three-part stage for the theatre of Imola (1780)—see Figure 10. He compared the width of the proscenium of an elliptic-shaped plan (segment GH in Figure 10) with that of the proscenium in the bell-shaped plan (segment EF) and the horse-shoe shaped one (segment DD). Indeed, the proscenium would be wider using the elliptical-shaped design. According to the architect, this configuration allowed also a larger stage for the scenes, a better visibility from the boxes, and a higher occupancy in the stalls.

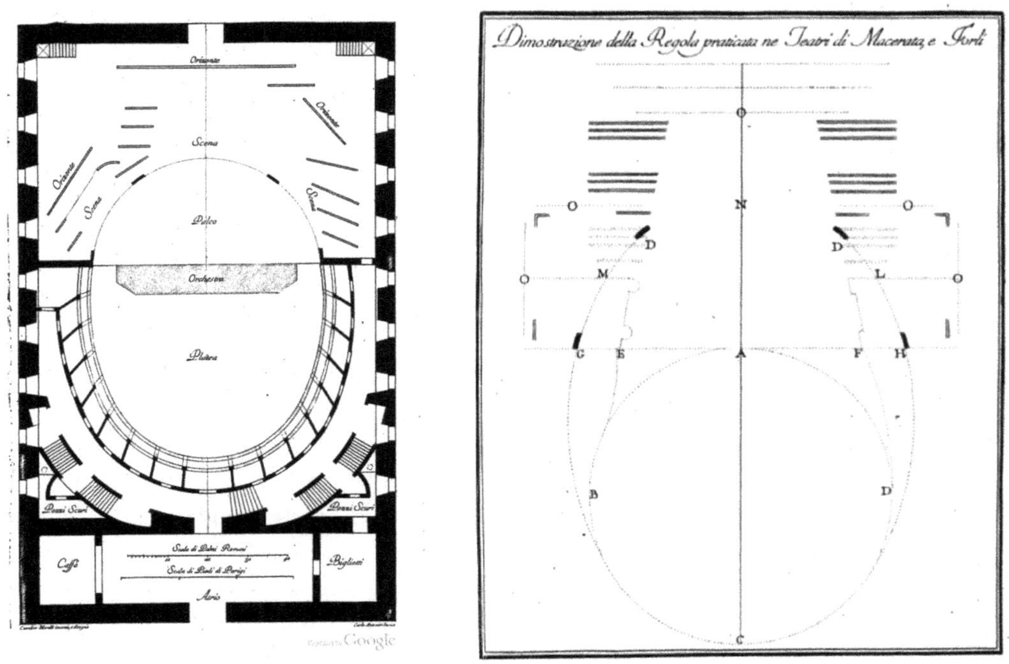

Figure 10. Plan of the Theatre in Imola designed by Cosimo Morelli [61].

The essay by the Venetian author and nobleman Andrea Memmo (1729–1793), titled "Semplici lumi tendenti a render cauti i soli interessati nel teatro da erigersi nella parocchia di S. Fantino in Venezia" (Simple Hints for Those Who Are Interested in the Theatre to Build in the Parish of S. Fantino in Venice, 1790) is an interesting account of the scientific knowledge on the theatrical construction in those years [63]. Far from the political and social ideologies that had been surrounding the idea of theatre in the second half of the 18th Century, Memmo focused on the building characteristics of a space for music and performances. He first pointed out the directivity of the sound sources. In his opinion, the Vitriuvian design of circular-shaped theatres was not optimised to the voice directivity, which decreases with the increase of the angle from the principal direction. Furthermore, the timbre of the voice depends both on the direct sound and the reverberated field. The author thus noted that reflected acoustic rays should converge to a central line and not to a point. In this case, a phenomenon of resonance (circumsonare) can be avoided. He focused on the necessity of balancing sound reflexion and diffusion, with the proper control of concave and convex shapes. At the same time, the direct and the reverberated fields had to be balanced: in particular, the latter could be equalised with a proper choice of the materials. In fact, Memmo was aware of the different acoustic absorption of porous and rigid materials, and he also referred to the problem of the different kinds of wood and their ageing. Furthermore, he stressed the importance of vaults made of light materials, in order to have low frequency absorption due to membrane-resonance effects. Quoting the essay of the Neapolitan engineer Vincenzo Lamberti (17??–18??), who dedicated his work to Memmo himself [64], Memmo referred to the bell shape—named pelcinona by Lamberti—as the optimal shape for a theatre. It is also interesting to notice that the author made no specific distinction between the acoustic and the lighting technique, which on the contrary were usually divided between the concerns of the physicists and those of the scenographers. By reading this essay, it is possible to claim that the technical and scientific knowledge around the building of theatres was extremely high in late 18th-Century publications, despite the lack of written works by the architects who were active in the field. This confirms the refined results by contemporary theatres, such as Teatro alla Scala in Milan (1776–1778) by Giuseppe Piermarini (1734–1808) and the already mentioned theatres by the architects Galli Bibbiena. A similar knowledge was established also by George Saunders (1762–1839), who wrote A Treatise on Theatres (1790) [65]. In this book, Saunders designed a theatre whose dimensions were based on audible threshold of a human voice and prescribed the use of wood in order to absorb the right amount of sound. Saunder's design of theatre was semicircular and showed some similarities with Enea Arnaldi's treatise (see the next section).

These last two sections have dealt with still existing Italian Opera houses, but it is important to keep in mind that several others were demolished or burned down. These events contributed, as stated in the Introduction, to a "Darwinian" evolutionary process of the optimisation of the shape and of the technical solutions of the theatres that were rebuilt after their demolition. For a complete chronology of the Opera houses of this period, see [66].

5. The Ideal Theatre

Throughout the 18th-Century, a widespread theoretical debate exploded on the social role of the theatre. Several French authors and theorists, such as Voltaire (1694–1778) and Diderot (1713–1784) [67] among the others [68–80], wrote on the educational importance of theatre performances. At the same time, the model of the classical, circular theatre was taken at the ultimate formal example [81,82], thus mirroring the necessity to erase the strict social division that characterised the partition between the audience and the boxes. Indeed, equality in listening and seeing had to match social and political equality, as a forerunner of the ideologies that led to the French Revolution. For a precise analysis of the connections between Italian theatres and French theoretical texts, see [83,84]. If in France the only architectural outcome of these theories was the theatre of Claude-Nicolas Ledoux (1736–1806) in Besançon (1775–1784) [81], these experimentations had no physical results in the Italian context. Nevertheless, architectural theorists like Enea Arnaldi (1716–1794) and Francesco Milizia (1725–1798) published on the topic of the "ideal theatre" and, most probably, they directly influenced some of the French designs [83,84].

In 1762, the architect Enea Arnaldi published in Vicenza the essay titled L'idea di un teatro nelle principali sue parti simile a' teatri antichi all'uso moderno accomodato (The Idea of a Theatre Being Similar in its Parts to the Ancient Theatres and Adjusted to the Modern Use) [85]. His ideal theatre merged the characteristics of the ancient models—of Vitruvian origins—together with the Italian typology. In fact, he added a series of semi-circular steps to the audience and gave a cylindrical shape to the orchestra pit (see Figure 11).

Figure 11. Section of the theatre proposed by Enea Arnaldi [85].

Shortly after, in 1773, the theorist Francesco Milizia wrote an essay about the Italian theatre [86]. He compared the so-called modern theatre with the ancient one, thus complaining about its degraded conditions. The author observed that modern theaters were built with flammable materials and they were affected by humidity problems, whereas the main features of such constructions were supposed to be robustness, comfort and convenience. Among all European theatres, he only mentioned as "solid" those of Turin, Naples, Bologna and Berlin. Examining the shape of the auditorium, Milizia also highlighted the inappropriate form adopted for the Italian theatre. He compared the semi-circular ancient theatre, where all the spectators were able to see and hear properly, with the shape of the modern theatre and its boxes, that did not allow satisfying visuals to the whole public. He wrote: "the geometry itself demonstrates that, in a circle, all the angles of the circumference, which have as their basis the same diameter, are equal" [86]. For this reason, the author recommended assuming a circular shape and to place the stage on the main diameter, as the only way to guarantee the same perspective on the performance to each person in the audience. In addition, Milizia criticised the presence of boxes, which, in his opinion, divided the volume of the theatre in too many small spaces. On this, he wrote that "the boxes cut and reverberate the resonant air", thus confusing the perceived sound. In order to strengthen the criticism of the modern theatre shape, Milizia necessarily quoted Vitruvius. In fact, the Latin author affirmed that the inclination of the ancient theatre was chosen following musical proportion and it permitted a visual contact between the listeners and the actors. On this issue, Milizia published the project of an ideal theatre, called Teatro Ideale (1773). Taking example from Arnaldi's theory, he designed a more radical project, introducing French revolutionary themes into theatre design. In the Teatro Ideale, the architect replaced the separated boxes, reminders of the class divisions, with open galleries. The ideal theatre was characterised by a semi-circular shape, semi-circular ceiling, cavities and resonators, and galleries. On one hand, Milizia's assumptions were not acoustically appropriate and this was shown by the numerical simulation of the Teatro Ideale, realised from the original drawings [12]. In fact, considering the overall values of the acoustic descriptors the Teatro Ideale appears as a reverberant theatre—$T_{30} = 2.6\,\mathrm{s}$ at 1000 Hz in unoccupied conditions—more than the existing Italian theatres of that period [15,19,20]. On the other hand, Milizia's theoretical contribution was extremely important when it comes to the social role of the theatre, which was seen as a public space. Indeed, his struggle for a more democratic spatial configuration of the audience could be seen as a forerunner of Richard Wagner (1813–1883) and his "democratisation of acoustics" [87]. While the 18th-Century theatre was still a place for aristocracy, Milizia saw in theatre performances a means to convey the social image of a more equal and educated society, able to mirror the idea of a nation—the Italian one—still scattered into different states (see Figure 12).

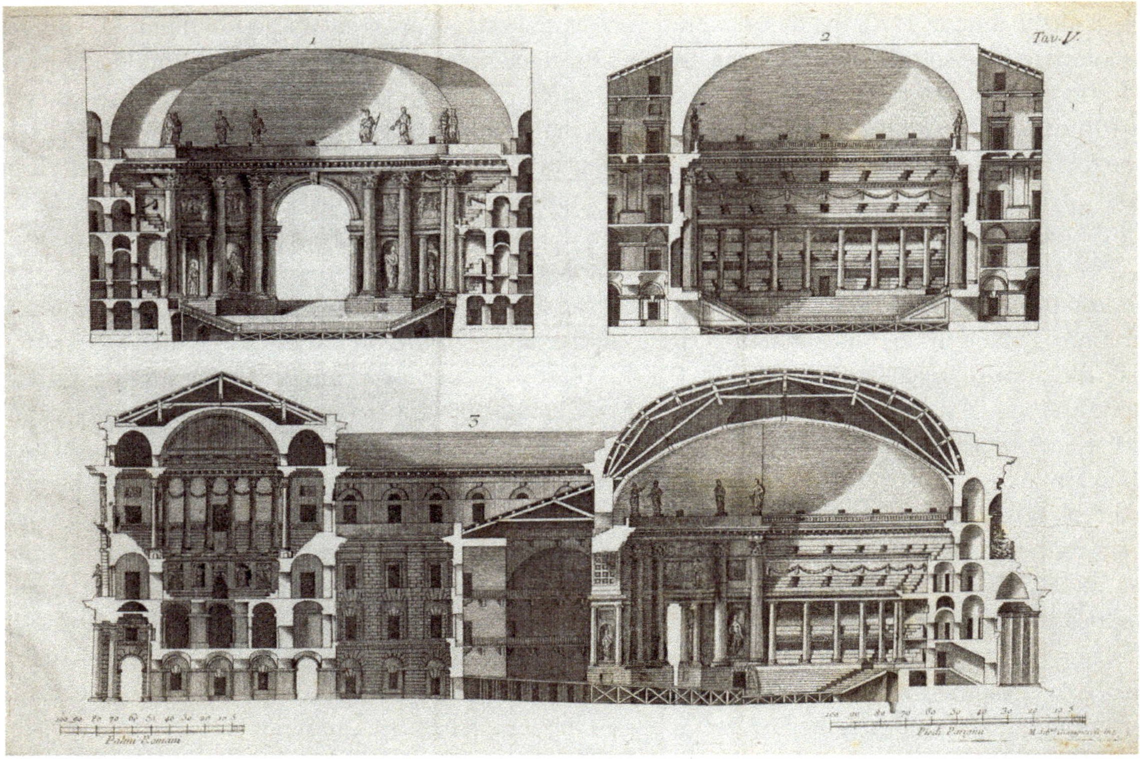

Figure 12. Design of the "Ideal Theatre" by Francesco Milizia and Vincenzo Ferrarese [86].

Milizia's writings strongly influenced the work of the architect Giovanni Antonio Antolini (1753–1841) who designed the grand "Foro Bonaparte" (1801) in Milan, in the area surrounding the Sforza Castle. Within the unbuilt architectural complex of the "Foro", Antolini introduced Milizia-Ferrarese circular-shaped theatre—that he named the "school for morals" [88]—equally divided between the stage and the semi-circular steps for the audience (see Figure 13).

Another follower of Milizia's theories was Tommaso Carlo Beccega (17??–18??), who published in 1817 an essay titled Sull'architettura greco-romana applicata alla costruzione del teatro moderno italiano e sulle macchine teatrali (On the Application of the Graeco-Roman Architecture to the Construction of the Modern Italian Theatre and to the Theatrical Machines) [89]. The basis of this essay was Vitruvius's theory and, at the same time, he tried to develop a theatre standard, whose shape was a semi-circle with elongated extremities (see Figure 14). Recalling the Vitruvian principle of comfort, he deducted the stage inclination, the orchestra pit position—lower than audience in the stalls—the proscenium arch shape and proportion, the construction of the dividers between boxes. He also claimed that the theatre height should have been equal to the cavea width, in order to avoid excessively deep halls. He fixed the maximal length of the audience, which should not exceed 45 m, the recognized audibility threshold.

Figure 13. Giovanni Antonio Antolini's design of a theatre in the Foro Bonaparte in Milan (after [83]).

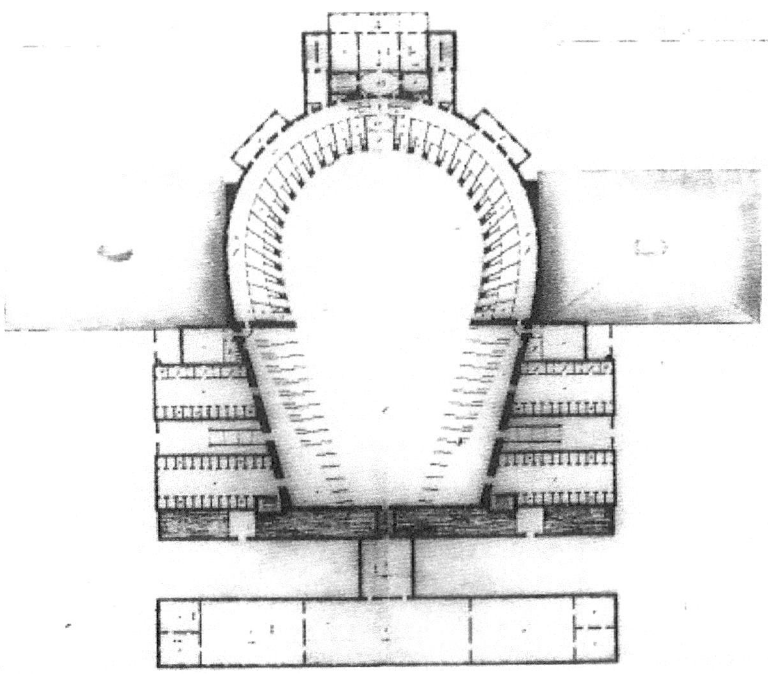

Figure 14. Ideal theatre shape according to Tommaso Carlo Beccega (1817) (after [83]).

6. The 19th-Century Theatre

The evolution of the 18th-Century into the 19th-Century theatre has not been diffusely discussed in architectural essays. However, it is clear that the theaters built during the 19th Century presented similar characteristics to the buildings belonging to the preceding century. By the mid-19th Century, the form of theatre was standardised and the vast majority of Italian opera houses was built after this period, in order to satisfy the request of each small town to have its theatre and the request of bigger cities to build increasingly larger ones.

At the beginning of the 19th Century, Antonio Niccolini (1772–1850) published a pamphlet entitled Alcune idee sulla risonanza del teatro (Some Ideas on Theatre Resonance) [90], in which he attributed the theatre deafness to its proportions and, exactly as Milizia did, to the use the boxes [83]. Differently from Milizia, Niccolini understood that the modern theatre characteristics derived from social and economical needs; therefore, he elaborated solutions in order to improve the acoustic of these theatres, taking into account the circumstantial necessities. Niccolini's research followed a completely different path than those of his predecessors: he investigated the stage dimensions—and compared to the dimensions of the audience—and the vault continuity as deafness causes. In his opinion, the shape of the theatre and the used materials were irrelevant. The first element significantly contributing to the theatre resonance was a wide stage opening, which conveyed an air flow from the stage area to the audience (see the Niccolini's plan of Teatro San Carlo in Figure 15)—where the air was more rarefied because of the presence of public—thus reinforcing the sound propagation. The second element was the vault which in his opinion should ensure the uniformity of the sound path.

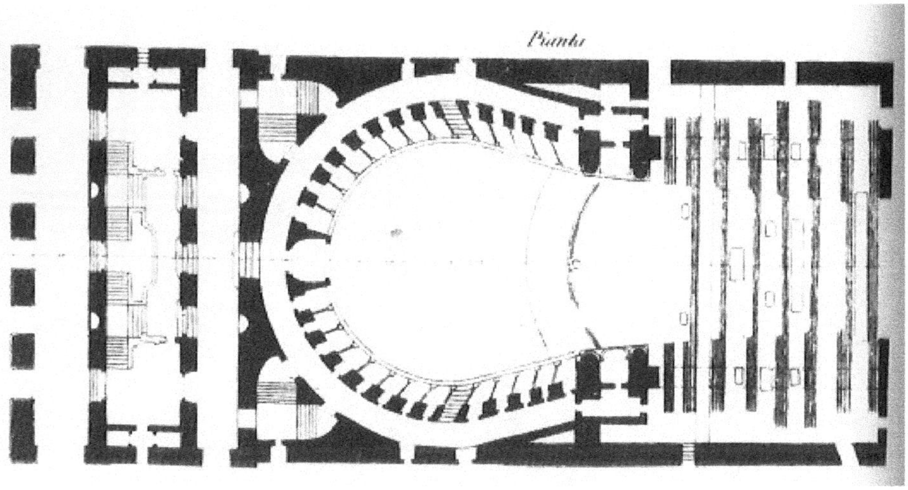

Figure 15. Antonio Niccolini's design of San Carlo theatre in Naples (1816) [90].

The architect and scenic painter Paolo Landriani (1757–1839), in his Osservazioni sui difetti prodotti nei teatri dalla cattiva costruzione del palco scenico e su alcune inavvertenze nel dipingere le decorazioni (Observations on the Flaws Deriving by the Poor Construction of the Stage and on Some Oversights in Decoration Paintings) [91], provided a description of a common method used to determine the theatre shape in that period. The usual shape adopted for theatres was a curve composed by a semi-circle and two lines, with a focal point at a distance of two diameters of the aforementioned semi-circle. The whole length of the theatre was usually similar to the length of the circle diameter; the proscenium width was determined by the intersection of the aforementioned lines with a line tangent to the circle. It could also happen that the length of the audience was longer than the circle diameter: therefore, the proscenium width could be determined as explained in Figure 16. It was the author's firm belief that the acoustic quality of a theatre depended on: the horseshoe shape commonly adopted; the lack of interruption or projections in the cavea boundary; the wooden parapets of the boxes; the elliptical vault above the audience, which should be as flat as possible; the proscenium opening, which should be wider on the audience side; a slightly inclination of the proscenium arch ceiling towards the stage, so that the actors' voices could spread easily into the auditorium as a ray.

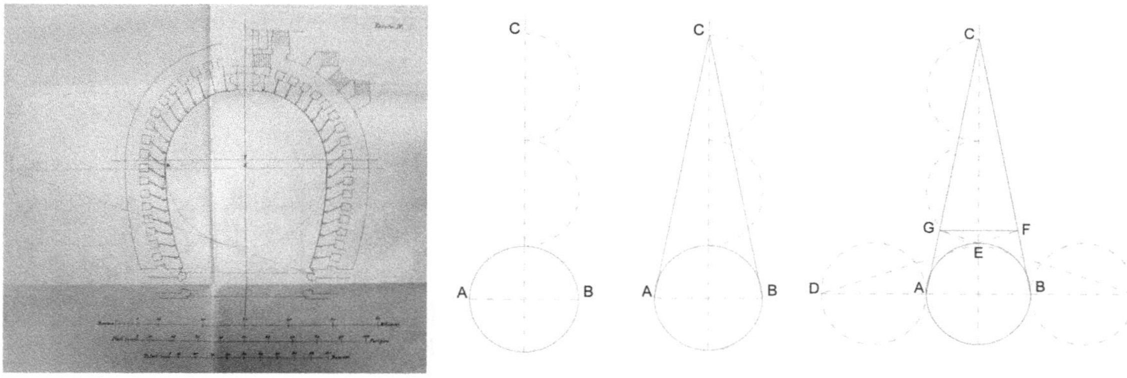

Figure 16. Paolo Landriani's guidelines to determine the proscenium width. Landriani's original drawing (1828) and redrawings [50]. See details in the text.

As stated before, theatres were prone to fires and therefore they were often rebuilt. Just like Niccolini, also the Venetian architect Giovanni Battista Meduna (1800–1886) and his brother Tommaso (1798–1880) had to deal with the reconstruction of a theatre. In their case, they rebuilt the theatre La Fenice in Venice, already debated by Memmo in his essay: the theatre was destroyed by fire in 1836 and it was reopened in 1837 [92]. Meduna's method is still "classic", regarding both the shape of the hall and the use of the materials. In particular, the Meduna brothers used the properties of timber in order to create resonant volumes under the stalls [93].

Already in the first decades of the 19th Century, the shape and the building techniques for theatre construction were almost consolidated: therefore, contemporary scholars focused on manuals or historical overviews [94–97], which progressively accounted for a theatre form with standard shapes and building typologies. It is not a coincidence if, in the same years, some Italian composers—such as Gaetano Donizetti (1791–1848), Vincenzo Bellini (1801–1835) and Giuseppe Verdi (1813–1901)—wrote some of the most iconic Italian operas [98]. The popular fame of these operas is strongly connected to the need of building theatres in every town, thus "scaling" the geometries on reduced volumes and occupancy [16–18].

From the second half of the 19th Century onwards, there was a progressive change in the building typologies of theatres. For example, the use of steel (such as in the Politeama theatre in Palermo, designed and built between 1865 and 1891) [99] and the use of cast-iron in the bearing structures, thus replacing timber or masonry, had resulting effects on the acoustic of such spaces [100,101].

The last moment in which Italian Opera houses were extensively built can be located in time just before the arrival of cinema. In those years, some great Sicilian theatres were built: like the Bellini theatre in Catania (inaugurated in 1890) and the Massimo theatre in Palermo (inaugurated in 1897), which is still today the biggest theatre of the Italian peninsula [102,103]. These last theatres were deeply influenced by geometries and spatial schemes that could not be described as merely Italian anymore, but "European", as they were affected by French (Opéra Garnier, 1875) and Austro-Hungarian (Wiener Staatsoper, 1869) designs. In fact, they do not only show boxes, but also galleries; they have large foyers that could host the new bourgeoisie, together with the high classes.

A reverse exemple—that is, an Italian architect working in an European architectural background, as the architects Galli Bibbiena had done in the former Century—was the architect Antonio Canoppi (1769–1832). In 1830, he published his Opinion d'A.C. sur l'architecture en général et en spécialité sur la construction des théâtres modernes (The Opinion of A.C. on Architecture in General and in Particular on the Construction of Modern Theatres) [104]. He was an Italian architect who had been working in Russia since 1805 as theatre director for the Russian Empire. He designed two theatres in St. Petersburg (see Figure 17), whose occupancy were, respectively, of 2200 and 1000 people. Thanks to studies on optics, perspective and acoustics, he fixed the length of the cavea—in the larger theatre—at

25.8 m and the proscenium width at 12.9 m. The dimensions of the stage were very large, in order to host monumental scenographies.

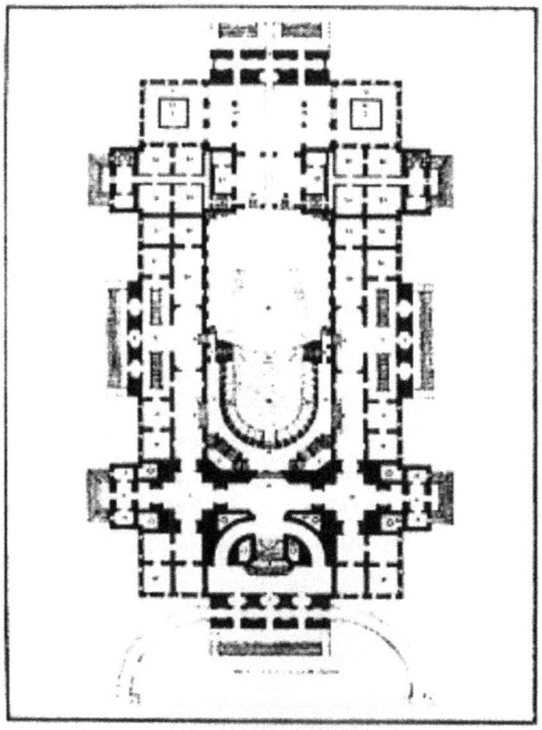

(**a**)large-sized theatre

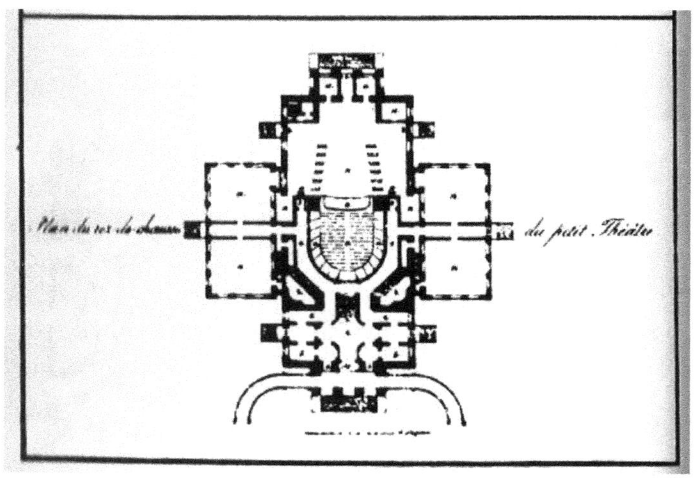

(**b**)small-sized theatre

Figure 17. Antonio Canoppi's drawings of two theatres in St. Petersburg (after [83]).

In order to better understand the chronological evolution of each hall shape, together with their analysis in literature, Figure 18 represents a timeline overlapping the construction year and shape of each mentioned theatre, together with the life spans of the debated authors.

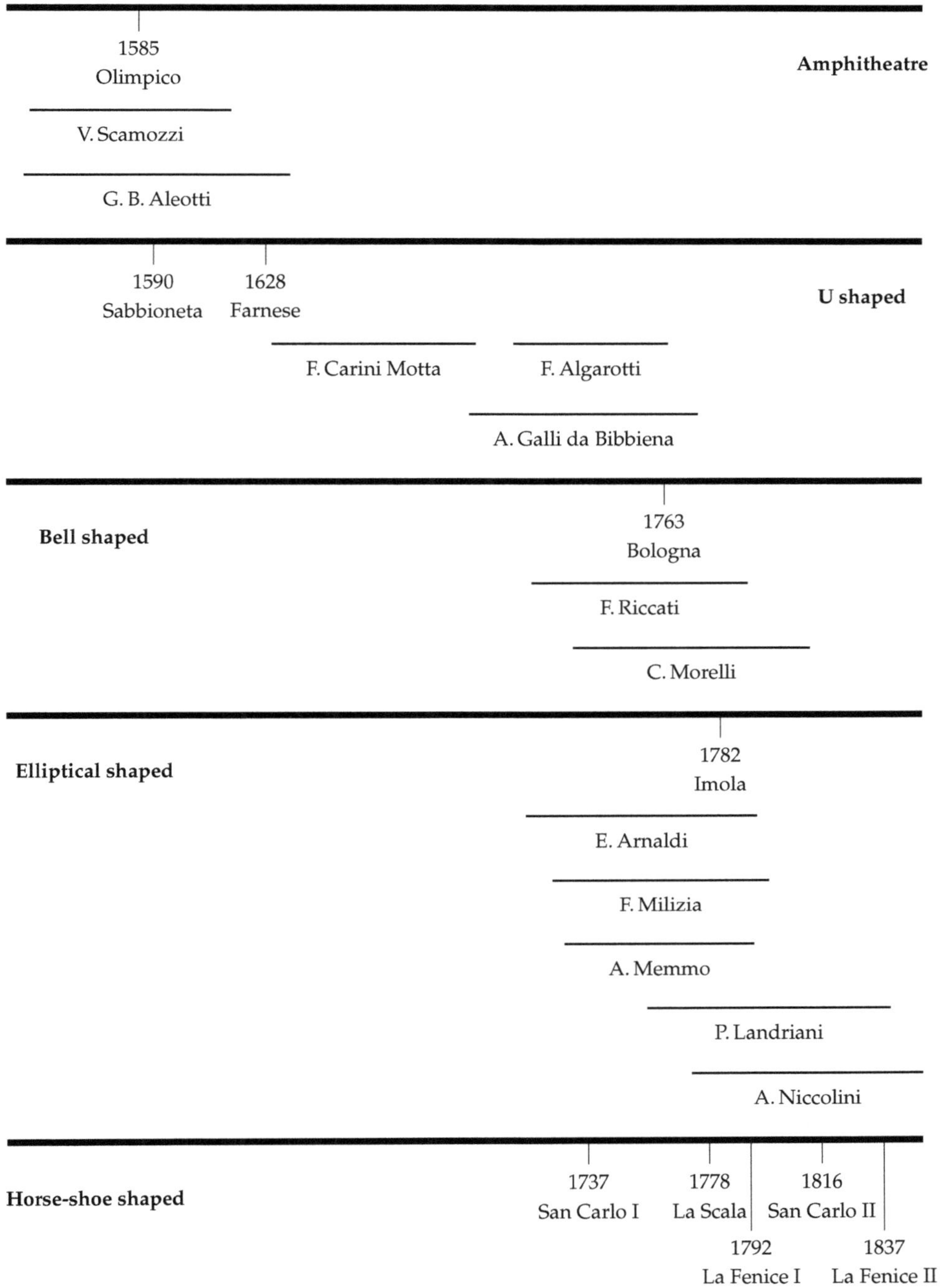

Figure 18. Evolution during time of the different shapes for Opera houses. In particular, some of theatres mentioned in the text have been marked on each timeline, together with the life span of mentioned authors.

7. Evidences in Small–Sized Theatres of the 19th Century

Five small-sized Italian Opera houses have been selected as case studies because of their similar dimensions and social function. All of these theatres were designed and built in the 19th Century: Municipal Theatre of Cervia (CER), Municipal Theatre of Cesenatico (CES), Municipal theatre of Russi (RUS), Dragoni theatre of Meldola (DRA), Petrella Theatre of Longiano (PET); see further details in Table 1. Such theatres were chosen due to the availability of the original designs in the city archives (see Figure 19) and a complete documentation about the restoration works of the last several decades [36,105–112].

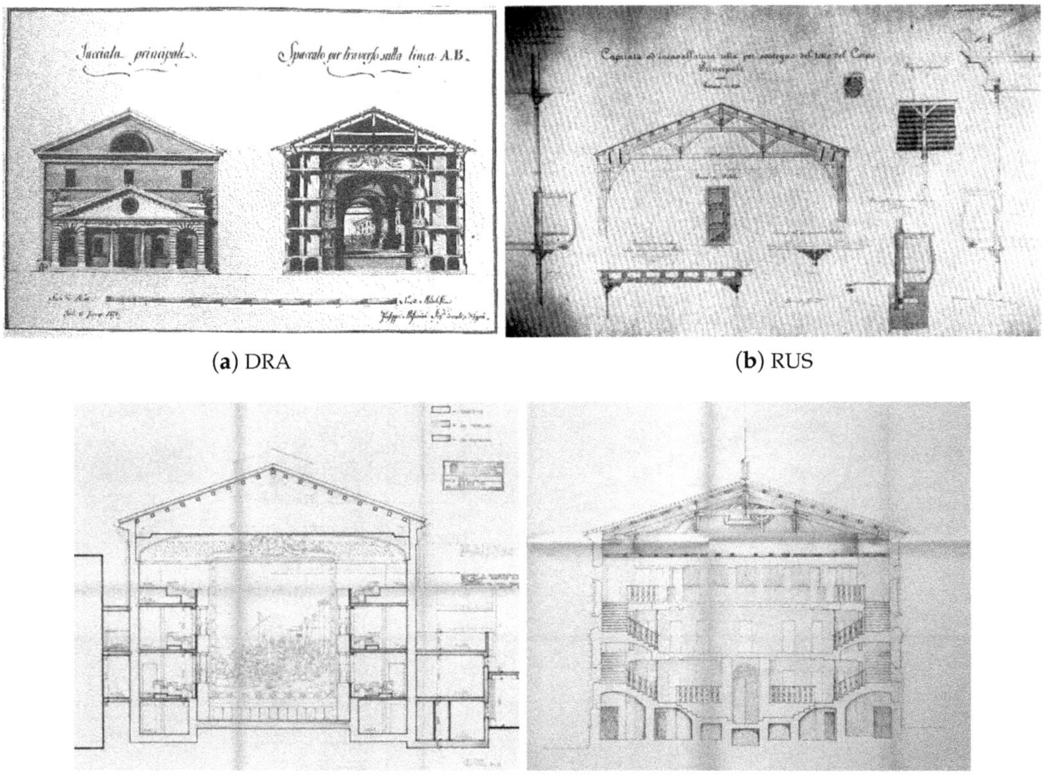

(**a**) DRA (**b**) RUS

(**c**) CES (**d**) PET

Figure 19. Some original drawings of the small–sized opera houses under study. Such drawings are available at the local city archives and local publications [36,105–112] and represent one of the key sources for the historical research on such buildings, collected in [50].

Table 1. Architectural features of the small Italian historical Opera houses under study. $T_{30,M}$ is the measured value of Reverberation time at mid-frequencies averaged over the whole audience. Other measured room criteria are fully available on [19].

	Theater, City	**Seats**	V_{hall}	V_{stage}	**Scenic Arch**	$T_{30,M}$ **(s)**	**year**
DRA	Dragoni, Meldola	318	1140	1080	6.8×7.3	0.83	1838
RUS	Comunale, Russi	305	900	1370	8×7.5	0.97	1887
CES	Comunale, Cesenatico	271	870	1320	7.4×8.6	0.90	1865
CER	Comunale, Cervia	224	730	1140	8.1×7.4	0.84	1862
PET	Petrella, Longiano	241	630	1390	7.5×9.1	1.07	1870

A subset of these theatres can be made considering their shape: for example, CES and MEL show an elongated shape. Nevertheless, the two theatres show a significant difference, as CES has two boxes included in the proscenium arch (one on each side). Its shape is thus characterised by a linear part linked with the horseshoe shape. Among the others, CER and RUS show exactly the same bending. PET, instead, is very similar in its bottom part, even if it is smaller than the other two, but with a linear evolution closer to the stage. Considering the stage area, all theatres show more or less a rectangular shape, with the same irregularity due to the presence of service rooms. CER is an exception: because of urban necessities, the stage area is characterised by a strong asymmetry and the rear wall of the stage is not parallel to the proscenium arch.

The aforementioned guidelines by Carini Motta, for the design of horse-shoe theatres, may be applied to these case studies. Following the author's instructions, an elongated shape with a linear part close to the proscenium arch can be obtained: therefore, this method has been applied to the theatres of CES and DRA, which present similar features Figure 20. The shape obtained following Carini Motta's method, with the dimensions of CES, is an offset of the real curve defining the audience boundary of the theatre. Considering the dimensions of DRA, the shape obtained is only a little too rounded on the sides of the audience, but, in general, it corresponds quite well to the real theatre shape. It also tried to apply Carini Motta's method to a different geometry, such as CER, yet with unsatisfactory results.

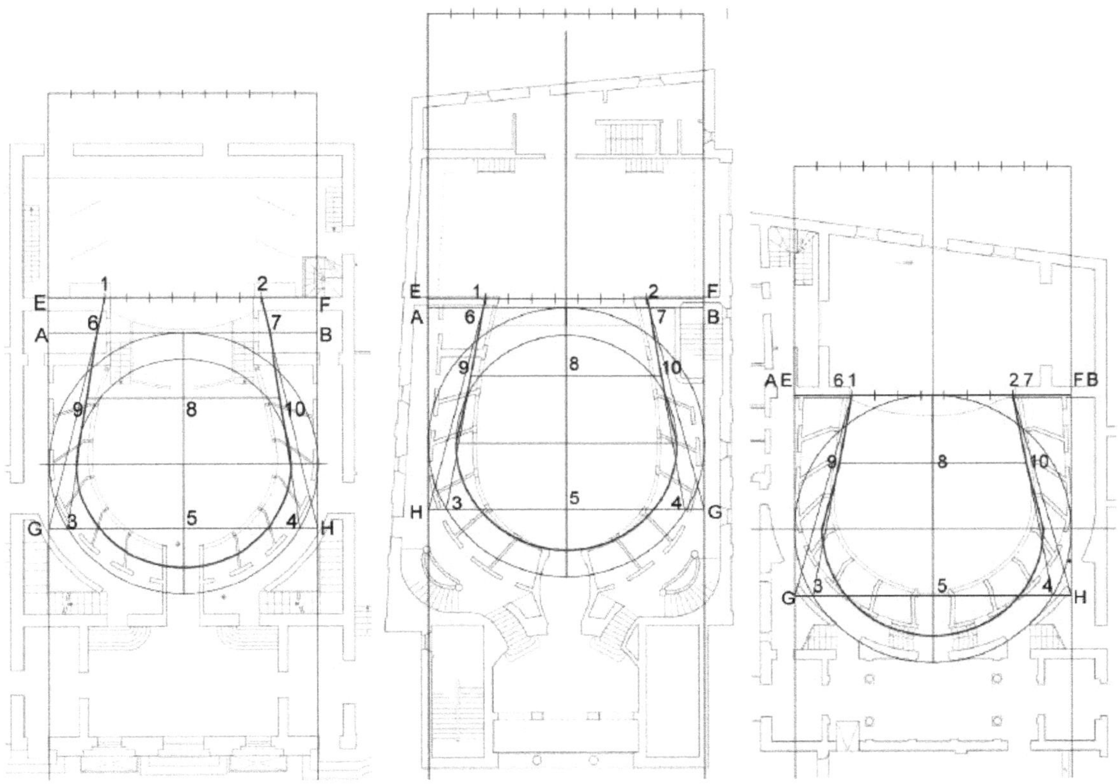

(a) CES (b) DRA (c) CER

Figure 20. Attempts to apply Carini Motta's guidelines for drawing horseshoe-shaped theatres.

In Figure 21, Landriani's method to design the proscenium arch was applied to the small-sized theatres under study. This method has been first applied to the theatres with a more rounded shape, such as CER and RUS. For both theatres, the shape obtained is quite close-fitting to the real one. However, Landriani's method to define the proscenium arch width did not give satisfying results. The figures of CER and RUS show that the shape obtained following Landriani's guidelines is bigger if compared to the boundary of CER audience, while it is a bit smaller in the part close to the stage in the case of RUS. This means that RUS is defined by a more rounded shape. The same method was also applied to the theatres with a more elongated shape (that is, DRA, CES, PET). In DRA, the shape obtained fits quite well with the rear part of the audience, but it becomes too narrow close to the stage. The same observation also counts for PET; instead, Landriani's method applied to CES produced a tight-fitting shape. As for the theatres of CER and RUS, Landriani's instructions to define the proscenium arch width did not give satisying results.

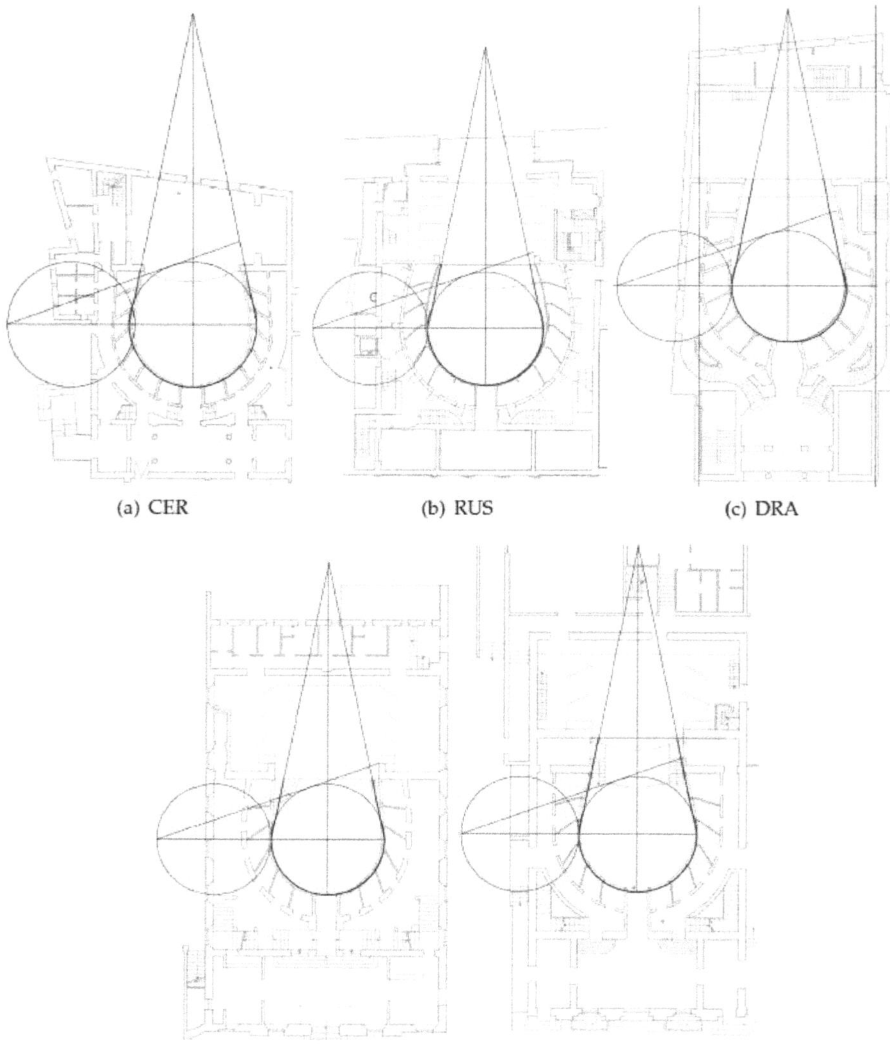

(a) CER (b) RUS (c) DRA

(d) PET (e) CES

Figure 21. Application of Paolo Landriani's guidelines for dimensioning the proscenium width.

Such design features may influence the acoustic criterion of the hall. Despite of refurbishment differences in each theatres—changes in velvet, chairs and other elements and materials—the acoustic behaviour at the present day may still be dependent from design. The theatres under study were measured in a previous campaign of theatres in Romagna [19]. The positions of the omnidirectional sound source [113] on the stage were chosen similarly in all theatres: the first position on the longitudinal axis of the stage at 1 m from the edge (fore-stage); the second one at the centre of the area closed by the curtains (centre-stage). ISO 3382 [114] criteria were extracted from the measured impulse responses [115].

The values of reverberation time are quite low, if compared with other Italian Opera houses of bigger dimensions—swinging between 1.2 s (La Scala Theatre in Milan), those classified as "type B" by Prodi et al. [20], and 1.7 s (Teatro Comunale in Bologna), those classified as "type A". As the reverberation time appears to be quite constant on all the audience, due to the complex connection between geometry and positions, it may be more useful to analyse a value that depends on the positions of the source and the receiver, such as the sound strength [114].

Table 2 summarises the measured values of G_M, using the two source positions described above (one in the proscenium and one at the centre of the stage), and also the averaged values on all the receivers of the same categories (all positions in the stalls, boxes and galleries). In general, results show how the position of the soloist on the stage influences the sound strength in the audience. G_M values in the fore-stage are higher that the ones in the centre-stage. This may be due to the higher acoustic coupling when the soloist is placed in the fore-stage. This situation was well known by Opera authors, who often stressed in their librettos the right place of each aria, in order to balance the sound level of each soloist, the choir and the orchestra [87].

Table 2. Measured values of Sound Strength at mid-frequencies G_M in the theatres under study (dB).

Sources:	Proscenium			Center-Stage		
Receivers:	Stalls	Boxes	Gallery	Stalls	Boxes	Gallery
DRA	10.2	7.3	5.5	8.1	6.2	4.5
RUS	11.0	8.6	7.3	8.7	7.4	6.8
CES	11.4	9.7	9.4	9.3	8.4	8.4
CER	16.0	12	12.8	13.0	12.0	11.9
PET	15.7	10.4	11.5	11.8	9.0	10.3

The use of a velvet band within the cavea of DRA—introduced by the renovation works of 1970s [36,105,108]—is for sure the cause of a low *G* value in the stalls. However, in this theatre, the shape factor of the hall is different from the others, as DRA is much higher than what the cavea surface would allow. This influences a low *G* value in the gallery.

CER and CES, as stated above, are very similar. The choice of drawing the stage further back, in CES, in order to insert a small orchestra pit, has a negative influence on the sound strength (about 3 dB are lost, for each couple source-receiver). In CES, the presence of the original design for the proscenium shows a sound strength homogeneity in the audience. In fact, apart from the values in the stalls for a source on the proscenium, that are necessarily influenced by its proximity, *G* values are more or less constant in all audience, gallery included. This behaviour mirrors the scope of a good design of a theatre built for a community, which is a city theatre.

If RUS is too wide if compared to Landriani's method, PET fits perfectly into these guidelines. The latter was restored by preserving the original materials, as the former was renovated into a small contemporary prose theatre. PET also shows a great result of acoustic coupling between the fly tower and the cavea. *G* values are constant, with the exception of those in the stalls, due to the proximity effect.

To conclude, the measures of sound strength and its spatial distribution show that the design of a theatre hall is mainly based on the shape of the theatre itself, more than on the materials used (this is valid for small-sized Opera houses with a horseshoe shape, such as the aforementioned case studies). Restoration works may influence negatively on the acoustic behaviour of the halls, yet the influence of the original design still plays an important role.

8. Conclusions

This study had the aim of summarising the copious literature on theatre design related to pre-Sabinian scholars, from the 15th- and 16th-Century rediscovery of Vitruvius's treatise to 19th-Century standardised applications of guidelines and instructions by several Italian authors. The aim was to draw evidence of such literature in the design of Italian Opera houses, with a particular attention to small-sized Opera houses to be found in minor Italian towns.

In a contemporary acoustic point of view, the pre-Sabinian Italian scholars pointed out all the critical aspects of theatre design: Aleotti the relevance of the scenic arch [9], Carini-Motta [48] the separation of the acoustic volumes of the stage and the audience, Milizia [86] the coupling of the boxes, Riccati [55] the homogeneity of direct sound to the listeners, Rizzetti [60] the diffusion of the sound, Niccolini [90] the energetic transmission between stage and cavea and the relationship between reverberation and acoustic mean free path, and Beccega [89] the geometric ratios of the theatre and the echo-flutter phenomena. The dualism optics-acoustics might have often been useful to these authors, the knowledge in optics at the times being more advanced than that in acoustics.

Some of the topics that emerged from this study are at the core of the acoustic properties of Italian Opera houses. The majority of existent and active Italian theatres was bult between 1750 and 1880 and is characterised, for example, by a vast use of timber, whose effects have only partially been studied so far. Recent renovation works and the adaptation to fire-escape regulations have reduced the occupancy of such theatres. Moreover, their acoustic quality does not mirror the acoustic conditions at the time of their construction and full functionality in the 19th Century. In the authors' opinion, this may not necessarily be seen as a problem as the theatre has always been a flexible building, adaptable to the changes of the audience and to the functional necessities of the performances. The knowledge of such literature could instead be of help in the case of refurbishment of a historical theatre or of a new building with new materials, which still aims at preserving the acoustic characteristics of Italian Opera houses.

The birth of architectural acoustics as an independent science dates back to Wallace Clement Sabine's work. Interestingly, some Vitruvian categories were still used by Sabine himself [1], regarding speech intelligibility. Nevertheless, some of the main issues of architectural acoustics had already been consciously tackled by pre-Sabinian Italian scholars in their treatises. Although the concept of reverberation was not explicitly known, this literature shows a particular attention to the effects of materials, related to both intelligibility and abatement. In fact, timber was used not only for its ability to radiate sound energy (the analogy with the musical instruments is often recalled), but also for its capacity of absorbing part of the sound energy and thus equalising the sound. Moreover, issues related to the echo-flutter were tackled, both focusing on the elliptical shape and on the maximum dimensions of the hall. Furthermore, the Italian Opera house is a system of acoustic coupling of volumes. The fly tower was born for scenographic reasons—it is missing in the first performance spaces, but it soon became a critical element for the correct acoustic coupling of sound sources on the stage. On one hand, the fly tower cannot be too big for scenographic issues; on the other hand, some geometries like the bell and the ellipse enlarge the dimension of the scenic arch in order to host a sufficient number of people in the audience. This need is perhaps the reason why the horseshoe-shape eventually took hold.

To conclude, by quoting one of the authors analysed in the paper, this sentence by Antonio Niccolini—designer of the new hall of the Teatro San Carlo in Naples (1817)—may be still valid today: "L'indole della propagazione del suono è tuttora nel nostro teatro reputata da molti come un mistero ignoto, e da molti altri erroneamente definita" (Until the present day, the nature of sound propagation is considered by many as an unknown mystery, and it is erroneously defined by many others).

Author Contributions: Conceptualization and methodology, D.D.; Investigation, D.D. and S.N.; Writing—Original Draft, D.D. and S.N.

Funding: This research received no external funding.

Acknowledgments: The authors acknowledge Elena Bo for her contribution about ancient theatre and the work of Vitruvius, Lorenzo Fecchio for his suggestions on the works by Sebastiano Serlio and Andrea Palladio, and Valentina Sartini for bibliographic research and for the use of Figures 20 and 21.

Conflicts of Interest: The authors declare no conflict of interest.

References and Notes

1. Sabine, W. *Collected Papers on Acoustics*; Harvard University Press: Cambridge, MA, USA, 1922.
2. Kwon, Y.; Siebein, G.W. Chronological analysis of architectural and acoustical indices in music performance halls. *J. Acoust. Soc. Am.* **2006**, *128*, 654–663.
3. Barbieri, P. The acoustics of Italian opera houses and auditoriums. *Recercare* **1998**, *X*, 263–328.
4. Barbieri, P. The state of architectural acoustics in the late renaissance. In *Architettura e Musica Nella Venezia del Rinascimento*; Howard, D., Moretti, L., Eds.; B. Mondadori: Milan, Italy, 2006.

5. Postma, B.N.J.; Jouan, S.; Katz, B.F.G. Pre-Sabine room acoustic design guidelines based on human voice directivity. *J. Acoust. Soc. Am.* **2018**, *143*, 2428. [CrossRef] [PubMed]
6. Vitruvius. *I dieci libri dell'architettura [...] tradotti e commentati da Mons. Daniel Barbaro*; Ten Books of Architecture); de' Franceschi: Venezia, Italy, 1567. (In Italian)
7. Biancani, G. *Sphaera Mundi seu Cosmographia, ac Facili Methodo Tradita*; Sebastiano Bonomi for Geronimo Tamburini: Bologna, Italy, 1620. (In Latin)
8. Postma, B. A history of the use of time intervals after the direct sound in concert hall design before the reverberation formula of Sabine became generally accepted. *Build. Acoust.* **2013**, *20*, 157–176. [CrossRef]
9. Fabbri, P.; Farina, A.; Fausti, P.; Pompoli, R. The second life of the Teatro degli Intrepidi by Giovan Battista Aleotti through the new techniques of virtual acoustics. In Proceedings of the 2nd International Conference on Acoustic and Musical Research, Ferrara, Italy, 19–21 May 1995. (In Italian)
10. Cammarata, G.; Fichera, A.; Pagano, A.; Rizzo, G. Acoustical prediction in some Italian theatres. *Acoust. Res. Lett. Online* **2001**, *2*, 61. [CrossRef]
11. Prodi, N.; Pompoli, R. The acoustics of three Italian historical theatres: The Early days of modern performance spaces. In Proceedings of the Tecniacustica 2000, Madrid, Spain, 16–20 October 2000.
12. Tronchin, L. Francesco Milizia (1725–1798) and the Acoustics of his Teatro Ideale (1773). *Acta Acust. United Acust.* **2013**, *99*, 91–97. [CrossRef]
13. Weinzierl, S.; Sanvito, P.; Schultz, F.; Büttner, C. The acoustics of renaissance theatres in Italy. *Acta Acust.* **2015**, *101*, 632–641. [CrossRef]
14. Iannace, G.; Ianniello, C.; Maffei, L.; Romano, R. Objective measurement of the listening condition in the old Italian opera house "Teatro di San Carlo". *J. Sound Vib.* **2000**, *232*, 239–249. [CrossRef]
15. Farina, A. Acoustic quality of theatres: Correlations between experimental measures and subjective evaluations. *Appl. Acoust.* **2001**, *62*, 899–916. [CrossRef]
16. Pompoli, R. (Ed.) *Proc. of Teatri d'opera dell'Unità d'Italia (Proc. of Opera houses of the Italian Unification)*; Associazione Italiana di Acustica: Venezia, Italy, 2011.
17. Cirillo, E.; d'Alba, M.; Martellotta, F. Acoustic characterisation of Apulian historical theatres. In Proceedings of the 38 Congress of Italian Acoustic Association, Rimini, Italy, 8–10 June 2011. (In Italian)
18. Ceniccola, G. Architetture in scena. Teatri storici in Campania tra XVIII e XX secolo: Conoscenza e nodi critici nel progetto di conservazione (Performance Architecture. Historical Theatres in Campania between 18th and 20th Century: Knowledge and Criticalities in the Conservation Project). Ph.D. Thesis, University of Naples "Federico II", Naples, Italy, 2011.
19. Garai, M.; Morandi, F.; D'Orazio, D.; De Cesaris, S.; Loreti, L. Acoustic measurements in eleven Italian opera houses: Correlations between room criteria and considerations on the local evolution of a typology. *Build. Environ.* **2015**, *94*, 900–912. [CrossRef]
20. Prodi, L.; Pompoli, R.; Martellotta, F.; Sato, S. Acoustics of Italian Historical Opera Houses. *J. Acoust. Soc. Am.* **2015**, *138*, 769–781. [CrossRef] [PubMed]
21. De Cesaris, S.; Morandi, F.; Loreti, L.; D'Orazio, D.; Garai, M. Notes about the early to late transition in Italian theatres. In Proceedings of the 22nd International Congress on Sound and Vibration (ICSV22), Florence, Italy, 12–16 July 2015.
22. Morandi, F.; De Cesaris, S.; D'Orazio, D.; Garai, M. Energy criteria in Italian historical opera houses: A survey over 11 theatres. In Proceedings of the 9th International Conference on Auditorium Acoustics, Paris, France, 29–31 October 2015; Volume 37.
23. Garai, M.; De Cesaris, S.; Morandi, F.; D'Orazio, D. Sound energy distribution in Italian opera houses. *Proc. Mtgs. Acoust.* **2016**, *28*, 015019.
24. Serlio, S. *Sette libri dell'architettura di Sebastiano Serlio Bolognese (Seven Books of Architecture)*; Francesco de'Franceschi senese: Paris, France, 1545. (In Italian)
25. Mazzucato, T. Idea del espacio escénico y lugares para la representación teatral entre los siglos XV y XVI. Modelos de teatro a la manera de Italia (Idea of the Scenic Space and Places for Theatre Performance between the 15th and the 16th Century. Models for the Italian Theatre). *Studia Aurea* **2009**, *3*, 139–172. (In Spanish)
26. Canac, F. *L'acoustique des Theatres Antiques: Ses Insegnements (The Acoustics of Ancient Theatres: Its Lessons)*; Edition du Centre National de la Recherche Scientifique: Paris, France, 1967. (In French)
27. Maconie, R. Musical Acoustics in the Age of Vitruvius. *Music. Times* **2005**, *146*, 75–82. [CrossRef]

28. Farnetani, A.; Prodi, N.; Pompoli, R. On the acoustic of ancient Greek and Roman Theatres. *J. Acoust. Soc. Am.* **2008**, *124*, 157–167. [CrossRef] [PubMed]
29. Chourmouziadou, K.; Kang, J. Acoustic evolution of ancient Greek and Roman theatres. *Appl. Acoust.* **2008**, *69*, 514–529. [CrossRef]
30. Mo, F.; Wang, J. The Conventional RT is Not Applicable for Testing the Acoustical Quality of Unroofed Theatres. *Build. Acoust.* **2013**, *20*, 81–86. [CrossRef]
31. Amadei, D. L'ordine e la geometria nel teatro antico. Diffusione e fortuna del De Architectura di Vitruvio. Caso studio: Il teatro romano di Fanum Fortunae (Order and Geometry in the Ancient Theatre. Diffusion and Fame of Vitruvius's De Architectura. ase Study: The Roman Theatre of Fanum Fortunae). Ph.D. Thesis, Polytechnic University of Marche, Ancona, Italy, 2015. (In Italian)
32. Mullin, D.C. The Influence of Vitruvius on Theatre Architecture. *Educ. Theatre J.* **1966**, *18*, 27–33. [CrossRef]
33. Torello–Hill, G. The exegesis of Vitruvius and the creation of theatrical spaces in Renaissance Ferrara. *Renaiss. Stud.* **2014**, *29*. [CrossRef]
34. Sanvito, P.; Weinzierl, S. L'acustica del Teatro Olimpico di Vicenza (The Acoustics of the Teatro Olimpico in VIcenza). *Odeo Olimpico* **2013**, *23*, 463–492. (In Italian)
35. Pigafetta, F. *Due Lettere descrittive l'una dell'ingresso a Vicenza della Imperatrice Maria d'Austria dell'anno MDLXXXI l'altra della Recita nel Teatro Olimpico dell'Edippo di Sofocle nel MDLXXXV (Two Letters on the Arrival of Maria of Austria in 1581 and on the Performance at the Teatro Olimpico of Sofocles's Oedipus in 1585)*; Crescini, V., Ed.; Raccolta di Lettere Sulla Pittura, Scultura ed Architettura: Padova, Italy, 1830; pp. 25–31. (In Italian)
36. Farneti, F.; Van Riel, S. *L'architettura teatrale in Romagna 1757–1857 (Theatrical Architecture in Romagna)*; Uniedit: Firenze, Italy, 1975. (In Italian)
37. Scamozzi, O.B. *Le fabbriche e i disegni di Andrea Palladio (Andrea Palladio's Works)*; Giovanni Rossi: Vicenza, Italy, 1796. (In Italian)
38. Scamozzi, V. *L'idea della architettura universale (About of Universal Architecture)*; Expensis Auctoris: Venezia, Italy, 1615. (In Italian)
39. Stone Peters, J. *Theatre of the Book 1480–1880. Print, Text, and Performance in Europe*; Oxford University Press: Oxford, UK, 2000.
40. Crabtree, S.; Beudert, P. *Scenic Art for the Theatre*; Focal Press: London, UK, 1998.
41. Peacock, J. *The Stage Designs of Inigo Jones: The European Context*; Cambridge University Press: Cambridge, UK, 1995.
42. Bjurstrom, P. *Giacomo Torelli and Baroque Stage Design*; Almqvist & Wiksell: Stockholm, Sweden, 1961.
43. Sabbatini, M. *Pratica di Fabricar Scene e Machine Ne' teatri (On Making Scenes and Machines for Theatres)*; CNR IRCrES: Ravenna, Italy, 1638. (In Italian)
44. Lazardzig, J.; Rössler, H. (Eds.) *Technologies of Theatre: Joseph Furttenbach and the Transfer of Mechanical Knowledge in Early Moderne Theatre Cultures*; Verlag Vittorio Klostermann: Frankfurt, Germany, 2016.
45. Donati, P. *Descrizione del Gran Teatro Farnesiano di Parma e notizie storiche sul medesimo di Paolo Donati parmigiano architetto teatrale e accademico di Bologna e professore della Reale Accademia di Firenze (Description of the Farnese Theatre in Parma and Historical Information . . .)*; Blanchon: Parma, Italy, 1817.
46. Mariano, F. *Il Teatro Nelle Marche: Architettura, Scenografia e Spettacolo (The Theatre in the Marche: Architecture, Scenography and Show)*; Banca delle Marche: Ancona, Italy, 1997. (In Italian)
47. Quagliarini, E. *Costruzioni in Legno nei Teatri All'Italiana del '700 e '800: Il Patrimonio Nascosto Dell'architettura Teatrale Marchigiana (Wooden buildings in Italian Historical Opera Houses: The Hidden Heritage of Theatrical Architecture)*; Allinea: Firenze, Italy, 2008. (In Italian)
48. Carini Motta, F. *Treaty about the Structure of the Theatres and of the Scenes*; Guastalla, 1676; Republished; Il Polifilo: Milano, Italy, 1972. (In Italian)
49. Donnelly, M.C. Theaters in the Courts of Denmark and Sweden from Frederik II to Gustav III. *J. Soc. Arch. Hist.* **1984**, *43*, 328–340. [CrossRef]
50. Sartini, V. Comparative Analysis of Five Small Sized Italian Theatres of the 19th Century. Master's Thesis, University of Bologna, Bologna, Italy, 2012.
51. Galli Bibiena, F. *L'architettura Civile Preparata su la Geometria e Ridotta Alle Prospettive (Civil Architecture: Geometry and Perspective)*; Paolo Monti: Parma, Italy, 1711. (In Italian)

52. Algarotti, F. *Saggio Sopra L'opera in Musica (Essay on Opera)*, 2nd ed.; Library of Congress: Washington, DC, USA, 1764. (In Italian)
53. Giordano, L. Il Teatro dei Quattro Cavalieri e la presenza di Antonio Galli Bibiena a Pavia (The "Teatro dei Quattro Cavalieri" and the work of A. Galli Bibiena in Pavia). *Bollettino D'arte* **1975**, *60*, 88–102. (In Italian)
54. Planelli, A. *Dell'opera in Musica. Trattato del Cavaliere Antonio Planelli Dell'ordine Gerosolimitano (About the Opera)*; D. Campo: Napoli, Italy, 1772. (In Italian); re-published and edited by F. Degrada; Discanto edizioni: Fiesole, Italy, 1981. (In Italian)
55. Riccati, G. *Della Costruzione De' Teatri Secondo il Costume d'Italia: Vale a Dire Divisi in Piccole Logge (About the Construction of Theatres According to the Italian Style, I.e., With Small Boxes)*; Remondini from Venice: Bassano, Italy, 1790. (In Italian)
56. Barbieri, P. *Giordano Riccati Fisico Acustico e Teorico Musicale (Giordano Riccati Acoustician and Musical Theorist)*; Leo S. Olschki: Florence, Italy, 1990; pp. 279–304. (In Italian)
57. Bagni, G.T. *Vincenzo, Giordano, Francesco Riccati e la matematica del settecento (V., G.; F. Riccati and Mathematics in the 18th Century)*; Teorema: Treviso, Italy, 1993. (In Italian)
58. Bortolozzo, R. *L'universo ben Temperato Dei Riccati: Cosmologia e Musica in Una Famiglia di Illuministi Trevigiani (The Well Tempered Universe of the Riccati Family ...)*; Il Cardo: Venezia, Italy, 1995. (In Italian)
59. Rizzetti, F. Memoria intorno alla più perfetta costruzione di un Teatro. as quoted in Riccati, 1790.
60. Rizzetti, L. Risposta del sig. Conte Luigi Rizzetti alle accuse date al Teatro da lui proposto (Response [...] to the allegations made to the theatre he proposed). In *Collection of Scientific and Literary Brochures*; Stamperia Coleti: Ferrara, Italy, 1792. (In Italian)
61. Morelli, C. *Pianta e Spaccato Del Nuovo Teatro d'Imola (Plan and Section of the New Theatre of Imola)*; Casaletti: Roma, Italy, 1780. (In Italian)
62. Matteucci, A.M.; Lenzi, D. *Cosimo Morelli e L'architettura Delle Legazioni Pontificie (Cosimo Morelli and the Architecture of the Papal States)*; University Press Bologna: Imola, Italy, 1977. (In Italian)
63. Memmo, A. *Semplici lumi tendenti a render cauti i soli interessati nel teatro da erigersi nella parocchia di S Fantino in Venezia [...] (Simple Hints for Those Who Are Interested in the Theatre to Build in the Parish of S. Fantino in Venice)*. 1790. (In Italian)
64. Lamberti, V. *La regolata costruzion de' teatri secondo il costume d'Italia, vale a dire divisi in picciole logge ... (Italian Construction of Theatres, i.e., Divided in Boxes ...)*; V. Orsini: Napoli, Italy, 1782. (In Italian)
65. Saunders, G. *Treatise on Theaters*; I. and J. Taylor: London, UK, 1790.
66. Available online: https://www.theatre-architecture.eu (accessed on 8 February 2019).
67. D'Alembert, J.; Diderot, D. *"Theatre" in Encyclopédye ou Dictionaire raisonnè des Sciences, des Arts et del Métiers*; Briasson: Paris, France, 1751–1772. (In French)
68. Bondin, N. *Teatro Antico. Ragionamento Sopra la Forma e la Struttura (Ancient Theatre. Reasoning on Form and Structure)*; s.e.: Venezia, Italy, 1746. (In Italian)
69. Cochin, C.-N. *Lettre sur les salles de spectacle, Mercure de France (Letter on Musical Halls)*. 1760. (In French)
70. Dumont, G.M. *Suite de projects détaillés de salles de spectacle, avec del principes de construction, tant pour la mécanique del theatres que pour das décoration ... (Set of Detailed Projects of Performance Halls, with Construction Principles, both for Theatre Mechanics and Decoration)*. Paris, France, 1773. (In French)
71. De La Dixmerie, N.B. *Lettres sur l'etat present de nos spectacles, avec del vues nouvelles sur chacun d'eux, particulierement sur la Comedie Francaise et l'Opera (Letters on the Present Situation of Our Performances, with New Opinions, Especially on the Comedie Française and l'Opéra)*; Duchesne: Amsterdam, The Netherlands; Paris, France, 1765. (In French)
72. Chaumont, D.J. (*Veritable construction exterieure d'un theatre d'Opera, a l'usage de France, relative a celle donnee l'annee dernere, pour la Construction interieure (Exterior Construction of a French Style Opera House, Related to the One Published Last Year, Related to the Interior)*; De Lormel: Paris, France, 1766. (In French)
73. Monginot, G.M. *Exposition des principes qu'on doit suivre dans l'ordonnance des theatre modernes*; C.A. Jombert: Paris, France, 1769. (In French)
74. Damun, J. *Prospectus du noveau theatre tracé sur les principes des Grecs et des Romains (Project of a New Theatre Based on Greek and Roman Principles)*. Paris, France, 1772. (In French)
75. Roubo le Fils, A.J. *Traité de la construction des theatres et des machines theatrales (Treatise on the Construction of Theatres and Theatre Machines)*; Cellot et Jombert: Paris, France, 1777. (In French)

76. Patte, P. *Description du theatre de la ville de Vicence en Italie, chef-d'oeuvre d'Andrea Palladio (Description of the Theatre of Vicenza, in Italy, Work by Andrea Palladio)*. Paris, France, 1780. (In French)
77. Noverre, J.G. *Observations sur la construction d'une salle d'Opera (Observation on the Construction of an Opera House)*; Cellot: Paris, France, 1781. (In French)
78. Patte, P. *Essai sur l'architecture théatrale, ou, De l'ordonnance la plus avantageuse à une salle de spectacles, relativement aux principes de l'optique et de l'acoustique: Avec un examen des principaux téatres de l'Europe, et une analyse des écrits les plus importans sur cette matiere (Essay on Theatre Architecture, or on the Most Advantageous Design of a Performance Hall, and an Analysis of the Most Important Writings on This Topic)*; Moutard: Paris, France, 1782. (In French)
79. Bouellet. *Essai sur l'art de construire les theatres, leurs machines et leur mouvements (Essay on the Art of Building Theatres, their Machines and their Movements)*; Ballard: Paris, France, 1801. (In French)
80. Grobert, J.-F.-L. *Exécution dramatique en rapport avec le materiel de la Salle et de la Scene (Dramatic Execution rEgarding the Materials of the Hall and the Scene)*; Schoell: Paris, France, 1809. (In French)
81. Ledoux, C.N. *L'architecture considerée sour le rapport de l'Art, des Moeurs et de la Legislation ... (The Architecture Considered through the Connections with Art, Traditions and Law)*. Paris, France, 1804. (In French)
82. Boullée, E.L. *Architecture, Essai sur l'art (Architecture, Essay on Art)*. London, UK, 1953. (In French)
83. Tamburini, E. *Il luogo teatrale nella trattatistica italiana dell'800 (The Theatrical Space in Italian 19th-Century Treatises)*; Bulzoni Editore: Roma, Italy, 1984. (In Italian)
84. Barbieri, P.; Tronchin, L. L'impostazione acustica dei teatri nei progetti del primo neoclassicismo Italiano (1762–1772), (The Acoustical Structure of Theatres in the First Italian Neoclassical Projects). In *Francesco Milizia e il teatro del suo tempo Architettura, Musica, Scena, Acustica*; Russo, M., Ed.; Collana Studi e Ricerche n. 2: Trento, Italy, 2011; pp. 137–161, ISBN 978-88-8443-396-1. (In Italian)
85. Arnaldi, E. *Idea di un teatro nelle principali sue parti simile a' teatri antichi all'uso moderno accomodato*; Veronese: Vicenza, Italy, 1762. (In Italian)
86. Milizia, F. *Trattato completo, formale e materiale del teatro (About the Theatre)*; Pasquali: Venezia, Italy, 1773–1794. (In Italian)
87. D'Orazio, D.; De Cesaris, S.; Morandi, F.; Garai, M. The aesthetics of the Bayreuth Festspielhaus explained by means of acoustic measurements and simulations. *J. Cult. Herit.* **2018**, *34*, 151–158. [CrossRef]
88. Westfall, C.W. Antolini's Foro Bonaparte in Milan. *J. Warbg. Courtauld Inst.* **1969**, *32*, 366–385. [CrossRef]
89. Beccega, T. *Sull'architettura greco-romana applicata alla costruzione del teatro moderno italiano e sulle macchine teatrali (On the Application of the Graeco-Roman Architecture to the Construction of the Modern Italian Theatre and to the Theatrical Machines)*. Alvisopoli, Italy, 1817. (In Italian)
90. Niccolini, A. *Alcune idee sulla risonanza del teatro (Some Ideas on the Resonance of the Theatre)*; Masi: Napoli, Italy, 1816. (In Italian)
91. Landriani, P. *Osservazioni sui difetti prodotti nei teatri dalla cattiva costruzione del palco scenico e su alcune inavvertenze nel dipingere le decorazioni (Observations on the Flaws Deriving by the Poor Construction of the Stage and on Some Oversights in Decoration Paintings)*; Regia tipografia di Milano: Milan, Italy, 1828. (In Italian)
92. Meduna, T.; Meduna, G. *Il Teatro La Fenice in Venezia ... (On the La Fenice Theatre ...)*; Antonelli: Venezia, Italy, 1849 . (In Italian)
93. Cocchi, A.; Garai, M.; Tronchin, L. Influenza di cavità risonanti poste sotto la fossa orchestrale: Il caso del teatro Alighieri di Ravenna (The Influence of Resonating Cavities under the Orchestra Pit: The Case of the Alighieri Theatre in Ravenna). In *Teatri storici. Dal restauro allo spettacolo*; Nardini Editore: Firenze, Italy, 1997; pp. 135–153.
94. D'Apuzzo, N. *Cenno intorno ai teatri moderni e sopra gli archi di trionfo degli antichi (Notes on Modern Theatres and Ancient Triumphal Arches)*. 1817. (In Italian)
95. De Grazia, V. *Discorso sull'architettura del teatro moderno (On the Architecture of the Modern Theatre)*. 1825. (In Italian)
96. Ferrario, G. *Storia e descrizione de' principali teatri antichi e moderni (History and Description of the Main Ancient and Modern Theatres)*. Forni, Italy, 1830. (In Italian)
97. de Cesare, F. *La scienza dell'architettura applicata alla costruzione, alla distribuzione, alla decorazione degli edifici civili (The Science of Architecture on the Construction, Distribution and Decoration of Civil Buildings)*. 1885. (In Italian)

98. D'Orazio, D.; De Cesaris, S.; Garai, M. Recordings of Italian opera orchestra and soloists in a silent room. *Proc. Meet. Acoust.* **2016**, *28*, 015014. [CrossRef]
99. Barbera, P. *Giuseppe Damiani Almeyda artista architetto ingegnere (Giuseppe Damiani Almeyda Artist, Architect, Engineer)*; Pielle Edizioni: Palermo, Italy, 2008.
100. Boniotto, E.; Bovo, M.E.; Di Bella, A.; Frinzi, G.; Granzotto, N.; Rinaldi, C.; Zecchin, R. L'acustica nel restauro dei teatri storici: Il caso del Teatro Civico Di Schio (The Acoustics in the Restoration of Historical Theatres. The Case of the City Theatre in Schio). In Proceedings of the Congress of Italian Association of Acoustics, Ischia, Italy, 10–12 May 2006.
101. Silingardi, V.; Rinaldi, C.; Granzotto, N.; Barbaresi, L.; di Bella, A. A restoration based on the result of a public debate: The case of Civic Theatre of Schio. *Rivista Italiana di Acustica* **2017**, *41*, 1–14.
102. Basile, G.B.F. *Sulla costruzione del Teatro Massimo Vittorio Emanuele (On the Construction of Teatro Massimo Vittorio Emanuele)*; Tip. Dello Statuto: Palermo, Italy, 1883. (In Italian)
103. Fundaró, A.M. *Il Concorso per il Teatro Massimo di Palermo—Storia e progettazione (Competition for the Teatro Massimo in Palermo: History and Design)*; STASS: Palermo, Italy, 1974
104. Canoppi, A. *Opinion ... sul l'architecture en général et en spécialité sul la construction des Théatres modernes (On Modern Theatres and Ancient Triumphal Arches)*. 1830. (In French)
105. Bortolotti, L.; Masetti, L. *Teatri storici. Dal restauro allo spettacolo (Historical Theatres. From Renovation to Performance)*; Nardini: Fiesole, Italy, 1977. (In Italian)
106. Savini, F. *Cenni storici e vicende del Teatro Comunale di Cesenatico (Historical Information on the City Theatre of Cesenatic)*. Cesenatico, Italy, 1979.
107. Fabbri, P. *Teatri di Russi, dal vecchio al nuovo Comunale (The City Theatre in Russi, from the Old to the New)*; Longo Editore: Ravenna, Italy, 1979.
108. Van Riel, S. (Ed.) *Il Teatro di Meldola: Storia e Restauro (The Theatre in Meldola: History and Renovation)*; Alinea: Firenze, Italy, 1982.
109. Castagnoli, S. Il teatro "E. Petrella" di Longiano dalla fabbrica al restauro (The Theatre Petrella in Longiano, from Design to Renovation). In *Studi Romagnoli*; XXXVI; La Fotocromo Emiliana: Bologna, Italy, 1985. (In Italian)
110. Bortolotti, L. *(a cura di), Le stagioni del teatro. Le sedi storiche dello spettacolo in Emilia Romagna*; Grafis Industrie Grafiche: Bologna, Italy, 1995.
111. Ceredi, C.; Piraccini, O. (Eds.) *Teatro Comunale di Cesenatico (City Theatre in Cesenatico)*. Comune di Cesenatico, Italy, 2004.
112. Vasumi Roveri, E. *I teatri di Romagna. Un sistema complesso (The Theatres of Romagna: A Complex System)*; Compositori Editore: Bologna, Italy, 2005. (In Italian)
113. D'Orazio, D.; De Cesaris, S.; Guidorzi, P.; Barbaresi, L.; Garai, M.; Magalotti, R. Room acoustic measurements using a high SPL. In Proceedings of the 140th AES Convention, Paris, France, 4–7 June 2016.
114. ISO 3382-1. *Acoustics—Measurement of Room Acoustic Parameters. Part 1: Performance Spaces*; ISO: Geneva, Switzerland, 2009.
115. Guidorzi, P.; Barbaresi, L.; D'Orazio, D.; Garai, M. Impulse responses measured with MLS or Swept–Sine signals applied to architectural acoustics: An in-depth analysis of the two methods and some case studies of measurements inside theaters. *Energy Procedia* **2015**, *78*, 1611–1616. [CrossRef]

11

The Acoustics of the Choir in Spanish Cathedrals

Alicia Alonso *, Rafael Suárez and Juan J. Sendra

Instituto Universitario de Arquitectura y Ciencias de la Construcción, Escuela Técnica Superior de Arquitectura, Universidad de Sevilla, Seville 41012, Spain; rsuarez@us.es (R.S.); jsendra@us.es (J.J.S.)
* Correspondence: aliciaalonso@us.es

Abstract: One of the most significant enclosures in worship spaces is that of the choir. Generally, from a historical point of view, the choir is a semi-enclosed and privileged area reserved for the clergy, whose position and configuration gives it a private character. Regarding the generation and transformation of ecclesial interior spaces, the choir commands a role of the first magnitude. Its shape and location produce, on occasions, major modifications that significantly affect the acoustics of these indoor spaces. In the case of Spanish cathedrals, whose design responds to the so-called "Spanish type", the central position of the choir, enclosed by high stonework walls on three of its sides and with numerous wooden stalls inside, breaks up the space in the main nave, thereby generating other new spaces, such as the trascoro. The aim of this work was to analyse the acoustic evolution of the choir as one of the main elements that configure the sound space of Spanish cathedrals. By means of in situ measurements and simulation models, the main acoustic parameters were evaluated, both in their current state and in their original configurations that have since disappeared. This analysis enabled the various acoustic conditions existing between the choir itself and the area of the faithful to be verified, and the significant improvement of the acoustic quality in the choir space to become apparent. The effect on the acoustic parameters is highly significant, with slight differences in the choir, where the values are appropriate for Gregorian chants, and suitable intelligibility of sung text. High values are also obtained in the area of the faithful, which lacked specific acoustic requirements at the time of construction.

Keywords: worship acoustics; Spanish cathedrals; choir space

1. Introduction

Throughout history, the development and construction of worship spaces in Europe has been adapted to the different artistic styles of each era. Although the same style can be dated for different centuries in each European country, there is a number of common characteristics as well as formal and constructive aspects that define each typology. It is evident that from the acoustic point of view, the materials, the geometry, and the variety of particularities belonging to each style can exert a significant effect on the evaluation of the sound field of the temple. In this regard, several studies have analysed the acoustic evolution of worship spaces and have focused on the main qualities of each era and characterised the sound field through the analysis of various acoustic parameters [1,2]. However, not only do dimensions, materials, and stylistic considerations determine the acoustic behaviour of a space, but the spatial conception of its interior also constitutes another main aspect that determines the acoustics of the temple. The variation of the use of the temple in terms of the requirements of each epoch has given rise to certain spatial changes undergone over the years, that is to say, the organisation of its component zones: Naves, transept, ambulatory, and choir.

Occasionally, acoustic needs have influenced the formal, artistic, and spatial configuration of the temples. Knowledge on the origin and situation of elements that have influenced the acoustics of the temple allow the way of solving the interior liturgical space to be understood more clearly, thereby enabling the various configurations adopted in each area to be outlined in accordance with the time or type of event developed. The variation in the location of the choral space, apart from its stylistic, constructive, and chronological considerations, has changed the configuration of worship spaces throughout history [3]. Due to its ability to create and transform the indoor space, the choir would become the main enclosure to configure and articulate the spatial relationship of the temple and the location of the clergy and the faithful. Therefore, it is confirmed that the position of this semi-enclosed and privileged area reserved for the clergy occasionally produces major modifications of these indoor spaces.

In Europe, there are several types of churches and cathedrals that establish the choir space in different positions. In this article, the evolution and acoustic impact of the case study of Spanish cathedrals is analysed, in which the central position of the choir stalls fragments the space in the main nave where, unlike other models elsewhere in Europe, the congregation is not permitted to participate in all types of ceremonies. This work concerns the reconstruction of the process by which Spanish cathedrals have a strong personality within the European panorama, by virtue of the location of the choir.

2. The Choral Space in Churches and Cathedrals

From the architectural point of view, the choir enclosure commands a role of the first magnitude in the configuration of ecclesial indoor spaces and thereby constitutes one of the elements that configure the acoustic issue of worship spaces. The cardinal element of the liturgy can also be considered in its spatial development, and the conception of the choir can only be understood in the space for which it was conceived. The hierarchy of the choir constitutes the most faithful portrait of the clergy and responds to a different personality in each diocese [4]. In the case of cathedrals, the choir position has a significant condition in the architectural floor plan, since liturgical functions are developed around the choir. The fact that it is enclosed by high stonework walls on three of its sides, with numerous wooden stalls inside, has a noticeable effect on the sound field.

In the following sections, the historical evolution and the main typologies of European choirs are defined according to how their location affects the interior of the ecclesial space.

2.1. *European Typologies of European Cathedrals*

Ever since the construction period of the first Christian churches, the presence of a space reserved for the clergy has stood out, which would be used for the recital of the word in the form of singing. At first, it was located at the head of the temple and later, in its central nave, facing the presbytery, which then led to the schola cantorum, or choir enclosure, in early Christian basilicas [1].

The first position of the choir was later altered in European cathedrals, whereby that central area disappeared and the grouping of spaces for the clergy, altar, and choir were established, thereby creating a deep head in the temple. In Europe, various models and typologies, which are shown in Figure 1, were established, depending on the location of the choir:

- French typology. At first, the choral space was established at the head of the temple around the High Altar, changing its name to choeur. In this regard, the French model had a deep head of the temple to lodge the High Altar and the choir, which, in certain cases, was later moved to the centre for functional reasons. This model corresponds to the initial state of great French cathedrals, such as those of Reims, Amiens, Chartres, and Notre Dame. This opacity and the choir–presbytery represented, to a certain extent, an enclosed canonical church within the cathedral space, and hence the solemn liturgy could not be seen but could be heard by the faithful. As mentioned previously, this typology was an innovation with respect to the Romanesque churches, whose position of the choral space was in the centre of the nave following the model of Cluny Abbey [5]. In the central nave, the sequence followed was that of altar–choir–faithful.

- English typology. English choirs, such as those in the cathedrals of York and Lincoln, created a large private enclosure and were totally closed off to the faithful, and were located within a very long header. Among many other aspects in relation to the arrangement of the floor plan, form, and use of the cathedral, the processional entrance to the choir led through the door that opened into the retrochoir (trascoro). The sequence followed was that of altar–choir–faithful.
- Italian typology—Trentino model. The Schola cantorum of early Christian basilicas preserved throughout the Middle Ages has its origin in Italy. In accordance with major reforms enacted in the Council of Trent (16th century), decisive actions affecting the interior space and the liturgy of the cathedral were carried out. For this reason, renovation work based mainly on the transfer of choral space was promoted in certain European cathedrals, in order to strengthen this estranged relationship between the clergy and the faithful. This typology responds to the solution funded by Carlos Borromeo in the cathedral of Milan, in which the choir is located behind the main altar, taking this intermediate position between the choral space and the faithful. In the central nave, the sequence followed was that of choir–altar–faithful.

These European models (Figure 1) allow the congregation to be accommodated throughout the central nave in all kinds of celebrations.

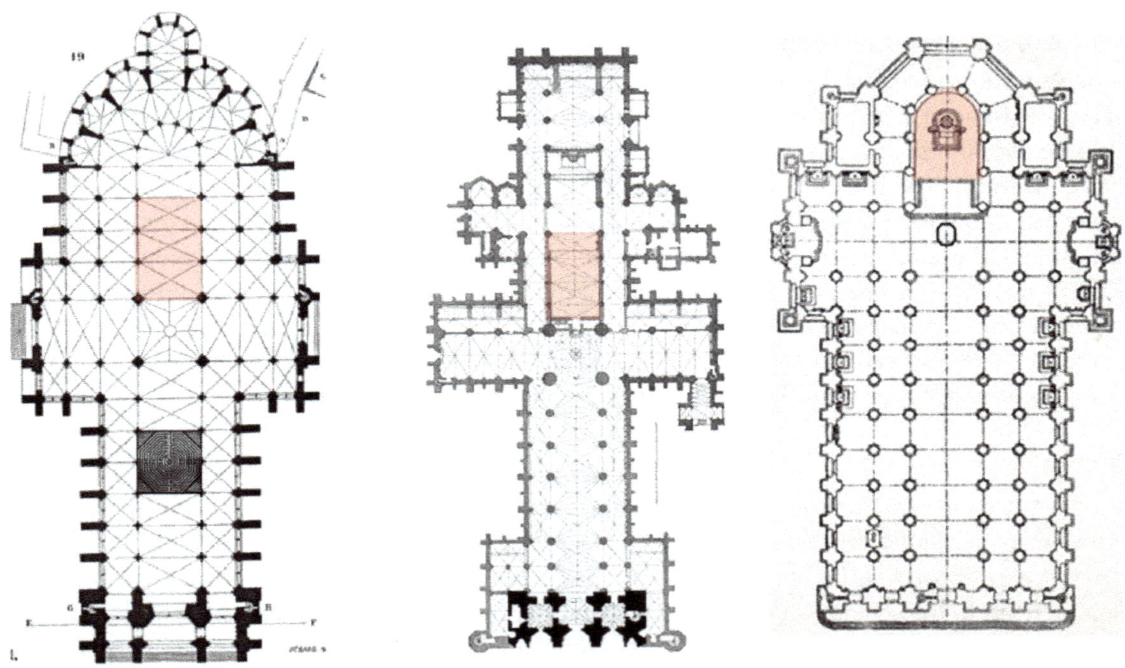

(a) (b) (c)

Figure 1. European models. (**a**) French typology, cathedral of Reims (13th century); (**b**) English typology, cathedral of Lincoln (14th century); (**c**) Italian typology—Trentino model, cathedral of Milan (16th century). Red zone: Choir stalls.

2.2. Historical Evolution of the Choir Space in Spanish Cathedrals

In Spain, the choir is usually placed in the middle of the central nave, thereby responding to the old tradition of monasteries, and these, in turn, respond to the schola cantorum of the early Christian basilicas [1,3]. Thus, the chancel and the choir, closely united, became a church for the clergy within the cathedral itself, where the congregation was excluded. In the Spanish typology, the position of the choir generated a division of the central nave into two differentiated zones: On the one hand, the solemn space of the great liturgy and, on the other hand, the space where the ordinary religious ceremonies were developed. This configuration delimited a space for preaching in front of the presbytery, to which the voice could be reached from the pulpits. Moreover, the central position of the choir stalls fragmented the space in the main nave and generated other new spaces, such as the trascoro. As previously indicated, this arrangement of the choir responds to the old tradition of Early Christian basilicas in which the schola cantorum had a similar disposition. In this case, the sequence adopted in the central nave was that of altar–faithful–choir–trascoro–faithful. In the Spanish model, the development of both the word and the musical liturgy is supported by a series of celebratory sources, such as the choir. Longitudinally, the choir space has the dimension of two intercolumniations. Throughout history, the celebration of certain events has required the transfer of this space to other areas of the interior, such as the trascoro, a space that becomes a new stage for music, thanks to its spatial amplitude.

A brief description is made of the course of the evolution of the choir space in Spain, whereby a description is given of the spatial configuration of some of the main cathedrals, both for their outstanding heritage character and the influence of their architectural configuration, transferred to other Spanish and New World cathedrals. Figure 2 shows various floor plans of Spanish cathedrals. Table 1 summarises the data of the main dimensions of the cathedrals under study and their choir spaces.

The tour of historical evolution of the choir space starts with the Latin cross floor plan of the cathedral of Santiago de Compostela (12th century), which is based on monastic buildings such as Cluny Abbey, with a shallow presbytery and a wide stone choir in the central nave [5,6]. This choir space is located in the first three sections of the nave from the transept, with a fourth section assigned to the leedoiro, a high tribune used for sermons in ceremonies. It was in the cathedral of Santiago where the location of the choir between the altar and the faithful in the centre of the nave was first made in Spain. The choir typology of this cathedral played a role of obligatory reference for future Spanish temples. It should be noted this typology generates a problem that has been analysed in other studies, which is the need to look for a configuration that accommodates a large congregation, since its position prevents the use of the depth of the central nave [7]. Although in many Spanish cathedrals, the architectural project raised the location of the choir in the main nave, it should be noted that in certain cathedrals, such as the cathedrals of Burgos and of León, the French model was initially followed. However, later work was carried out in order to restructure the model and move the choir to the central location. These proposals for the transfer of the choir to the nave obey no aesthetic reasoning, but rather a desire for ecclesial renewal.

In the cathedral of Toledo (14th century), the model of the cathedral of Santiago is followed, where the choir, named compostelano [4], is positioned in the centre of the temple, in this case at a double height. Its artistic style is varied and cannot be associated to a single period, since it was built in various stages.

In the case of the cathedral of Seville (15th century), following the previously established model, the choir space is also located in the middle of the central nave, facing the presbytery. It should be noted that the floor plan of the cathedral of Seville was the cathedral model exported to Ibero–America, that of a rectangular church with a straight ambulatory, thus ensuring the continuity of the processional space.

Unlike the other cathedrals, in the cathedral of Granada (16th century), there have been major interventions that have involved the modification of the situation of the choral space. At first, the choral space was located at the centre of the main nave, used mainly for the performance of instrumental music and singing. In the early 20th century, as in the case of the cathedral of Santiago, the choir was moved to the head of the temple, thereby enabling an optimal space-functional relationship by recovering its position at the High Altar. The last intervention, carried out at the end of the 20th century, completely suppressed the choir by distributing the stalls throughout the temple.

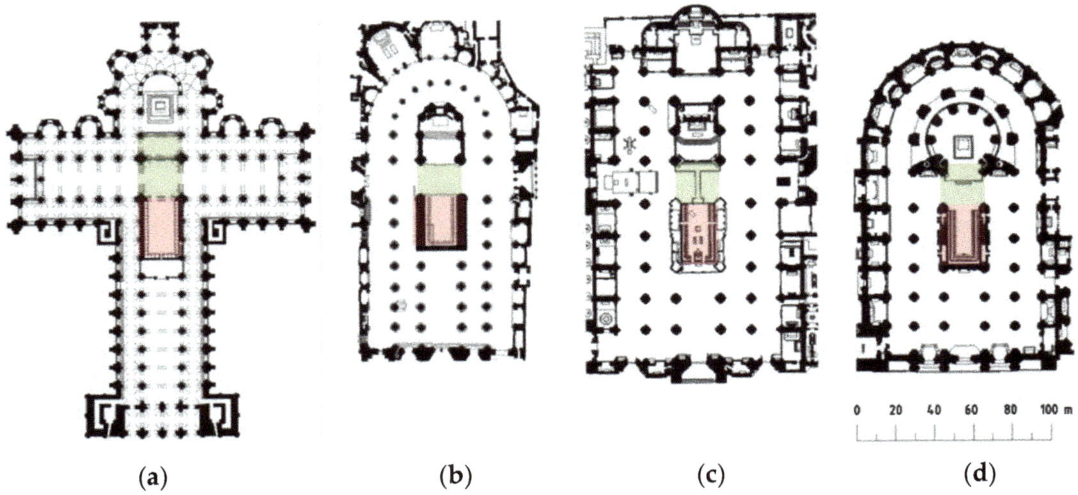

Figure 2. Spanish typology. Floor plans of: (**a**) Cathedral of Santiago de Compostela (12th century); (**b**) cathedral of Toledo (13th century); (**c**) cathedral of Seville (15th century); (**d**) cathedral of Granada (16th century). Red zone: Choir stalls; Green zone: Area of the faithful.

Table 1. Dimensions of Spanish cathedrals under study and their choir spaces.

Cathedral	Dimensions			
	Long (m)	Width (m)	Height (m)	Volume (m^3)
Santiago	174	40 (130 transept)	20 to 43	48,000
Choir	30	16	6.5	-
Toledo	108	55	19 to 31	125,000
Choir	19	11.5	6	-
Seville	116	76	35 to 40	205,000
Choir	22	13	8	-
Granada	106	63	25 to 32	160,000
Choir	20	11.5	7	-

3. Methodology

In order to analyse the acoustic environment of the Spanish cathedrals when the sound source is located in the choir space, two different techniques were carried out, depending on the case study: On the one hand, onsite acoustic measurements were carried out in two of the cathedrals that are preserved as they were built centuries ago (Toledo and Seville); and on the other hand, an archaeoacoustics procedure was developed, that is to say, a virtual acoustic reconstruction of the configuration of the spaces in the past using acoustic simulation tools (Santiago and Granada). However, the cathedral of Granada was also acoustically measured, since this procedure enabled the calibration of the virtual model that faithfully represents the current indoor acoustic environment, a procedure that is studied in Alonso et al. [8]. Once the model was calibrated, spatial transformations were incorporated that reproduce the acoustic sound field of the cathedral in the past with a different ancient choir. This technique, in which onsite measurements and computer simulation are used as a method of evaluation of the acoustic quality of worship spaces, is evidenced by other previous studies [2,7–9].

3.1. Onsite Measurements

An experimental technique was carried out based on onsite measurements to obtain room impulse responses (RIRs) of the cathedrals of Toledo, Seville, and Granada [7–9]. The measurements were carried out within two research projects, each of which lasted for 3 years, financed by the National Plan, in order to meet the objectives that were set out in the project, in which the recovery of the intangible heritage was proposed, as was the addition of possible solutions for the improvement of the acoustic behaviour of the cathedrals. The results of these research projects are published in several studies [7–10]. The RIR reflects the behaviour of a space between the point of emission and that of reception and can be used to calculate the acoustic parameters. Measurements were conducted at night in the unoccupied temple and followed the standard ISO 3382 [11] and guidelines established by Martellotta [12]. Temperature values and relative humidity were monitored, and the values of each cathedral are shown in Table 2.

Table 2. Data monitored during measurements in each cathedral.

Cathedral	Temperature (°C)	Relative Humidity (%)
Toledo	15.8–18	53–66.7
Seville	25–25.5	55–58
Granada	11.7–12.5	36–43

High impulse-to-noise ratio (INR) values require the adjustment of the level and duration of the measured signal and determine a suitable quality of the signal recorded, which provides reliable results of the main acoustic parameters. Since background noise during the measurements was low, it was possible to obtain INR values above 45 dB.

The measurement process started with the excitation response of the cathedrals by issuing sine sweep signals, emitted by an omnidirectional dodecahedron sound source (AVM do-12), raised up to 1.5 m from the floor or support surface, with frequency increasing exponentially over time. The duration of the sweep was set to 20 s and covered the octave bands from 63 to 16 kHz with a power amplifier (B&K 2734). In the case of the cathedral of Seville, a self-amplified Beringher Eurolive B1800D-Pro subwoofer was incorporated in order to improve the low-frequency results. A single output channel was used, since the signal is sent to the subwoofer, which is self-powered and has a high-pass filter (cut-off frequency at 90 Hz) that allows the signal to be sent up to the dodecahedron [13].

The signal generation, acquisition, and analysis were performed using the WinMLS2004 software tool together with an audio interface EDIROL UA-101 in the case of Seville and Granada, and the commercial software EASERA v1.2 with AUBION x8 multichannel sound card in the case of Toledo. Consequently, RIRs were acquired using a multipattern microphone Audio-Technica AT4050/CM5 that can be easily changed from omnidirectional to figure-of-eight in order to simulate the spatial impression of the listener through the lateral acoustic energy perceived, and whose amplifier and polarisation source is Earthworks Audio LAB1. In this study, acoustic parameters were considered when obtained only from omnidirectional directivity characteristics. The microphone was placed at a height of 1.2 m from the floor, which is the approximate ear level of a seated person, at each of the reception points throughout the audience area, with direct vision from the sound source. For the binaural RIRs, a Head Acoustics HMS III (Code 1323) dummy head was used. A moment of the measurement procedure in the choir space of the cathedrals of Seville and Toledo is captured in Figure 3a,b.

3.2. Simulation

Acoustic simulation is a highly useful tool, both for the prediction of the acoustic behaviour of an enclosure in the design stage and for the recovery of the sound of the past in spaces of worship that no longer exist. In this regard, simulation software also enables the characterisation of ancient acoustic conditions by offering the possibility of listening to the acoustics of the space of another era [13]. This is the case of the cathedral of Santiago since the old stone choir space, executed by Master Architect Mateo, was removed from the cathedral in the 17th century to be replaced by a wooden choir in the central nave. Since in the 20th century, it was also decided to dismantle the wooden choir, there is a need to use acoustical prediction tools to analyse the acoustic environment of the temple in past eras. Virtual acoustic models of the cathedral of Santiago and the cathedral of Granada were created in order to reconstruct these spaces from the past.

In the case of Santiago, documented floor plans researched by Conant [6] were used and the model contained 5600 planes and a volume of 48,202 m^3, and in the case of Granada, almost 5000 planes were used to create a volume of 160,000 m^3. After having created the models using SketchUp 3D modelling software, they were then exported to CATT-Acoustic v9.0c [14]. The predicted values were provided using the calculation engine TUCT v1.1a, incorporated in the acoustic software. On the one hand, the data collection of sound absorption/scattering coefficients was carried out by categorising the existing materials in the space after a visual inspection [15]; and on the other hand, coefficients of materials of inexistent enclosures, such as the choir space, were based on documented bibliography. The sound source was located in the choir space, corresponding to the liturgical use that is the object of study, and the receivers were positioned according to occupation. The acoustic model simulations provided the RIRs, which led to the calculation of the most significant acoustic parameters of the cathedral. Virtual models of the cathedral of Granada and Santiago, into which the inexistent choir space was incorporated, are shown in Figure 3c,d.

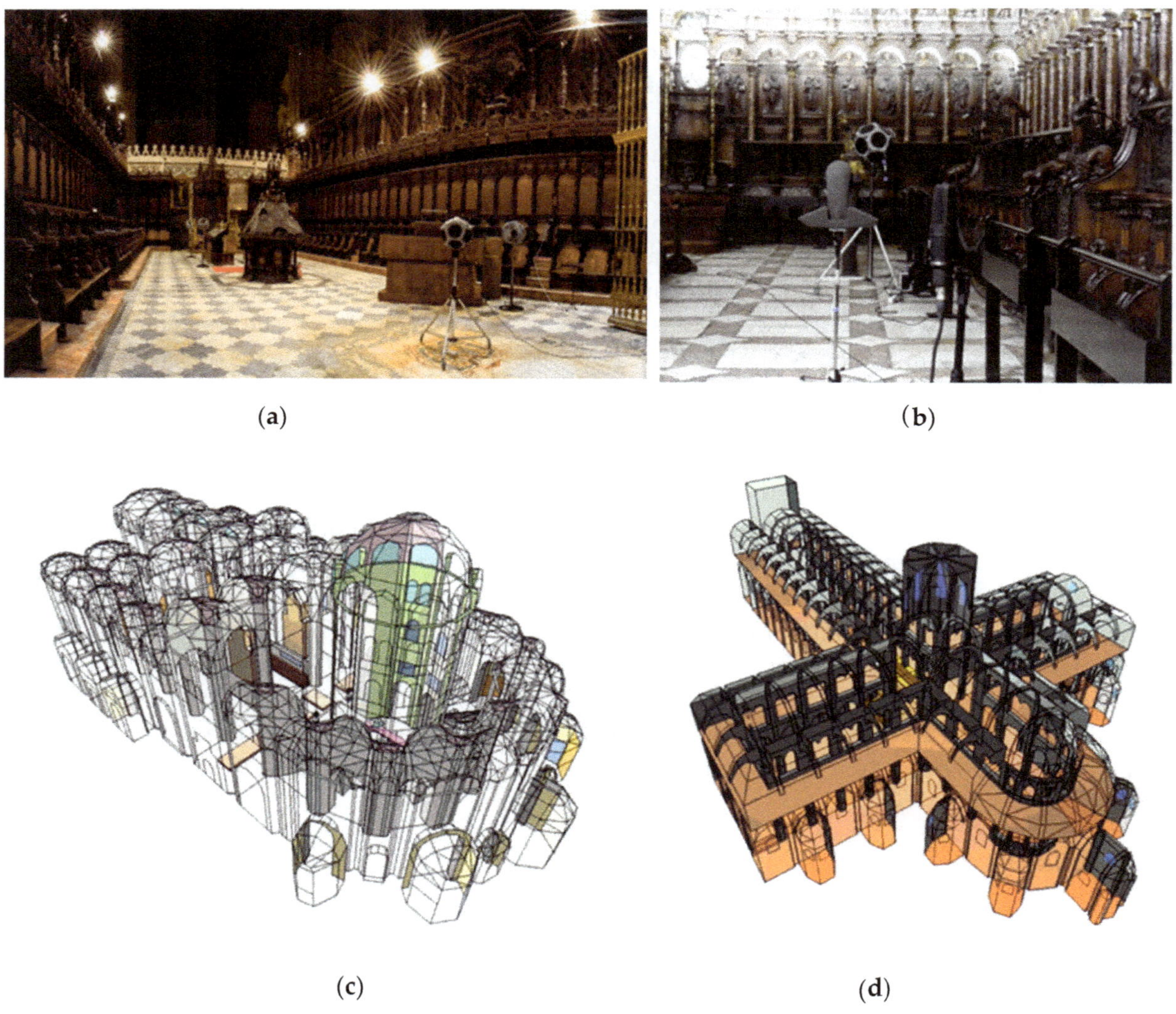

Figure 3. Acoustic measurements conducted inside cathedrals: (**a**) Choir stalls of the cathedral of Seville; (**b**) choir stalls of the cathedral of Toledo. Virtual models in which the choir space was incorporated: (**c**) cathedral of Granada; (**d**) cathedral of Santiago.

4. Acoustical Assessment of the Choir Space in Spanish Cathedrals

This study analysed the acoustic environment of Spanish cathedrals when the sound source is located in the choir stalls. The sound field of the Spanish cathedrals under study was evaluated through the analysis of the RIRs measured or predicted at the receiver points with direct sound from sources located in the choir and in the main altar. In this regard, a comparison of the values obtained was established in order to evaluate the acoustic impact of locating the sound source in the choir space, while considering different groups of receiver positions: One group located in the choir space itself, and another group located in the area of the faithful. This latter group was located in the area between the main altar and choir stalls, with a distribution of between 3–6 receiver points, depending on the cathedral studied. In order to correctly obtain the acoustic parameters, it was ensured that all the receiver points analysed guaranteed the arrival of direct sound. As previously stated, in relation to the acoustics, it is known that the choir constitutes one of the elements that configures the acoustic issue of cathedrals. This analysis enables the different acoustic conditions existing between the choir itself and the area of the faithful to be verified, while also noting the significant improvement of the acoustic quality in the choir space. The experimental results are shown in the various forms explained below.

Figure 4a–d depicts the acoustic parameter values, spatially averaged, versus frequency octave-band when the sound source is located in the choir and the receiver points are located in both positions of the choir and the area of the faithful. Table 3 provides a summary of spatial average values (500–1000 Hz) of the acoustic parameters analysed in the study, in accordance with Table A.1 ISO 3382 [11]. On the other hand, Table 4 shows the comparison in terms of the Just Noticeable difference (JND) of acoustic parameters between values of points located in the choir space and in the area of the faithful, when the sound source is located in the choir.

It can be observed from Figure 4a that T_{30} values are independent of the position inside the cathedral, since the reverberation time is obtained by calculating the time the signal needs to drop by 30 dB. In terms of early decay time (EDT), the calculation is performed using the first 10 dB, thereby raising the importance of the variation of the first reflections. This fact means that there is probably a difference in the results when the sound receivers vary the position and, therefore, vary the existence of the closest surfaces. Moreover, the choir enclosure is surrounded by high stonework walls on three of its sides and with numerous wooden stalls inside, which varies both the absorption and scattering coefficients. This study also centres on the values provided by parameters such as clarity (C_{80}), which measures the range of the listening perception for musical use, the centre time (Ts), which is related to the balance between early and late energy reaching the receiver, and definition (D_{50}), which shows the capacity of the enclosure to generate precision in the articulation of vocal sounds. All these factors lead to a noticeable effect in the sound field when the receiver points are located inside the choir space, as can be seen in Figure 4b–d and Table 4. It becomes apparent that the spectral values of EDT at mid-frequencies, spatially averaged, present a significant difference, greater than 10 JNDs in all cases, and more than 2 JNDs for C_{80}.

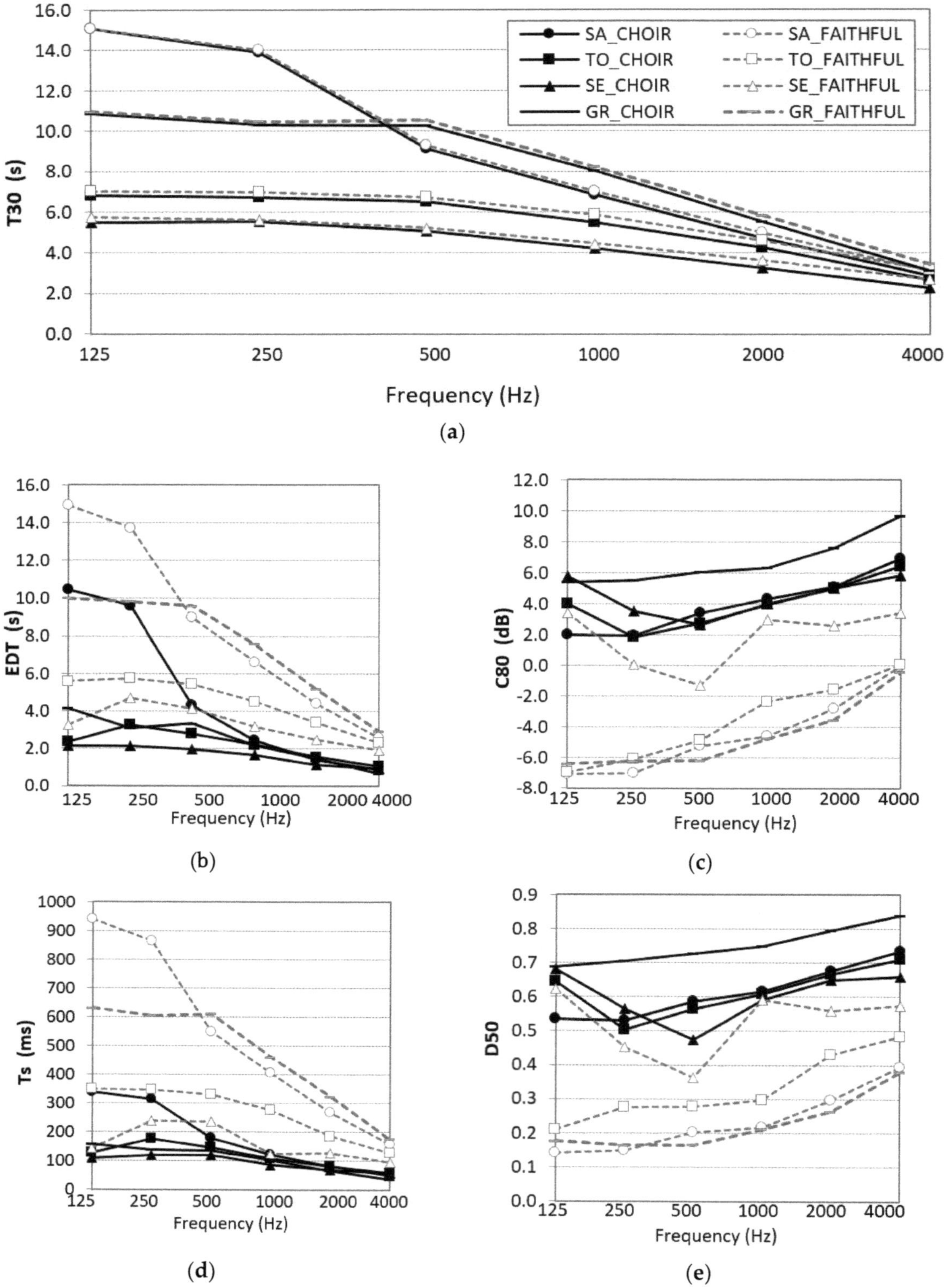

Figure 4. Acoustic parameter values, spatially averaged, versus frequency octave in the unoccupied cathedrals studied when the sound source is located in the choir and in the altar: (**a**) T30 (s); (**b**) EDT (s); (**c**) C_{80} (dB); (**d**) Ts (ms); (**e**) D_{50}. SA = Santiago, TO = Toledo, SE = Seville. EDT: Early decay time.

Table 3. Summary of average values (500–1000 Hz) of acoustic parameters analysed, in accordance with Table A.1 ISO 3382 [11]. Spatial average values by zone: Choir space and area of the faithful.

	T_{30} (s)	EDT (s)	C_{80} (dB)	T_S (ms)	D_{50}
SA_CHOIR	7.99	3.37	4.28	150	0.60
SA_FAITHFUL	8.15	7.81	−4.21	477	0.21
TO_CHOIR	6.01	2.50	3.89	128	0.59
TO_FAITHFUL	6.31	4.97	−2.93	302	0.29
SE_CHOIR	4.65	1.83	3.88	102	0.53
SE_FAITHFUL	4.84	3.66	1.42	181	0.48
GR_CHOIR	9.14	2.75	6.67	119	0.74
GR_FAITHFUL	9.37	8.59	−4.83	535	0.19

Table 4. Differences in JND of acoustic parameters between values of receiver points located in the choir space and in the area of the faithful, when the sound source is located in the choir.

	T_{30} (s)	EDT (s)	C_{80} (dB)	T_S (ms)	D_{50}
SA	0.4	26.3	5.7	29.2	3.9
TO	1.0	19.8	3.8	18.2	3.0
SE	0.8	11.2	2.6	6.4	0.6
GR	0.5	42.4	46.7	7.7	5.5

The variations of the EDT parameter are highly significant, with slight differences in the choir, where the values are appropriate for the Gregorian chant, and high values in the area of the faithful. The Gregorian chant is characterised as being a prayer sung with a prosodic rhythm, centred on intonation. This is a plainchant and a purely melodic, monodic song into which more complex developments, such as organum parallelum, have been incorporated. The relevance of the Gregorian chant is justified due to the importance of ecclesiastic chant at the liturgy and owing to the significant role played by choirs in cathedrals. In the case of D_{50}, the situation is similar, whereby adequate intelligibility is attained of the text sung in the choir.

Another influential factor is that of the acoustic quality of the cathedral, since the differences observed are lower when the acoustic conditions are better and the reverberation is lower. In the case of the cathedral of Seville (4.79 s), whose values of T_{30} are drastically lower than those of Granada (9.14 s), the differences in certain parameters are extremely small when the source is located on the altar and in the choir. In the cathedral of Seville, mostly rough and highly porous limestone, as well as the ornate decoration, exert considerable influence on the sound absorption by the venue, and, by extension, on its acoustic parameters [7]. It is interesting to observe that with the exception of the cathedral of Seville, the results reflect very similar sound conditions in the choir area, identified as the church of the clergy, with greater acoustic requirements for the chant. In the same way, the significant differences between the results of the two areas are shown, as influenced by the conditions of the enclosure. The area of the faithful represents the church without the many current acoustic requirements.

In addition, it should be noted that the JNDs in the cathedral of Granada are highly significant; however, this fact is justified by the great reverberation and the diffuse sound of the cathedral in the area of the faithful, which contrasts with the excellent acoustic conditions of its choral space. With the exception of the cathedral of Santiago, the subjective reverberation values in the choir are below 3 s, and the clarity values are in the range 2–4 dB, which is considered optimal for large-scale spaces [16].

5. Conclusions

The choir enclosure is one of the elements that structures and influences the acoustic issue of worship spaces, since it plays a role of the first magnitude in the configuration of ecclesial indoor spaces. From a historical point of view, the choir is a private, semi-enclosed, and privileged area reserved for the clergy. The spatial complexity of cathedrals responds to liturgical requirements. In this regard, in Europe, different models and typologies are established, depending on the location of the choir. In the case of most Spanish cathedrals, the central position of the choir stalls fragments the space in the main nave.

In this article, a grand tour is developed regarding the acoustic impact of the case study of Spanish cathedrals: Santiago, Toledo, Seville, and Granada. A noticeable effect is perceived in the sound field when the source is located inside the choir space since, in relation to the acoustics, it is known that the position of the choir is one of the elements that influences the acoustic issue of cathedrals.

When analysing the evolution of the acoustic impact of the choir in the Spanish cathedrals, it is not possible to establish an improvement in acoustic quality over time, since factors, such as the dimensions of the enclosure and the height of the walls, can influence the results. This analysis has successfully confirmed the various acoustic conditions existing between the choir itself and the area of the faithful, whereby the significant improvement of the acoustic quality in the choir space, identified as the church of the clergy, can be noted. The variations in the acoustic parameters are highly significant, with slight differences in the choir, where the values are appropriate for the Gregorian chant, with EDT values in the range of 1.8–3.3 s, and suitable intelligibility of sung text, with D_{50} values higher than 0.60 and C_{80} in the range of 3.8–6.6 dB. On the other hand, high values are obtained in the area of the faithful, which represents the congregation of the church, without the many current acoustic requirements.

Author Contributions: Conceptualization, A.A., R.S. and J.J.S.; Methodology, A.A., R.S. and J.J.S.; Software, A.A.; Validation, A.A., R.S. and J.J.S.; Investigation, A.A., R.S. and J.J.S.; Resources, A.A., R.S. and J.J.S.; Data Curation, A.A.; Writing-Original Draft Preparation, A.A.; Writing-Review & Editing, R.S. and J.J.S.; Visualization, A.A.; Supervision, R.S. and J.J.S.; Project Administration, R.S. and J.J.S.; Funding Acquisition, R.S. and J.J.S.

Funding: This work has been financially supported by FEDER funds and the Spanish Government, with reference BIA2014-56755-P.

Conflicts of Interest: The authors declare that there are no conflicts of interest. The founding sponsors played no role in the design of the study; in the collection, analyses, or interpretation of data; in the writing of the manuscript; and in the decision to publish the results.

References

1. Suárez, R.; Sendra, J.J.; Alonso, A. Acoustics liturgy and architecture in the Early Christian Church. From the Domus ecclesiae to the Basilica. *Acta Acust. United Acust.* **2013**, *99*, 292–301. [CrossRef]
2. Pedrero, A.; Ruiz, R.; Díaz-Chyla, A.; Díaz, C. Acoustical study of Toledo Cathedral according to its liturgical uses. *Appl. Acoust.* **2014**, *85*, 23–33. [CrossRef]
3. Navascués, P. *La Catedral en España: Arquitectura y Liturgia*, 1st ed.; Lunwerg: Barcelona, Spain, 2004; ISBN 8497851080.
4. Franco-Mata, A. El coro de la cathedral de Toledo. *Abrente* **2010**, *42*, 113–165.
5. Suárez, R.; Alonso, A.; Sendra, J.J. Archaeoacoustics of intangible cultural heritage: The sound of the Maior Ecclesia of Cluny. *J. Cult. Herit.* **2016**, *19*, 567–572. [CrossRef]
6. Conant, K.J. *Cluny, les Églises et la Maison du Chef D'ordre*, 1st ed.; Mâcon: Cambridge, MA, USA, 1968.
7. Alonso, A.; Sendra, J.J.; Suárez, R.; Zamarreño, T. Acoustic evaluation of the cathedral of Seville as a concert hall and proposals for improving the acoustic quality perceived by listeners. *J. Build. Perform. Simul.* **2014**, *7*, 360–378. [CrossRef]
8. Alonso, A.; Suárez, R.; Sendra, J.J. Virtual reconstruction of indoor acoustics in cathedrals: The case of the Cathedral of Granada. *Build. Simul.* **2017**, *10*, 431–446. [CrossRef]
9. Girón, S.; Álvarez-Morales, L.; Galindo, M.; Zamarreño, T. Acoustic evaluation of the cathedrals of Murcia and Toledo, Spain. In Proceedings of the International Congress of Sound and Vibration (ICSV24), London, UK, 23–27 July 2017.
10. Martellotta, F.; Álvarez-Morales, L.; Girón, S.; Zamarreño, T. An investigation of multi-rate sound decay under strongly non-diffuse conditions: The crypt of the cathedral of Cadiz. *J. Sound. Vib.* **2018**, *70*, 261–274. [CrossRef]

11. ISO 3382-1. *Acoustics—Measurement of Room Acoustic Parameters. Part1: Performance Rooms*; International Organization for Standardization: Geneva, Switzerland, 2009.
12. Martellotta, F.; Cirillo, E.; Carbonari, A.; Ricciardi, P. Guidelines for acoustical measurements in churches. *Appl. Acoust.* **2009**, *70*, 378–388. [CrossRef]
13. Suárez, R.; Alonso, A.; Sendra, J.J. Intangible cultural heritage: The sound of the Romanesque cathedral of Santiago de Compostela. *J. Cult. Herit.* **2015**, *16*, 239–243. [CrossRef]
14. Dalenbäck, B.I.L. *CATT-Acoustic v9 Powered by TUCT. User's Manual*; CATT: Gothenburg, Sweeden, 2011.
15. Vorländer, M. *Auralization*; Springer: Berlin, Germany, 2008.
16. Gade, A.C. Acoustics in halls for speech and music. In *Springer Handbook of Acoustics*, 2nd ed.; Rossing, T.D., Ed.; Springer: New York, NY, USA, 2007; pp. 301–350. ISBN 978-1-4939-0755-7.

Permissions

All chapters in this book were first published in ACOUSTICS, by MDPI; hereby published with permission under the Creative Commons Attribution License or equivalent. Every chapter published in this book has been scrutinized by our experts. Their significance has been extensively debated. The topics covered herein carry significant findings which will fuel the growth of the discipline. They may even be implemented as practical applications or may be referred to as a beginning point for another development.

The contributors of this book come from diverse backgrounds, making this book a truly international effort. This book will bring forth new frontiers with its revolutionizing research information and detailed analysis of the nascent developments around the world.

We would like to thank all the contributing authors for lending their expertise to make the book truly unique. They have played a crucial role in the development of this book. Without their invaluable contributions this book wouldn't have been possible. They have made vital efforts to compile up to date information on the varied aspects of this subject to make this book a valuable addition to the collection of many professionals and students.

This book was conceptualized with the vision of imparting up-to-date information and advanced data in this field. To ensure the same, a matchless editorial board was set up. Every individual on the board went through rigorous rounds of assessment to prove their worth. After which they invested a large part of their time researching and compiling the most relevant data for our readers.

The editorial board has been involved in producing this book since its inception. They have spent rigorous hours researching and exploring the diverse topics which have resulted in the successful publishing of this book. They have passed on their knowledge of decades through this book. To expedite this challenging task, the publisher supported the team at every step. A small team of assistant editors was also appointed to further simplify the editing procedure and attain best results for the readers.

Apart from the editorial board, the designing team has also invested a significant amount of their time in understanding the subject and creating the most relevant covers. They scrutinized every image to scout for the most suitable representation of the subject and create an appropriate cover for the book.

The publishing team has been an ardent support to the editorial, designing and production team. Their endless efforts to recruit the best for this project, has resulted in the accomplishment of this book. They are a veteran in the field of academics and their pool of knowledge is as vast as their experience in printing. Their expertise and guidance has proved useful at every step. Their uncompromising quality standards have made this book an exceptional effort. Their encouragement from time to time has been an inspiration for everyone.

The publisher and the editorial board hope that this book will prove to be a valuable piece of knowledge for researchers, students, practitioners and scholars across the globe.

List of Contributors

Braxton Boren
Audio Technology Program, Department of Performing Arts, College of Arts & Sciences, American University, Washington, DC 20016, USA

Zühre Sü Gül
Department of Architecture, Bilkent University, 06800 Ankara, Turkey

Chenhui Zhao and Marehalli G. Prasad
Department of Mechanical Engineering, Stevens Institute of Technology, Hoboken, NJ 07030, USA

Alessandro Franco and Carlo Bartoli
Department of Energy, Systems, Territory and Constructions Engineering (DESTEC), University of Pisa, 56126 Pisa, Italy

Zorana Đorđević
Institute for Multidisciplinary Research, University of Belgrade, 11000 Belgrade, Serbia

Dragan Novković
The School of Electrical and Computer Engineering of Applied Studies, 11000 Belgrade, Serbia

Uroš Andrić
Front Element Media, 11000 Belgrade, Serbia

Oliver Chmiel, Ivo Baselt and Andreas Malcherek
Institute of Hydromechanics and Hydraulic Engineering, University of the German Armed Forces Munich, 85579 Neubiberg, Germany

Hassan Azad and Gary Siebein
School of Architecture, University of Florida, Gainesville, FL 32611, USA

Julie Meyer and Tapio Lokki
Department of Computer Science, Aalto University, FI-02150 Espoo, Finland

Heow Pueh Lee
Department of Mechanical Engineering, National University of Singapore, 9 Engineering Drive 1, Singapore 117575, Singapore

Linus Yinn Leng Ang
Department of Mechanical Engineering, National University of Singapore, 9 Engineering Drive, Singapore 117575, Singapore
Kinetics Design and Manufacturing, ST Engineering Land Systems Ltd., 249 Jalan Boon Lay, Singapore 619523

Yong Khiang Koh
Kinetics Design and Manufacturing, ST Engineering Land Systems Ltd., 249 Jalan Boon Lay, Singapore 619523

Rupert Till
Department of Music and Drama, University of Huddersfield, Huddersfield HD1 3DH, UK

Dario D'Orazio
Department of Industrial Engineering, DIN University of Bologna, 40126 Bologna, Italy

Sofia Nannini
Department of Architecture and Design, DAD Politecnico di Torino, 10125 Torino, Italy

Alicia Alonso, Rafael Suárez and Juan J. Sendra
Instituto Universitario de Arquitectura y Ciencias de la Construcción, Escuela Técnica Superior de Arquitectura, Universidad de Sevilla, Seville 41012, Spain

Index

CPSIA information can be obtained
at www.ICGtesting.com
Printed in the USA
LVHW062008100122
708212LV00002B/103

9 781639 892518